Ecology and Applications of Benthic Foraminifera

In this volume John Murray investigates the ecological processes that control the distribution, abundance, and species diversity of benthic foraminifera in environments ranging from marsh to the deepest ocean. To interpret the fossil record it is necessary to have an understanding of the ecology of modern foraminifera and the processes operating after death leading to burial and fossilisation. This book presents the ecological background required to explain how fossil forms are used in dating rocks and reconstructing past environmental features including changes of sea level. It demonstrates how living foraminifera can be used to monitor modern-day environmental change.

Ecology and Applications of Benthic Foraminifera presents a comprehensive and global coverage of the subject using all the available literature. It is supported by a website hosting a large database of additional ecological information (www.cambridge.org/9780521828390) and will form an important reference for academic researchers and graduate students in Earth and Environmental Sciences.

JOHN MURRAY is Emeritus Professor at the School of Ocean and Earth Science, University of Southampton. He was awarded a D.Sc. by the University of London and has held honorary positions with the Palaeontological Association, the Micropalaeontological Society and the Geological Society of London. Professor Murray has over 40 years of experience in studying foraminifera ecology and has written several books on the subject.

CAMBRIDGE UNIVERSITY PRESS
Cambridge, New York, Melbourne, Madrid, Cape Town, Singapore, São Paulo

Cambridge University Press
The Edinburgh Building, Cambridge CB2 8RU, UK

Published in the United States of America by Cambridge University Press, New York

www.cambridge.org
Information on this title: www.cambridge.org/9780521828390

First published 2006
This digitally printed version 2008

A catalogue record for this publication is available from the British Library

ISBN 978-0-521-82839-0 hardback
ISBN 978-0-521-07009-6 paperback

Contents

Preface

Why write another book on foraminiferal ecology when I have already previously written two (1973, 1991)? Writing those was stimulating because it brought together all the information currently available making it possible to see where progress had been made and where there were still gaps in knowledge. The same philosophy applies to this book. But the simple answer is that a vast amount of literature has appeared since 1991 and ideas have changed. Also, I know that my previous books are cited much more often than any of my papers. Two events encouraged me to undertake this major task. The 1991 book went out of print in the late 1990s but there continues to be a demand for it so the publishers re-introduced it in 2002 as a 'print-on-demand' edition. Also, I was invited to give a keynote lecture at the Forams 2002 meeting in Perth and while writing that the need for a new book became glaringly obvious. Another factor is that my circumstances have changed. I started research on foraminiferal ecology in 1959 and became an academic in 1962 so for 41 years research was carried alongside teaching ever increasing numbers of students, administration (including being head of a geology department for 14 years, and an exponential increase in bureaucracy over the last decade) and serving the scientific community on the councils of scientific societies and as editor. I started to write this book in 2003 during my final year of employment. I retired in September 2003 and since then I have had the advantage of more time for reading and research and freedom from most other duties.

My 1973 book was based on living distributions and I believe that worked well. However, in the 1991 book in addition to data on living I included data from dead and total (live plus dead) assemblages. I now think that was a mistake for ecological relationships can satisfactorily be based only on the distribution of live organisms. Foraminifera are sufficiently small that post-mortem transport and even reworking of tests from older deposits can introduce distributional anomalies that go undetected if live and dead are not separated. Here I have based the ecological discussion solely on data on living (stained) forms. Geologists wish to know what differences arise between the living assemblages and the ones that become fossilised in the sediment. That is most easily addressed by studying the taphonomic processes that affect tests after the death of the organism and those are discussed in Chapter 9. Web Appendix 1 lists references on taxonomy, dead and total assemblages.

An additional aim has been to compile spreadsheets of data on living assemblages and to make them available to other workers and this is being done via the web. Published data from older literature have been entered into spreadsheets. Many authors have kindly responded to my request for electronic versions of their spreadsheets from recently published papers. One thing I soon realised when I started examining the published data tables is how many contain mistakes – particularly of totals not adding up to 100%. No doubt there are some mistakes in the tables presented here but I have tried to cross check everything as much as possible (web Tables WA-1 to WA-219). I have also used a standard set of generic and species names (web Appendix 2). With such a large amount of data, it has been difficult to know how to divide it into manageable portions. Whatever way divisions are introduced, there are always problems because the boundaries are artificial. I have followed the environmental route because it is already known that there are big differences between the major environments. Within specific environments, I have followed a geographical trail from north to south and then from the Atlantic to the Pacific and Indian oceans (see the website for extended versions of some distribution charts in Chapters 4 and 5). Coverage of the literature is up to early summer 2005.

The main focus of the book is on ecology and providing the essential database against which the fossil record can be compared. Because benthic foraminifera are strongly controlled by their environments, all applications of foraminiferal data must involve an understanding of their ecology. In Chapter 10 my aim has been to outline the methodology of the applications and to provide some case studies particularly from the Quaternary and Holocene. It is beyond the scope of the book to include a wealth of examples or to extend back into the Cenozoic or Mesozoic. Those are tasks for others to undertake in the future.

I have had great support from the foraminiferal community which I gratefully acknowledge. Everyone I asked for information, data or samples did their best to help: Elisabeth Alve (Norway), Joan Bernhard (USA), Martin Brasier (England), Alex Cearreta (Spain), Thomas Cronin (USA), Jean-Pierre Debenay (France), Henko de Stigter (The Netherlands), Paula Diz (Spain), Robin Edwards (England), Sue Goldstein (USA), Bruce Hayward (New Zealand), Silvia

Hess (Germany), Ben Horton (USA), Katrine Husum (Norway), Anne Jennings (USA), Frans Jorissen (France), Hiroshi Kitazato (Japan), Karen-Luise Knudsen (Denmark), Sergei Korsun (Russia), Wolfgang Kuhnt (Germany), Laetitia Licari (France), Jeremy Lloyd (England), Andreas Mackensen (Germany), Rosalie Maddocks (USA), Kjell Nordberg (Sweden), Emil Platon (USA), Wylie Poag (USA), Sally Radford (England), Rosanna Serandrei-Barbero (Italy), Ralf Schiebel (England), Gerhard Schmiedl (Germany), Joachem Schönfeld (Germany), James Scourse (Wales), Mike Scrutton (Wales), Barun Sen Gupta (USA), Scott Snyder (USA), Simon Sturrock (Australia), Jane Swallow (England), Donata Violanti (Italy). I apologise to anyone whose name I have inadvertently omitted.

In addition, I have received great help from the Natural History Museum, London, for access to collections (Clive Jones) and libraries (John Whittaker). The National Oceanography Centre, Southampton, has kindly provided office space, technical support (Barry Marsh, Richard Pierce, Elaine Watson, Daphne Woods) and access to IT and library facilities.

Finally, it is a pleasure to record my gratitude to my father, William Murray, Elisabeth Alve and John Whittaker for their continued support and encouragement to write this book. Elisabeth and John have provided both stimulating discussion and valuable advice on structure and content. Juliette Topping has been immensely helpful with entering data into spreadsheets, checking citations of references, figures, tables, etc. Elisabeth and Juliette also read the entire draft and offered valuable comment. Small sections were also read by Ben Horton (USA), Marty Buzas (USA), Bob Jones (BP) and John Whittaker (UK). I also gratefully acknowledge a two-year Leverhulme Emeritus Fellowship in support of the research. I have thoroughly enjoyed writing this book and especially the luxury of doing so without carrying out a full-time job at the same time. I hope it will be useful to readers!

John Murray
Southampton

1

Introduction

1.1 Objectives and strategy

Ecology is the study of the causes of patterns of distribution and abundance of organisms. It is concerned with interactions between individuals and their physical and chemical environment, interactions between individuals of the same species and between species. Ecology may be investigated through field studies, laboratory experiments and mathematical modelling. Foraminifera are generally small (<1 mm) although some exceed 1 cm and most have a shell or test which may be preserved in the fossil record. These attributes make foraminifera extremely valuable as they provide not only a contemporary but also a historical record of previous environments. They are therefore of interest both to biologists and geologists.

The primary objective of this book is to present a state-of-the-art synthesis of ideas and data on foraminiferal ecology that will be of value to those carrying out new studies or wishing to interpret new data from modern or ancient environments. In this book similarities are stressed because it is very easy to overlook the broad picture if the focus is on small differences. All applications of benthic foraminifera involve an understanding of their ecology (Chapter 10).

1.2 Taxonomic scope of foraminifera

The foraminifera form an order (Foraminiferida) in the Phylum Protista: 'Cytoplasmic body enclosed in a test or shell of one or more interconnected chambers...' (Loeblich and Tappan, 1987, p. 7). It has recently been argued that there are naked forms (without a test) (Pawlowski et al., 1998) but such forms do not leave a fossil record. Molecular geneticists are investigating

1

both the antiquity of the group and whether or not it is monophyletic but no definitive answers have yet been reached (see review by Pawlowksi, 2000).

Although biological species are defined on the ability successfully to reproduce sexually, foraminiferal species are defined primarily on wall structure, chamber and test shape, and the position of the aperture(s); hence they are morphospecies. The life cycle is known for no more than 30 species and several patterns have been observed with the basic cycle appearing very ancient (Goldstein, 1997, 1999). One of the problems faced by those studying foraminifera is inconsistent use of species and generic names (Boltovskoy, 1990). Generic terminology is stabilised by using taxonomic treatises such as that of Loeblich and Tappan (1987); however, that is now almost two decades old so inevitably there have been revisions. Some highly variable species that are difficult to separate on morphology seem even more complex from a molecular genetic perspective as there are cryptic species that have no morphological expression (e.g., *Ammonia*, Holzmann, 2000; Hayward *et al.*, 2004). There is clearly a need to integrate studies of morphospecies, life cycles and molecular genetics in order to determine true taxonomic relationships. So far this has been done only for a small group of glabratellids (Tsuchiya *et al.*, 2000). At present, foraminiferal ecology is based entirely on morphospecies but no doubt this will change in the future as the impact of molecular genetics becomes greater.

1.3   Historical development of ecological studies

The scientific discoveries of one generation are built upon the foundations laid by earlier workers. Although most of the ideas and data discussed in this book are from recent decades we should not overlook the contributions made by those who started our subject, many of whom earned their living in business or in other fields of science or medicine. Although fossil foraminifera had been recorded in the fifth century BC they were regarded as small molluscs or worms. The first modern foraminifera were described by Beccarius in 1731 but the collective term 'foraminifera' was not introduced until 1830 (Loeblich and Tappan, 1964). The study of foraminifera started as a hobby for eighteenth century gentlemen who examined various small objects under the newly introduced microscopes. In Britain, the first were Boys, a surgeon and naturalist, and Montagu. In the nineteenth century major contributions were made by Williamson, Professor of Natural History at Manchester; Carpenter, Professor of Medical Jurisprudence and Registrar of the University of London; and H.B. Brady, pharmacist who studied the Challenger expedition foraminifera in retirement. In the twentieth century Heron-Allen, a solicitor and polymath, and Earland, civil servant, formed a team producing numerous publications in the period 1913–43. In Austria, two keen collectors (Fichtel and Moll) described material from the Mediterranean and the Indian Ocean. In France, similar pioneering studies were made by d'Orbigny from various Atlantic localities, and in Germany by Ehrenberg and Rhumbler. In the USA, Flint was a pioneer but it was particularly Cushman and his co-workers who described numerous new taxa from a wide range of environments in the period 1909–48. All these authors concentrated on taxonomy, a necessity that had to precede true ecological studies; however they all reported on the occurrence of forms in different environments and habitats. The British school (Williamson and Carpenter) believed in a broad species concept, allowing for greater morphological variation than d'Orbigny and later workers such as Brady and Cushman. However, it has subsequently been shown that there is considerable morphological variation spanning several genera in *Cibicides lobatulus* (Nyholm, 1961) so perhaps species concepts should not be too narrow.

Observations on the soft parts of *Elphidium crispum* led to the determination of the life cycle (Lister, 1895) and since then around 30 species have been studied (Goldstein, 1999). Modern foraminiferal ecology started in the late 1930s to the 1950s with the works of Rhumbler (1935), who tried staining methods, Myers (1942), who introduced the first ideas on population dynamics, and Boltovskoy (1964), Bandy (1956) and Phleger (1951), who carried out field studies. In 1952, Walton introduced the now widely used rose Bengal staining method to differentiate live from dead in preserved samples. The introduction of the scanning electron microscope in the 1960s revolutionised our ability to illustrate foraminifera and the introduction of computers during the same period allowed the introduction of multivariate methods of data analysis. Whereas stable isotope studies of fossil foraminifera became commonplace from the 1960s, it is only in relatively recent years that they have been more widely applied to foraminiferal ecology.

Previous syntheses of foraminiferal ecology have shown the progressive increase in information and changes in which variables are thought to be significant: depth distributions (Phleger, 1960a); biogeographic distributions (Boltovskoy, 1965; Boltovskoy and Wright, 1976); the use of diversity indices and ternary plots of wall structure as features summarising the attributes of assemblages (Murray, 1973); associations related to water temperature, salinity and substrate (Murray, 1991). There have been no complete syntheses of ecology since then. Biology and ecology are briefly reviewed in Lee and Anderson (1991a). In Sen Gupta (1999) there are three chapters that discuss the role of oxygen and flux of organic material as ecological controls, and chapters on biogeography and symbiosis. Supplement 1 of *Micropaleontology* on the theme 'Advances in the biology of foraminifera' (Lee and Hallock, 2000) is especially useful. The practical application of using foraminifera to monitor pollution is discussed in Sen Gupta (1999), Lee and Hallock (2000), Martin (2000) and Scott *et al.* (2001), while the latter also consider their use in determining past sea level and in sediment transport. Foraminiferal ecology is commonly applied in the interpretation of the Quaternary palaeoecology (Haslett, 2002).

National museums play an important role in housing collections of type and reference material as well as maintaining comprehensive libraries of relevant literature, all of which are freely available for consultation by researchers. Key foraminiferal collections are housed in the Smithsonian Institution, Washington, USA (Cushman), the Natural History Museum, London, UK (Williamson, Brady, Millett, Heron-Allen and Earland), and the Museum National d'Histoire Naturelle, Paris, France (d'Orbigny). These are essential reference collections for taxonomic purposes. As micropalaeontological studies have grown in importance, scientific societies have been established to promote the subject, especially through their publications and by arranging symposia (Cushman Foundation, Gryzbowski Foundation, The Micropalaeontological Society). All these elements are important for the education of the next generation of micropalaeontologists. Over recent decades there has been a change in employment of micropalaeontologists from oil geology and basic geological surveying to interpreting palaeoecology and palaeoceanography (especially in connection with the deep sea) and in environmental monitoring (pollution, sea level, climate change). This trend is likely to continue.

1.4   Major developments over the past decade

No aspect of science develops in isolation. The development of new instrumentation and the greater involvement of biologists have led to the introduction of new techniques, e.g., biomarker analysis (lipids, sterols), molecular genetics, fluorogenic probes for detecting live individuals without causing them harm, and remote sensing images of sea-surface chlorophyll as an index of primary production. Improvements in analytical equipment make it possible to analyse ever smaller samples with greater accuracy for stable isotopes and trace-element geochemistry. There has been more experimentation: microcosm and mesocosm experiments to address specific questions (faunal response to changing oxygen conditions or to the introduction of specified pollutants); *in situ* sea-floor experiments (colonisation of barren substrates; exclusion of predators). Time-series studies showing the response to natural variability have been carried out in environments ranging from intertidal to the deep ocean. There is a greater understanding of sediment inorganic geochemistry especially in the top few centimetres where the majority of foraminifera actually live (redox boundary, nitrate boundary, etc.) because of the ability to make *in situ* measurements using probes. Also, it is now possible to quantify the biopolymers produced by bacteria and algae (potential food resources and important for stabilising the sediment surface against erosion). From a geological perspective, specific radiation events (^{137}Cs from nuclear-bomb testing which peaked in 1963–64; Chernobyl, 1986) and ^{210}Pb can be used to provide a timescale down core for the historical record of faunal and environmental change. All these developments are revolutionising our understanding of foraminiferal ecology and are discussed in the appropriate chapters.

The major patterns of distribution have been determined by field studies. We can readily distinguish between brackish, marine and hypersaline assemblages, and also between environments such as marsh, lagoon/estuary, continental shelf, bathyal and abyssal; and, for shallow-water environments, whether they are from cold, temperate or warm regimes. This database is readily applicable to the fossil record but with decreasing certainty of interpretation in passing from the Quaternary back through the Cenozoic and Mesozoic to the Palaeozoic (due to evolutionary changes in faunas through time and lack of environmental analogues). Apart from field surveys there are controlled experiments, where the impact of single variables can be isolated, and these give fairly definitive answers. We can look forward to an expansion of this type of study in the years ahead. This will help to explain some of the observed field relationships and help to bridge the gap between biological and geological approaches. Indeed, interdisciplinary collaboration is essential to advance the subject. However, as observed by Hilborn and Mangel (1997, p. 13) 'In both ecology and geology, experiments may be difficult to perform and so we must rely on observation, inference, good thinking, and models to guide our understanding of the world.' Finally, modelling is a way of considering how observations or data may be explained. Two broad types of model are recognised: *mathematical models*, where attempts are made to explain processes through calculation (e.g., the impact of variables on numbers of individuals as in population dynamics); *conceptual models*, which are essentially 'what if' questions: what would happen to species x if the value of variable y is changed? Mathematical models may be deterministic (where all components have known values, so for fixed starting values the model will always produce the same results) or stochastic (where some components are random, so several outcomes are possible depending on the values used). To do good science we need to consider whether we are asking the right questions and, if so, whether we are attempting to answer them using the best possible methods. It is important to test new ideas and to reject those that lead up blind alleys. Mathematical and conceptual models play a role in this and in advancing understanding in reality, but models should never be confused with reality itself (Hilborn and Mangel, 1997, p. 32).

1.5   The future

No doubt foraminiferal ecology will continue to play an important role in the interpretation of the geological record. Whereas geologists have the view that 'the present is the key to the past' it is now clear that this can also be turned around to help predict future events. Thus the study of past marine transgressions may help to determine the impact of future sea-level rise; past climatic warming may help to predict future responses. There are certainly lots of challenges for foraminiferal workers in the years to come and the future of foraminiferal ecology looks exciting.

2

Methods

2.1   Planning general field surveys

A key consideration is that it must be possible for the results to be compared with those obtained in other studies. As it is desirable to compare like with like, this means that there must be some standardisation of approach (e.g., using the same sieve size). The planning stage should include reviewing any previous studies on the geographic area: on foraminifera, ecology, oceanography, etc. Biologists like to take three to four replicates from each sampling area to determine patchiness and for valid statistical analyses. Most geologically oriented ecological surveys are based on single samples from each station. If possible, a preliminary survey should be carried out in order to decide on the following points.

- Take samples of adequate size; e.g., for standing crop, large enough to give more than 100 stained individuals.
- Is the study to include soft-shelled foraminifera or just those that might withstand fossilisation? If the former, then the samples will need to be examined wet. The method of examination will control the number of samples to be collected as well as the type of data collected.
- Which type of sampling equipment should be used (see below)?
- How many (or how few) sample sites should be chosen? Should replicates be taken. If so, how many? Their positioning needs very careful consideration as they should be random. For multicores and boxcores, replicates would be separate cores. Taking one or more subsamples from a single core is not true replication but pseudoreplication.

7

- What spatial and temporal patterns of sampling will be undertaken?
- Should sediment slices be taken down core in order to investigate abundance changes and infaunal depth stratification?
- How will the samples be preserved and, how they will be transported back to the laboratory; how long will it be before they are processed?
- Which environmental variables should be measured and how should this be done? Some require additional sediment samples (e.g., for grain-size analysis, lipid analysis or other geochemistry).
- The mathematical methods used to analyse the data should be decided at the time the sampling plan is drawn up.

Sometimes ecological studies of foraminiferal studies are part of a bigger programme so that choices of sampling pattern, number of samples, replicates, etc. are restricted. With few exceptions, the data discussed in this book have come from surveys that have not been planned according to these modern concepts.

2.2   Planning surveys to address a specific question

These should follow preliminary studies so that it is certain that the question can be addressed. Care must be taken not to prejudice the outcome. Many of the considerations listed above will apply. Additional points to consider include:

- It is essential to take replicates in order to determine variability.
- The environmental variable of prime interest should not co-vary with the only other variables measured (e.g., if oxygen is the variable of interest, it is likely to co-vary with clay content and total organic carbon (TOC) so other variables should also be measured).
- It may be necessary to collect time-series data. If so, consider short-term variability when planning the sampling interval. For instance, whether reproduction is periodic or continuous; if the former, the length of the life cycle is relevant to the sampling interval.

2.3   Types of sampler, taking and handling samples

A wide range of sediment samplers is illustrated and described in Mudroch and McKnight (1994). It is essential that the sampler used will take a representative sample without the loss of the surface veneer of sediment. For samples taken under water, this means that the sample must be sealed in the sampler on the sea floor otherwise there will be loss through 'washing' as the sampler is raised through the water column and from the water to the boat. For this reason, dredges, grabs and gravity corers are not very satisfactory as there may be loss of the surface layer on impact, and they do not seal once the sample has been taken. The ideal samplers for cohesive sediments are corers that have sealing devices and are lowered into the sediment gently so as to cause minimum disturbance and loss. Examples are the multiple corer (Barnett *et al.*, 1984), box corer (Mudroch and McKnight, 1994), and the Craib corer (Craib, 1965). These collect cores of sediment with the overlying few centimetres of bottom water. The cores can be subsampled at selected intervals (as described below) to study distributions below the sediment surface. However, corers do not always work well in non-cohesive (sandy) sediments. A sampling device that slices off the top 1 cm of sediment and seals it on the sea floor was designed to overcome this (Murray and Murray, 1987). In the intertidal zone, samples can be taken with a plastic ring of chosen height or with a core tube. The ring can be pressed into the sediment so that its upper surface is level with the sediment surface, a plate slid underneath and then the sample can be lifted out. Alternatively, the core tube can be pushed into the sediment and the base sealed with the plate.

To subdivide a core into sections the following equipment is needed: a clear plastic core tube, a shorter section of tube graduated with marks to determine sample thickness, a piston that fits the tube and a stand to hold the piston in a vertical position beneath the core (Figure 2.1). The core can be extruded into the graduated section to the desired height and the sediment sliced off using the plate. By repeating this procedure, a core can be sectioned into slices of chosen thickness. To avoid contamination it is essential to wash the plate and the graduated section of tube before each slice is taken.

2.4   Collecting live individuals

It is very easy to collect living foraminifera from the intertidal zone, especially from muddy sediments. Scrape off the top few millimetres of sediment making sure that no underlying anoxic material is included. The choice of sieve size will depend on the objectives of the study. While in the field, use ambient water with a pressure spray to wash out most of the mud. Place the residue in a container with some ambient water and keep it at the appropriate temperature during transit to the laboratory. Once there, spread the material thinly in a petri dish with water of ambient temperature and salinity and examine it under a binocular microscope. It will be easy to see tests with coloured contents and these are most likely to be alive. They can be removed

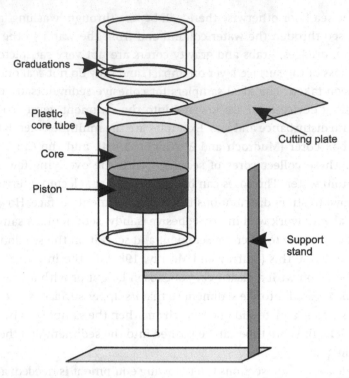

Figure 2.1. A schematic representation of the method used to subdivide a sediment core into slices.

with a pipette and placed in a small dish for examination. More detailed descriptions of field methods, and for the setting up of containers for maintaining foraminifera in the laboratory, are given in reviews by Arnold (1974) and Anderson *et al.* (1991).

2.5   Distinguishing live from dead foraminifera

A wide variety of methods of distinguishing living from dead foraminifera have been devised. In an excellent review, Bernhard (2000) differentiated between non-terminal and terminal methods and discussed their advantages and disadvantages. These techniques are listed below but only those commonly used are discussed in any detail and not all the arguments presented by Bernhard are reproduced here. The choice of technique depends on the objectives of the study. For experiments, it is clearly necessary to use non-terminal techniques. For most of the distributional studies reviewed in this book, sediment samples were collected and preserved. Forms that were alive at the time of collection were distinguished from those that were dead using staining methods (rose Bengal in almost all cases). Ideally, a second method should be used to validate the main method but in practice this is rarely done.

Non-terminal methods. The aim is to determine whether the organism is alive without killing it. This requires keeping it under suitable ambient environmental conditions. Such techniques are appropriate for isolating individuals to be used in experiments.

- Natural colour: due to the presence of symbionts, husbanded chloroplasts or from their food. This method should always be supplemented by other methods such as observing pseudopodial activity.
- Apertural bolus (a mass of food/detritus): not a very reliable method.
- Negative geotaxis/positive phototaxis: some individuals will crawl up the side of the tank or onto glass slides which have been pressed into the sediment.
- Cytoplasmic streaming in the test and pseudopodial movement outside the test. The method is suitable for isolating individual specimens but not for assemblage studies. It requires a microscope equipped with phase contrast or Nomarski optics.
- Fluorogenic probes: cause active cells to become fluorescent. Once the probe has been taken up and fixed, the individual can be examined either by epifluorescence microscopy (with suitable optics, specimens fluoresce but this is not easy to see in those with opaque tests) or by spectrofluorimetry (a membrane-permeating fluorogenic probe allows those with either transparent or opaque tests to fluoresce; the intensity of the fluorescence is measured with a luminescence spectrometer). The former method is slow but the latter is relatively fast and Bernhard considers that with practice at least 75 individuals can be isolated in one hour. These techniques are especially useful for ensuring that all specimens used in an experiment are alive at the start (see also Bernhard *et al.*, 1995).
- Calcein: this can be used in cultures of living foraminifera to label the newly-formed calcite chambers, which fluoresce in contrast to those formed prior to the experiment (Bernhard *et al.*, 2004).

Terminal methods. With the exception of the ATP method, the sediment is mixed with ethanol or formalin so the individuals are killed. The methods used are designed to determine those that were alive at the time of collection. At present, this is the only realistic approach to assemblage studies involving numerous samples.

- Non-vital stains: protoplasm is coloured regardless of whether or not it is alive at the time of staining. Stains that have been used are eosin (Rhumbler, 1935), rose Bengal (Walton, 1952) and Sudan Black B (Walker *et al.*, 1974). Only the rose Bengal technique is widely used. Protoplasm stains red while empty tests (dead) remain unstained or have a slight superficial pinkness. It is thus easy to distinguish stained from unstained. Even opaque tests become translucent when wetted, which is routinely done with a brush while picking dry samples so the colour can still be seen. It is not uncommon for staining to be concentrated in the final chambers. Recognition of stained individuals is perhaps slightly more problematic in deep-sea samples as the proportion of test filled with protoplasm is often much smaller than that in forms from upper-slope to coastal environments (see Altenbach, 1987). The advantages are that no special equipment is required and it is quick. It is thus ideal for dealing with large numbers of samples. The premise is that if protoplasm is present the individual was alive at the time of collection. Some authors are concerned that clumps of bacteria inside the test or cyanobacteria or fungi on the outside may be mistaken for protoplasm (Martin and Steinker, 1973) or that the stain may not be taken up if the individual is starving (Lutze and Altenbach, 1991). A major controversy is whether the protoplasm of dead individuals decays in a short time period (see Murray and Bowser, 2000) or persists in the test for weeks or months (see Bernhard, 2000, and references therein). Even if the latter is true, and for oxic environments that must be doubted, it is likely that the normal life-cycle processes affecting foraminifera do not lead to many individuals dying (Murray and Bowser, 2000): during reproduction the protoplasm of the parents is utilised to make the gametes or the offspring, and pre-reproductive deaths are most likely to be due to being eaten. Experiments to test the longevity of dead protoplasm use heat to kill the foraminifera, which makes it a non-analogue exercise (e.g., Bernhard, 1988). Nevertheless, it is undeniably true that where protoplasm is still present dead individuals will stain with rose Bengal and be counted as 'live'; but Murray and Bowser believe the problem to be overstated. Even Bernhard (2000) accepts that rose Bengal staining is the most appropriate method for broad-scale distributional studies.
- Life-position studies: sediments are impregnated with epoxy resin to make them hard so that they can be sectioned. It is possible to combine this method with a method of detecting live individuals to determine their position of life within the sediment. Rose Bengal was used by

Frankel (1970) but more recently fluorogenic probes have been used (Bernhard and Bowser, 1996). Care must be taken not to disturb the sample in any way prior to impregnation. The method is particularly suited to microhabitat studies but has not yet been widely used.

- ATP assay: adenosine triphosphate (ATP) is a nucleotide present only in live individuals as it rapidly dephosphorylates after death. This is a specialist technique that has been employed in a few studies (e.g., Bernhard and Reimers, 1991; Bernhard and Alve, 1996) but it must be carried out within a few hours of sample collection.
- Cell ultrastructure: this is another specialist technique based on the fact that cell organelles degrade after death. Bernhard (2000) suggests that this is a good technique to use to determine the status of individuals at the end of an experiment.

2.6   Preservation and fixing

As soon as possible after collection, samples need to be preserved to prevent decay of the protoplasm. There are three commonly used methods:

- Place the sediment in a container, add an equal volume of 70% ethanol, seal and shake to mix thoroughly. Ethanol is a preservative but not a fixative.
- Use formalin buffered with borax (4% solution; preservative and fixative). Formalin is toxic so protective gloves should be worn and it should be used only in a well-ventilated place. For details of how to avoid dissolution of calcareous foraminifera in samples kept for long periods see Maybury and Gwynn (1993).
- Freeze the sample and before processing thaw in an equal volume of 70% ethanol.

2.7   Processing modern sediments for foraminifera

Most ecological studies are based on assemblages > 63 µm, but other sizes are sometimes used. The number of individuals and the number of species will decrease with increasing sieve size and this may give misleading information especially in deep-sea environments (Schröder et al., 1987; Rathburn and Corliss, 1994; Wollenburg and Mackensen, 1998a, b; Alve, 2003).

Always wash the sieve carefully prior to use on each sample. If the sample is of known area/volume and preserved in ethanol or formalin, then:

1   Tip into a 230-mesh (63 µm) sieve, gently wash with tap water to remove clay and silt, pour the residue into a container and **label with**

the sample number (use a paper label and write with pencil as it is inert).

2   Add an equal volume of rose Bengal stain (1 g stain to 1 l distilled water), stir, and allow to stand for one hour.

3   Tip into a 230-mesh sieve and gently wash to remove excess stain.

4   Pour the residue into a container.

Cedhagen (1989) recommends soaking the sample in an alkaline detergent solution at 80–90 °C for 20–40 min. This breaks up the clay aggregates and, if the sample has been preserved in formalin, it eliminates the formalin odour. However, Lehman and Röttger (1997) advise against this as it may destroy some delicate tests.

There is then a choice of picking the sample wet or dry. For the latter:

1   Dry in an oven at <50 °C. If the sample contains a lot of organic detritus, spread it thinly over a large surface area. This can be brushed off when dry (Lehmann and Röttger, 1997).

2   When cool, check the sample under a microscope to determine whether it can be picked or whether it is necessary to concentrate the foraminifera from the sediment using trichloroethylene.

2.8   Foraminiferal separation using trichloroethylene

This procedure must be carried out in a fume chamber and the operator should wear rubber gloves. If trichloroethylene is spilled outside the fume chamber, leave the laboratory immediately and seek help.

Set up a filter funnel on a stand, with a beaker below. Label a filter paper with the sample number (use pencil) and place in the filter funnel. Put the sample into a separate beaker and keep the paper sample label with it:

1   Add trichloroethylene.

2   Stir and allow detrital grains to settle.

3   Decant the scum of foraminifera into the filter paper (make sure there is no scum on the sides of the beaker).

4   Repeat steps 2 and 3, once or twice until no further scum is present; on the final occasion drain off as much trichloroethylene as possible (so that it can be recycled).

5   Place the sample beaker on its side to dry.

6   Remove the filter paper from the funnel and leave to dry; I use a paperclip to hold it together.

7   Return the trichloroethylene to the bottle.

Neither the filter paper nor the beaker with the sediment residue should be removed from the fume chamber until they are completely dry. These two parts of the sample can be stored in labelled glass vials.

2.9   Splitting samples

Wet samples may be split using various devices (Elmgren, 1973; Jensen, 1982; Scott and Hermelin, 1993). Dry samples may be poured through a sediment splitter or divided by quartering.

2.10   Census data, standing crop and biomass

Ecological studies should be based on census data from living (usually stained) assemblages. The dead assemblage in the same sample has the potential to yield information both on the long-term contribution of foraminiferal tests to the sediment and the effects of taphonomic alteration. Therefore, the live and dead assemblages should be treated separately. Some authors use total assemblages in the belief that these more accurately represent what will become fossilised (Scott and Medioli, 1980a). This author believes this is false logic for the total assemblage consists of an arbitrary mixture of living and dead, depending on sample thickness; also the dead assemblage has already undergone taphonomic change whereas the live component has yet to do so (see Murray, 1982, 2000a; Horton, 1999).

For samples of known area and volume, quantitative counts of the live assemblage can be made. Either the whole sample must be picked or a known portion (measured by weight or proportion). Patterson and Fishbein (1989) consider that a count of 300 individuals is necessary for species that comprise ~10% of the sample. A much larger count is required to determine accurately the abundance of rarer species.

- For the dead assemblage and for non-quantitative live assemblages, most authors make a count of >250 individuals.
- It is essential that the *sum of individuals* counted in each assemblage should be stated in data tables regardless of whether the faunal data are given as numbers or per cent, and in addition to any other calculated values such as standing crop per unit volume. This information is essential both to judge the statistical validity of the count and for determining species diversity by methods such as the Fisher alpha index.

Assemblage counts are generally made on dried samples. If a separation has been made using trichloroethylene, it is important to check the sediment residue to see whether any foraminifera remain. If there are more than a few, then it will be necessary to pick them. The sample can be split either with a sample splitter or by quartering. Use a spatula to take a portion of the sample and gently spread it thinly over a gridded picking slide. Then systematically pick the foraminifera with a moistened brush, working cell-by-cell. Some authors prefer to count size fractions separately. If that is done, then the same proportion should be counted from each fraction in order not to introduce bias into the numerical results. The picked individuals are mounted on lightly glued cardboard slides. It seems that glass covers are better than celluloid as the latter may release acid which causes dissolution of calcareous tests (Barbero and Toffoletto, 1996).

For counts made on wet samples preserved in formalin the latter should be removed by sieving and the sample placed in water before examination under the microscope. The individuals may be merely counted or picked out using a pipette, to be stored in vials or well slides. Analysis of wet samples is made if there is a wish to document fragile forms such as allogromiids and poorly cemented agglutinated taxa or to avoid organic-rich samples from drying into paper-like sheets (e.g., Scott and Medioli, 1978). Compared with the dry method, examining samples wet is slow and laborious but it does give the most comprehensive results.

For quantitative samples, the standing crop (of the whole living assemblage or of individual species) is expressed as the number of individuals per unit volume (or area). The proportions of species are expressed as relative abundance (%). It is important to realise that absolute and relative abundance give different information. Peaks of abundance recorded by one method may not correspond with those of the other method. Percentage data are subject to the constant-sum constraint – where the increase in abundance of one species leads to a decrease in others (i.e., they are not able to vary in abundance independently). In Table 2.1 species A has uniform numbers but decreasing

Table 2.1. Relative and absolute abundance.

Species	Number			Percentage		
	Sample 1	Sample 2	Sample 3	Sample 1	Sample 2	Sample 3
A	80	80	80	52	40	33
B	38	50	60	25	25	25
C	35	70	101	23	35	42
Standing crop	153	200	241			

percentages because the standing crop increases from sample 1 to 3. Species B increases in number at the same rate as the standing crop so its relative abundance is uniform throughout. Only species 3 shows an increase in both numbers and percentages. Where many species are present in a sample, the effects are limited, but where only a few species are present the effects may be significant and cause misinterpretation. One way of overcoming the problem is to make a log ratio transformation of the data (Kucera and Malmgren, 1998).

Biomass can be expressed as the wet weight of organic matter, ash-free dry weight, weight of organic carbon, weight of ATP or volume of organic matter (Murray and Alve, 2000a; see Glossary). It is not easy to weigh the soft parts of foraminifera but it is possible to estimate the volume by approximating the shape to an oblate or prolate sphaeroid or a cone (Murray, 1973). However, the protoplasm does not always fill all the chamber space and there are vacuoles.

Although samples may be taken at random, the distribution of organisms is aggregated. Therefore, in order to determine local variability more than one sample must be taken. Confidence limits can then be calculated. For counts of 300 individuals abundances of <1% are close to confidence limits and therefore the latter cease to be meaningful (Buzas, 1990). It has been argued that for dead assemblages, which represent time-averaged accumulations, local variability is smoothed out; therefore a count of even 100 individuals is adequate to record all the important species, although it may not be adequate to decide which of the two main species is dominant. This is because if error bars are added to the calculated abundances they are likely to overlap (Fatella and Taborda, 2002). These authors recommend that all abundances should include the error estimate measured by the binomial standard error. In practice, although most authors would consider the species with the highest abundance to be dominant (without taking error estimates into consideration) few consider variations in abundances of occurrences of less than 5% to be meaningful.

2.11   Analysis of field data

This involves both faunal and environmental data. Although it is customary to undertake some form of quantitative (univariate or multivariate) data analysis, my own view is that it is wise to attempt to interpret the data using brainpower first. That way, anomalous computer-based interpretations can be avoided. In univariate analysis, the attributes of a sample are expressed as a single figure, e.g., measures of species diversity. Multivariate statistics is the analysis of multiple variables, such as species abundance and environmental factors, which may have been measured at several times, locations, etc. It is therefore possible simultaneously to analyse multiple dependent and multiple independent variables. A dependent variable is one that varies with an independent variable (e.g., for age and size, age is independent and size is dependent). Independent variables predict dependent variables; age is a good predictor of size (Kingsford and Battershill, 1998). However, the ease with which complex analyses can now be made makes it ever more important to ensure that the choice of method is appropriate to the problem being solved.

Species richness, species diversity, biodiversity or biological diversity. In many studies an aim is to compare species diversity from one sample to another but it is also of fundamental importance to know something of the pattern of diversity on a much larger scale. It has long been apparent that the number of species present in an area is partly related to the number of individuals sampled. The challenge is how to compare in a meaningful way two samples each having a different number of species and a different number of individuals. If all samples had the same number of individuals then the comparison would be easy. However, knowing the change in the number of species would not provide any information about how the individuals were divided between them. This can be overcome by using an appropriate univariate measure. Several such measures are available (see Hayek and Buzas, 1997) but only some have been applied to foraminiferal data.

The number of species in a sample or study area is termed species richness (S). Diversity indices relate S to the number of individuals (N). In the case of the Fisher alpha index (Fisher *et al.*, 1943) a log series distribution is used to predict the number of species represented by one individual, two individuals, etc. To calculate the log series, it is necessary to estimate values for α and x: $\alpha = \frac{n^1}{x}$, where x is a constant having a value <1 (this can be read from Figure 125 of Williams, 1964) and n^1 can be calculated from $N(1-x)$. With information on S and N, values of α can be determined from Appendix 4 of Hayek and Buzas (1997) and they can also be read from Figure 2.2. Samples of different size are readily compared; if it is desired to know how many species would be present in a smaller or larger sample, the number can be read off Figure 2.2 by tracking the relevant α contour to the chosen number of individuals. For example, a sample of $S=9$, $N=200$, Fisher alpha $=2$ is equivalent to another sample where $S=11$ and $N=600$. Another measure of species diversity is the information function H (Shannon, 1948; Hayek and Buzas, 1997):

$$H = -\sum_i p_i \ln(p_i),$$

where p_i is the proportion of the *i*th species ($p = \%/100$) and ln is the natural logarithm. The contribution to H of each species depends on its proportion p_i

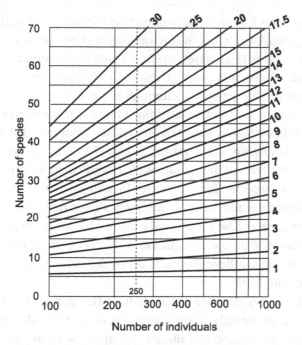

Figure 2.2. Graph to determine values of Fisher α diversity (numbers in bold) from the number of species and number of individuals in an assemblage.

so rare species make little contribution. If all species in the sample are equally abundant (which is never the case), then the maximum value of H will be attained:

$$H_{max} = \ln S.$$

The term evenness (E or J) is used to describe how individuals are divided between species; the greater the dominance of one species, the lower the value of evenness. Several different ways of calculating evenness have been proposed:

$$J = \frac{H}{\ln S} \quad \text{Pielou (1966)};$$

$$E = \frac{e^H}{S} \quad \text{Hayek and Buzas (1997)} \quad \text{where } e = \sim 2.718.$$

The relationship between H and E has now been resolved. The value of H recorded for an assemblage is equal to H_{max} minus the amount of evenness. Buzas and Hayek (1996) expressed this as

$$H = \ln S + \ln E$$

Note that the value of $\ln E$ is always negative.

Not only can the measures S, H and E be applied to single samples but also to the understanding of groups of samples. Buzas and Hayek (1996) proposed that by taking a data series, such as time series, making a cumulative sum of H and S, and calculating H and E for each increment, analysis of the pattern of species distribution could be made (SHE analysis). Although the results can be plotted on graphs they are most readily interpreted from tables. If $\ln E$ remains constant the distribution follows the broken-stick model (see Glossary); if H and α remain constant the distribution follows a log series; if $\ln E / \ln S$ is constant then the distribution is log normal. It is also possible to calculate whether the observed value H_{obs} is that expected for a log series $H_{ls} = \ln \alpha + 0.58$ (where 0.58 is Euler's constant). Therefore, SHE analysis uniquely shows how species distributions and their patterns can be recognised (Hayek and Buzas, 1997). This technique can also be applied to the recognition of biofacies along gradients such as traverses across the sea floor (Buzas and Hayek, 1998) or in a stratigraphic section (Wakefield, 2003).

It must be stressed that no measure of diversity is **the** index (Hayek and Buzas, 1997) because each reveals a different aspect of the relationship between the number of species and the number of individuals. In this book Fisher alpha and H are used individually and in combination (Figure 2.3).

Environmental stress index (Index of confinement).   In marginal marine environments there is a range of conditions from fully marine to either brackish or hypersaline. The foraminiferal assemblages are closely related to the local salinity conditions and this relationship can be used to measure stress on the assumption that it is minimal under marine conditions and increases as conditions deviate from this. The environmental stress index (Ic) was introduced for a situation where three associations were recognised: A, marine; B, more stressed (e.g., brackish); and C, maximum stress (e.g., marsh):

$$Ic = (C/B + C) - A/[(A + B) + 1]/2.$$

If the percentages for the three associations are entered into the equation the results are 0 for fully marine (only association A present) to 1 under the most restricted conditions (Debenay, 1990). The results can also be plotted on a ternary diagram where the three assemblages are the end-members.

Depth of life in the sediment.   Through the examination of subsurface sediment slices, it is possible to determine the average living depth (ALD):

$$ALDx = \sum_{i=0,x} (n_i \times D_i i)/N$$

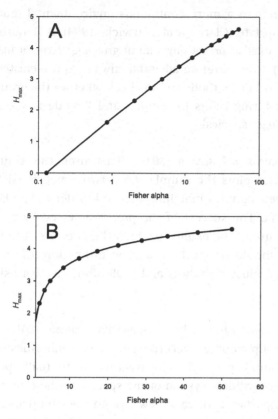

Figure 2.3. The relationship between Fisher alpha and H_{max}. A. When Fisher alpha is plotted on a logarithmic scale it shows a straight line relationship with H_{max}. B. It is more convenient to use an arithmetic scale so that values may be easily read from the graph.

where x is the lower boundary of the deepest sample, n_i is the number of specimens in interval i, D_i the midpoint of interval i, and N is the sum of individuals in all intervals (Jorissen *et al.*, 1995).

2.12 Multivariate analysis

Examples of the application of these methods are given in Dale and Dale (2002).

Cluster analysis. This uses a matrix of similarity coefficients computed between every pair of samples to construct a dendrogram. This method is appropriate where the samples are expected to fall into natural groups. Where the faunal pattern responds to a more continuous environmental gradient, ordination is more appropriate (Clarke and Warwick, 1994). A dendrogram always produces clusters *whether or not any natural groupings occur in the data* (Parker and Arnold, 1999). It is therefore advisable always to use cluster analysis in combination with other methods and to check whether the groupings are 'natural'. Q-mode clustering refers to samples and R-mode to measured environmental variables (e.g., species).

Non-metric multidimensional scaling (MDS). This uses the similarity matrix described above and plots the samples on a two-dimensional 'map' with the distances between samples matching the rank order of similarities from the similarity matrix. The success of the plot is assessed by a 'stress coefficient' that is a measure of the extent to which the two sets of ranks do not agree. Ideally stress should be < 0.1 for a good ordination (Clarke and Warwick, 1994). In reality, cluster analysis and MDS often give very similar groupings.

Principal component analysis (PCA); correspondence analysis (CA). These techniques attempt to group samples, according to their similarities or differences, in multidimensional space. The eigenvectors are mutually perpendicular axes defining the coordinate system of the space and the eigenvalues give a measure of the 'importance' of each new axis to the data (Parker and Arnold, 1999). Principal component analysis is most applicable to R-mode studies (measured environmental variables) because there are relatively few abiotic variables and they are continuously scaled; it is not well-suited to species abundance data with its predominance of zero values (Clarke and Warwick, 1994). Factor analysis is a refinement of PCA that rotates the factors orthogonally or obliquely. Correspondence analysis is similar but the data matrix is transformed in such a way that Q- and R-mode solutions are both equivalent and obtained simultaneously (Parker and Arnold, 1999). Caution must be taken to avoid covariance of factors that may lead to misinterpretation of the foraminiferal data (Loubere and Qian, 1997). Some multivariate-defined groupings in the literature are clearly not natural assemblages.

2.13   Designing and running experiments in field and laboratory

Relatively few experiments on benthic foraminifera have been carried out in the field. Ecosystems are complex: the biotic and abiotic components are intimately associated as they exchange materials and energy; they are temporally and spatially variable, and possibly chaotic (Graney *et al.*,

1995). One of the key problems is to design field experiments that separate natural spatial and temporal variability from the effects of the activity being monitored. Indeed, it is virtually impossible to set up a field experiment where only the activity of interest shows variability as can be seen from the designs below.

Control–impact design. This gives comparison between a *control* site (unaffected by the activity) and an *impact* site (affected by the activity). Examples: predator exclusion (Buzas, 1978; 1982); colonisation of sterile substrate under elevated temperature conditions (Schafer *et al.*, 1996). In the latter there were three variables – colonisation, seasonal variation in temperature and elevated temperature (4 °C above ambient temperature); colonisation and seasonal temperature variations affected both the 'control' and the 'impact' sites whereas elevated temperature affected only the 'impact' sites.

Before–after design. The site is sampled *before* and *after* the activity has taken place thus avoiding problems caused by natural spatial variation. However, such a design runs the risk of recording natural temporal variation rather than impact due to the activity.

Before–after–control–impact (BACI) design. A *control* and an *impact* site are sampled once *before* and once *after* the activity. A modification to this is the BACI paired series (BACIPS). In BACI the test relies on a comparison of the *before* differences and the *after* differences. In BACIPS the mean *before* difference is added to the average state of the *control* site in the *after* period. This yields an estimate of the expected state of the *impact* site in the absence of an impact during the *after* period; this is the null hypothesis (Osenberg and Schmitt, 1996).

Laboratory experiments. A distinction is made between maintaining organisms in the laboratory (i.e., keep them alive for some days or weeks during which they will feed, grow and possibly reproduce) and culturing them (when they go through continuous growth and reproduction over several generations) (Anderson *et al.*, 1991). For laboratory experiments it is normally considered adequate to maintain live forms. No experimental microcosm (containers $< 1\,\text{m}^3$) or mesocosm ($1–1000\,\text{m}^3$) can ever be identical to natural ecosystems so laboratory experiments provide a guide rather than conclusive proof as to what may happen in nature.

All experiments are time-consuming to run. Designing and running realistic and meaningful laboratory experiments requires very careful preparation and planning. The objective needs to be clearly defined. The experimental procedures must satisfy the statistical requirements so that the outcome is valid. The results must be applicable to natural ecosystems otherwise the exercise is meaningless. Basically experiments attempt to isolate the effects of the variable of interest from all the other variables. The experiment should be designed so that there is a *control set* and a *treatment set* each comprised of several replicates. It is not correct to have one control and one treatment and to take multiple samples as 'replicates' from each; these are regarded as pseudoreplicates (Clarke and Warwick, 1994). All replicates should be kept under exactly the same conditions with the exception that the treatment set will have one variable changed by the operator. Covarying factors must be avoided. Examples include changing dissolved oxygen levels (Alve and Bernhard, 1995 one control, one treatment; Moodley *et al.*, 1997, 12 treatments but no control) and disturbance (Ernst *et al.*, 2000, 13 treatments but no controls). The interpretation of experimental data always involves statistical analysis to test whether the control and treatment faunas are the same or different. One of the primary methods used is analysis of variance (ANOVA) as described in standard statistics texts. It is also possible to use the analysis of similarity test (ANOSIM) based on a simple non-parametric permutation procedure, applied to the rank similarity matrix used in the ordination of samples (Clarke and Warwick, 1994).

2.14   Essential data in publications

A surprising number of papers refer to species without giving the author name.

All papers should include a faunal reference list (name used and original name). Now that many journals put supporting information on a linked website, there is no justification for authors not providing their original data. Abundance can be given as numbers or per cent but it is essential that the number of individuals counted in each sample is given. Where known, the sample volume and area should also be listed. With this information, readers can then carry out further data analysis if they so wish.

2.15   Selection of data used in this book

The ecological studies discussed in this book are based on living (stained) assemblages. In this way problems of taphonomic bias are avoided (see Chapter 9). Most data are for >63 μm assemblages unless otherwise stated. It was my original intention to include only samples based on a count of >100 living individuals but it soon became clear that a large part of the literature would be excluded if that was the case. Instead I have included only samples of >50 individuals in order to eliminate statistically unrepresentative data. Where authors have used total data (living plus dead), if the counts are large enough to give a living assemblage of at least 50 individuals, the living data have been extracted. Papers dealing only with data for dead assemblages or samples that have not been preserved in alcohol or formalin and stained for protoplasm are listed in Web Appendix 1 (www.cambridge.org/9780521828390) for completeness but are not otherwise discussed.

In order to make meaningful comparisons between datasets produced by different authors it has been necessary to use a single taxonomic scheme, i.e., by using a consistent binomial name for each species (Web Appendix 2). This list contains all the species mentioned in the book but it does not include synonyms. However, it has clearly not been possible to check the accuracy of identifications in every paper. Although many species and subspecies of *Ammonia* have been described, most authors use *Ammonia beccarii* as a group term but this is clearly unsatisfactory. Until 2004, the most recent review was that of Walton and Sloan (1990) who undertook a global study based on morphology. From this they suggested that *Ammonia beccarii* could be divided between three ecophenotypic morphotypes that they regarded as formae (*beccarii, tepida, parkinsoniana*). Now, using a combination of molecular genetics and morphological analysis (based on 37 external characters of the test), Hayward *et al.* (2004) recognise 13 molecular types, some of which can be related to traditionally named taxa. They estimate that there are probably 30–40 *Ammonia* extant species. They further suggest that the use of *Ammonia beccarii* should be restricted to the large forms typical of some localities in the Mediterranean. This important study identifies the problem but does not provide the basis on which published records of *Ammonia 'beccarii'* can be reassigned to another species. Therefore in this book such records are termed *Ammonia* group.

Faunal counts provide data on numbers of individuals (sometimes with reference to a standard area or volume of sea-floor sediment), and species richness. From these, various measures have been calculated: number of specimens counted, number of individuals $10\,cm^{-3}$ of sediment (usually for the top centimetre) to give the standing crop, per cent of each species in the live assemblage to identify the dominant (most abundant) and subsidiary ($\geq$10%) taxa, proportions of wall structures, proportions of epifaunal/infaunal taxa where appropriate, and species diversity indices (calculated only on samples with $\geq$100 individuals). The details of multivariate analysis of data sets have generally not been included as they often add little to the interpretation. The data and patterns of distribution on both local and global scales are presented in Chapters 4–7 and summarised in Chapter 8. The ecological details of selected species and genera are listed in the Appendix. Tables of data on live (stained) assemblages are given on the linked website and labelled with the prefix WA- (Chapter 4, WA-1 to WA-104; Chapter 5, WA-111 to WA-173; Chapter 6, WA-181 to WA-185; Chapter 7, WA-191 to WA-219). Dominant species are highlighted in yellow and subsidiary species with abundances >10% in green. For Chapters 4 and 5 there are more comprehensive versions on the linked website of the figures showing the distribution of taxa.

3

Aspects of biology and basic ecology

3.1   Introduction

The aim of this chapter is to set the scene for examining and interpreting ecological distribution patterns in the major environments. Therefore it is appropriate to discuss relevant aspects of biology and ecological concepts. For definitions of basic terms see the Glossary.

3.2   Biology

The presence of the test in foraminifera makes the group attractive to geologists because it provides a fossil record (Cambrian to Recent) but to biologists the test obscures most of the soft parts so relatively few studies have been carried out on foraminifera compared with naked protist groups. Nevertheless, important and interesting observations were made on their life activities, particularly from the late-nineteenth to mid-twentieth century, by a small group of people who had the patience to sit at the microscope and make observations for long periods (e.g., Lister, Schaudinn, Schultze, Heron-Allen, Jepps, Myers and Arnold). In recent decades there have been great advances in both observation and understanding due to improved technology (transmission and scanning electron microscopy, introduction of new fixation techniques, radiotracers, stable isotopes, cell physiology, microanalytical methods, etc.). However, although there is now a greater understanding of the way foraminiferal cells operate, *there is still very little understanding of the physiological response to varying environmental conditions.* That surely holds the key to understanding the niches of individual species and the interpretation of patterns of distribution and should be a major topic of research in the years ahead.

Figure 3.1. Digital image of *Marginopora vertebralis*, a larger foraminiferan with endosymbionts: an example of a giant cell.

Extensive reviews of aspects of biology are given in Lee and Anderson (1991a), Goldstein and Bernhard (1997), Sen Gupta (1999), Lee and Hallock (2000) and Cedhagen *et al.* (2002). These should be consulted for more complete bibliographies on this topic. In this chapter, only those aspects of biology most relevant to ecology are discussed.

3.2.1   The cell

Although benthic foraminifera are unicellular protistans some reach a size of several centimetres and these can therefore be considered as giant cells (Figure 3.1). A distinction is often made between two types of cytoplasm: cell-body cytoplasm (otherwise called endoplasm, intrathalamous cytoplasm or the sarcode) and reticulopodia composed of pseudopodia (Bowser and Travis, 2000); reticulopodia are otherwise called ectoplasm, extracellular cytoplasm or rhizopodia. Observation of cytoplasmic streaming shows that there is a constant exchange of smaller organelles between the cell body and the reticulopodia (Anderson and Lee, 1991) so there is no obviously fundamental difference between the two (Alexander and Banner, 1984) and if a foraminiferan is disturbed it withdraws the reticulopodia into the test. The granuloreticulose cytoplasm of foraminifera is fundamentally different from that of all other protozoans in that it contains granules and the pseudopodia form a net (hence reticulopodia). It provides the major biological basis for separating the foraminifera as a distinct taxonomic group of protozoans.

3.2.2   Cell-body cytoplasm

This is less clearly visible because of the test but in forms with thin hyaline walls it is possible to see active cytoplasmic streaming. When the organism is well nourished, the cytoplasm is dense but during periods of food shortage it becomes highly vacuolated. It is probably for this reason that deep-sea foraminifera take up rose Bengal stain less intensively than shallow-water forms.

The major organelles are nuclei, ribosomes, Golgi bodies, lysosomes, digestive vacuoles, peroxisomes and mitochondria (Lee et al., 1991b). Messenger RNA (ribosomes) synthesises proteins that directly or indirectly control the structure and function of the cell components. Rough endoplasmic reticulum (RER) contains ribosomes whereas smooth endoplasmic reticulum (SER) does not; RER is responsible for protein synthesis and SER for lipid synthesis and other biosynthetic activities. The nearly clear aqueous fluid surrounding the endoplasmic reticulum is rich in enzymes and small macro-molecules, which are essential elements of the biochemical reactions that sustain life. The Golgi bodies generate membrane-bound vesicles, some of which secrete primary digestive enzymes (primary lysosomes) and others mucoid substances composed of carbohydrates and proteins. The latter include adhesives that are used by the pseudopodia to gather up bacteria and algal cells during feeding. The engulfed food is held in digestive vacuoles (phagosomes) formed by the reticulopodia. As these vacuoles are moved towards the aperture, they are reduced in size by the expulsion of water, and the pH is lowered through fusion with acid-containing vesicles (acidosomes). Peroxisomes convert waste products into potentially useful metabolic compounds. The waste from feeding (stercomata, xanthosomes) is transported in vacuoles from the cell body into the reticulopodia and then is discharged (Travis and Bowser, 1991) although in some tubular agglutinated forms the stercomata are stored (Mullineaux, 1987). Xanthosomes are conspicuous because of their amber colour. Mitochondria are the major centres of energy conversion that lead to the reduction of oxygen to water and the generation of ATP (energy-storage molecule). They are widely distributed throughout the cytoplasm. In the low oxygen environment of the California (USA) borderland basins, mitochondria are low in density and clustered at the inner terminations of wall pores of hyaline foraminifera (Leutenegger and Hansen, 1979). However, the pseudo-podial mitochondria probably carry out most of the respiration.

3.2.3 Reticulopodia

The most obvious parts of the living animal are the thread-like cyto-plasmic pseudopodia extending from the aperture. Although these can be seen using optical microscopes, progress in understanding their form and function has come from advanced techniques of fixation and electron microscopy (Bowser and Travis, 2000). Pseudopodia are the means by which the organism is in contact with its surroundings. They carry out many of the life processes: respiration, attachment or movement, feeding, test construction, and they are involved in reproduction (Travis and Bowser, 1991). Pseudopodia are prone to breakage in currents (Lipps, 1982) and some foraminifera make use of spines or elongate detrital particles such as sponge spicules to support them. If portions of pseudopods become detached, the parent organism can coalesce with them (Cushman, 1922). The pseudopods of the reticulopodia form a network (reticulum) of constantly changing form, with strands dividing and joining, and they bear granules (commonly mitochondria) which show that cytoplasmic flow occurs both away from and towards the aperture. The rate of transport of mitochondria in the reticulopodia increases with temperature and is thought to indicate an increased metabolic rate (Cedhagen and Frimanson, 2002).

Near the aperture the pseudopodia may form thick bundles but in the reticulum the strands are fine (2 μm in diameter). There are three major types of movement: extension, bending to form the network and withdrawal; incessant reticulopodial movement is essential for the function of the network. The reorganisation of the reticulopodia is considered to be autonomous and it responds to environmental stimuli such as the adhesiveness of the substratum (Travis *et al.*, 2002). If the pseudopodia of two individuals come into contact, they immediately repel one another (Cushman, 1922). The reticulopodia produce mucopolysaccharides (glycosaminoglycans) that act as an adhesive and also protect the cell from toxins (Bresler and Yanko, 1995). Network activity scours the substrate for food, such as bacteria or algae, and detrital grains (Travis and Bowser, 1991). This material is transported as aggregates towards the aperture where it may form a bolus. In *Haynesina germanica* the aggregates are passed through the lines of tubercles in front of the aperture and disaggregated prior to selective ingestion (Banner and Culver, 1978). The mitochondrial granules are essential for oxygen respiration and may be responsible for transporting metabolic energy (ATP) within the reticulopodia (Goldstein, 1999).

3.2.4   Life cycle

Around 30 species have been studied (see reviews by Lee *et al.*, 1991b; Goldstein, 1999). The basic pattern is an alternation of sexual and asexual generations but there is considerable variation in detail and it is more varied than that of other protozoans. The production of biflagellate gametes is common to representatives of the suborders Textulariina, Miliolina and Rotaliina; from this Goldstein (1997) concluded that the basic pattern of reproduction has shown little evolution through the long history of the foraminiferal record. In

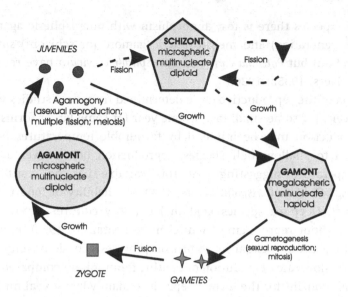

Figure 3.2. Schematic representation of the classic life cycle (alternation of generations) of benthic foraminifera (based on Lee et al., 1991b; Goldstein, 1999). Entire arrows: basic cycle; broken arrows: alternative multiple fission.

species where the alternation of generations is obligatory, reproduction in the agamont involves meiosis and the resulting young are uninucleate and haploid. The gamonts produce gametes that form zygotes and grow into agamonts (Figure 3.2, solid arrows). Common examples are *Elphidium crispum* (Lister, 1895; Schaudinn, 1895; Jepps, 1942) and *Patellina corrugata* (Myers, 1935a). Where the life cycle is a facultative alternation of generations, in some cases there are successive asexual cycles (reproduction by multiple fission as in *Ammonia tepida*, Bradshaw, 1957; Schnitker, 1974; Goldstein and Moodley, 1993; Stouff *et al.*, 1999c). The agamont starts as a diploid zygote. It may undergo meiosis and multiple fission to form haploid juveniles which grow into uninucleate gamonts; or it may undergo multiple fission to produce a second schizont that may produce several successive asexual generations (Figure 3.2, dashed arrows). This is called biological trimorphism and occurs in *Amphistegina gibbosa* (Dettmering *et al.*, 1998; Harney *et al.*, 1998) and *Heterostegina depressa* (Röttger, 1990; Röttger *et al.*, 1990a). Another strategy is to omit one generation (apogamy), e.g., the sexual phase in *Fissurina marginata* (Le Calvez, 1947) and *Spiroloculina hyalinea* (Arnold, 1964). The size of the proloculus varies between generations because in sexual reproduction there is just fusion of two gametes (small, microspheric, proloculus) whereas in asexual reproduction the juveniles inherit a nucleus, some of the parent cytoplasm and sometimes symbionts as well (larger, megalospheric proloculus; Hottinger,

2000). In some species there is test dimorphism with microspheric agamont larger tests (B generation) and megalospheric gamont (or schizont) smaller tests (A generation) but *Patellina corrugata* and *Spirillina vivipara* have reversed dimorphism (Myers, 1935b, 1936).

The duration of the reproductive cycle determined from field studies varies from a few weeks for some small taxa to one year for some larger forms. The onset of reproduction may be initiated by favourable temperatures but in some intertidal to shallow-shelf species, reproduction occurs over a wide range of temperature suggesting that this variable has no control on the process (e.g., *Elphidium crispum*, Myers, 1943). Availability of food may be more important. Deep-sea species kept in laboratory cultures survived for several years without reproducing (Hemleben and Kitazato, 1995). From an evolutionary aspect, only sexual reproduction provides genetic novelty. This may explain the slow rate of evolution of benthic foraminifera compared with their planktonic cousins for the former reproduce mainly by asexual multiple fission and the latter entirely by sexual reproduction. However, experimental studies of reproduction in four species of *Glabratella* suggest that each comprises small, reproductively isolated populations because individuals from distant localities (500 km) cannot interbreed (Kitazato *et al.*, 2000b).

From an ecological point of view, for those forms with symbionts asexual reproduction provides a means of passing symbionts to the next generation; the juveniles comprise not only a nucleus but also some cytoplasm with all its organelles (Röttger, 1974, 1981). Agamonts resulting from sexual reproduction have to acquire symbionts from the environment. Normally, the parent test effectively dies during reproduction as all the cell contents are used to make either gametes or juveniles. However, in a few cases the parent survives to reproduce again (Arnold, 1955; Sliter, 1965; Röttger, 1978; Stouff *et al.*, 1999c). In some instances, there is partial or total dissolution of the parent test to provide Ca^{2+} ions for the formation of the juvenile tests (e.g., Nyholm, 1961; Stouff *et al.*, 1999c). Thus, when corroded tests are found in the sediment, the damage may be due either to syn-reproductive dissolution or postmortem dissolution in the sediment.

3.2.5   Production of tests and population dynamics

The annual production of tests is the number of tests contributed to an area of sea floor during one year. It is controlled by the population dynamics (births and deaths) of each species. It will always be greater than the standing crop measured at any one time. It is difficult to estimate annual production if the species reproduce continuously or more than once per year so that growth stages (cohorts) cannot be separated. The method (of summing the peak standing crop values to give annual production) proposed by Murray (1967a) and applied to numerous shallow-water examples by Murray (1991, Table 4.6) has subsequently been shown to be unreliable, as the results depend on the frequency of sampling (Murray and Alve, 2000a). At present there does not seem to be a reliable way of estimating annual production through analysis of living assemblages. However, it is possible to estimate the rate of accumulation of tests in cores where reliable dates are available (e.g., Alve, 1996).

3.2.6 Dispersal

There are one active and three passive ways of dispersing benthic foraminifera (see review by Alve, 1999). Active dispersal is via self-locomotion but although this may operate over short distances the speed of movement and the lack of direction make it an unlikely method of dispersal on a wider scale. The passive methods involve getting material into suspension in the water column so that it can be transported. This can be as gametes, zygotes, propagules or juveniles; a temporary planktonic juvenile phase; or entrainment of individuals from the sediment by waves, currents or biotic activity. Although gametes and zygotes may not survive long periods of transport, propagules may be important because they appear to remain dormant for a few months and therefore could be transported over greater distances (Alve and Goldstein, 2002, 2003). Forms with a temporary planktonic phase are rare and the few examples have special float chambers (*Tretomphalus, Tretomphaloides, Millettiana, Cymbaloporetta, Rosalina*). There are some records of living benthic foraminifera in surface-water plankton samples (Arnold, 1964; Lidz, 1966; Loose, 1970) so this may be a realistic mechanism. However, it has been observed that foraminifera commonly attach themselves to the sea floor or to sediment grains with their reticulopodia (see above). Indeed, some forms are hydrodynamically shaped to withstand erosion from the sea floor (e.g., *Discobotellina biperforata*; Almers and Cedhagen, 1996). It is much easier to suspend empty tests than living individuals. Nevertheless, in a colonisation experiment in the Baltic Sea, new habitats suspended 1.5–5.0 m above the sea floor were colonised by individuals too big to be considered as juveniles (diameter 200 μm) so they may have been dispersed through resuspension (Wefer and Richter, 1976). In other experiments carried out at bathyal depths off Japan, two boxes with substrates for colonisation were placed on the sea floor and one was covered with a 90 μm mesh to exclude adults. The open jars were colonised but the netted jar was not. From this Kitazato (1995) concluded that the transported colonisers were larger forms transported either by benthic storms or through bioerosion by fish. Also, one individual of *Alveolophragmium advena* 1 mm in diameter colonised a jar containing glass beads and after one year had added chambers that incorporated glass beads. Thus this form was already large when it was transported into the jar. The importance of fish throwing foraminifera into suspension has been demonstrated in flume experiments (Palmer, 1988). The sediment is passed through the gills and disturbed by the movements of the tail and this causes rearward transport of foraminifera.

Larger marine animals and ships are frequently encrusted with smaller marine life and might provide a means of dispersal (Myers, 1936; Levin et al., 2001). It is thought that the muddy feet of wading birds may transport foraminifera from one estuary to another (Haake, 1982) or even into inland saline lakes (Resig, 1974; Haake, 1982; Almogi-Labin et al., 1992; 1995). The distances involved may be as great as 2000–3000 km (Patterson and McKillop, 1997; Boudreau et al., 2001). Ships discharging ballast water derived from a distant source may be responsible for the introduction of sediment-dwelling foraminiferal species into new areas (Witte, 1994). A remarkable occurrence of living foraminifera (miliolids, *Textularia* and *Nodosaria*) was recorded from saline groundwater in the Kara-Kum desert in Turkmenistan >350 km from the Caspian Sea (Brodsky, 1928).

3.2.7   Opportunists/generalists

Opportunistic species respond rapidly to short-term favourable circumstances by increasing their numbers whereas generalists do not respond in this way. A good example of an opportunistic species is infaunal *Stainforthia fusiformis* (Alve, 2003). It occurs in hydrographic frontal regions (Scott et al., 2003), physically disturbed sediment (Alve and Murray, 1997) and in ephemerally dysoxic/anoxic basins in fjords and seas over shelves and slopes. It can withstand short periods of anoxia, is a facultative anaerobe and husbands chloroplasts even though it lives deeper than the photic zone. It can rapidly recolonise formerly anoxic sediments (Alve, 1994; Finger and Lipps, 1981) and sediments containing $CuCl_2$ (Alve and Olsgard, 1999). The controlling factor seems to be its ability to cope with environmental stress in areas subject to rapidly changing conditions (Alve, 2003).

3.2.8   Mode and position of life

Active organisms are free living while inactive forms are attached to a substrate. The two positions of life are **epifaunal**, on the surface of the substrate, or **infaunal**, in the sediment. Epifaunal individuals live on soft substrates such as sediment or firm substrates such as animals, shells, rocks and plants (termed epiphytic or phytal). Epifaunal taxa may be sessile (*attached immobile* of Sturrock and Murray, 1981), clinging (*attached mobile* of Sturrock and Murray, 1981; A and B of Kitazato, 1981), or free living. Attachment is made using organic glue (glycosaminoglycan, a mucopolysaccharide; Langer, 1993, proposed calling this glycoglue). Some organisms not only secrete glycoglue beneath the test but also surround the test with it roughly to half their height. Video observation of movement has shown that some supposedly sessile taxa (*Gavelinopsis praegeri*, *Planorbulina mediterranensis* and *Rosalina floridana*) have periods of movement (Gross, 2000). Some authors have extended the term epifaunal to include infaunal taxa living in the top 1 cm of sediment (Corliss, 1991). *Rhabdammina* is a tubular form which has an erect orientation with the lower part buried in sediment and the upper part in the water column. It digs a burrow surrounded by a moat (Kaminski and Wetzel, 2004a). Some epifaunal taxa (*Gypsina globulus* and *Cribrobaggina reniformis*) are thought to be endolithic as they bore into carbonate and other material and therefore cause bioerosion (Vénec-Peyré, 1985a; 1993; 1996).

Infaunal taxa have been recorded living (stained) down to 60 cm (limit of sampling) below the sediment surface in marshes (Goldstein *et al.*, 1995) but in the majority of environments most live in the top few centimetres. Some surround themselves with a cocoon of detrital particles (Richter, 1961). Infaunal taxa may be attached, clinging or free. The free forms generate tiny burrows as they move through the sediment (Kitazato, 1994; Gross, 2002). It has been calculated that bioturbation by 100 infaunal foraminifera would turn over 420 cm^{-3} yr^{-1}, which is equivalent to the top centimetre of sediment four times per year (Gross, 2002). Even in laminated sediments from the dysoxic environment of the Santa Barbara Basin, Pacific USA, meiofaunal burrowing causes bio-irrigation of the near-surface sediments (Pike *et al.*, 2001). In laboratory tanks, epifaunal *Cibicidoides pachyderma* and *Ammodiscus anguillae* move in and out of the sediment (Bornmalm *et al.*, 1997). Infaunal bathyal taxa kept in experimental tanks migrated out of the sediment and up the tank walls to become epifaunal (Gross, 2002). *Elphidium crispum* normally lives on sand or on algae; in an experiment where it was placed on clay individuals showed an avoidance reaction (Murray, 1963). These observations show that the position of life is controlled by the ambient environmental conditions and that the terms infaunal and epifaunal apply only loosely in some instances.

Foraminifera can move ahead of a slowly advancing redox boundary as shown experimentally by Alve and Bernhard (1995) and Gross (2000) but if the environmental change is rapid and beyond their range of tolerance, they are in serious danger of being overwhelmed and killed off. A degree of tolerance to low oxygen and the presence of H_2S is a prerequisite for survival in soft sediments (Moodley *et al.*, 1998b). In some cases, they may be able to withstand adverse conditions (e.g., anoxia) for a period of weeks or months by remaining in the protection of their test either in a state of dormancy or as cysts (Gross,

2000; Moodley *et al.*, 1997). In disturbance experiments where the sediment was mixed and then allowed to stand for 22 days, the epifaunal taxa responded rapidly and returned to the surface whereas the infaunal taxa responded slowly or scarcely at all (Ernst *et al.*, 2000). This perhaps indicates, not unexpectedly, that epifaunal taxa are more sensitive than infaunal taxa to the redox front.

3.2.9   Behaviour

Movement. On a hard, flat surface a 'grip-and-tug' mechanism is employed (Travis and Bowser, 1991). The individual holds itself upright with its pseudopodia firmly attached to the substrate and with the aperture facing downwards. One set of pseudopodia contracts and pulls the test, the pseudopodia then move away, anchor themselves and repeat the process. In this way the individual moves in small increments but the process is normally very slow ($50 \, \mu m \, min^{-1}$, Winter, 1907; average 8–$82 \, \mu m \, min^{-1}$, Kitazato, 1988; 7–$17 \, \mu m$ min^{-1}, Wetmore, 1988; 1–$16 \, \mu m \, min^{-1}$, Bornmalm *et al.*, 1997; 8–$11 \, \mu m \, min^{-1}$ for bathyal *Laticarinina*, Weinberg, 1991; 20–$25 \, \mu m \, min^{-1}$ for bathyal forms, Gross, 2000) and there is not normally a sustained direction of movement. Also, the rate of movement is partly controlled by temperature and the need to search for food (Gross, 2000). The speed of *Quinqueloculina impressa* is inversely proportional to depth in the sediment and is controlled by the effective overburden pressure if individuals are larger than the interstitial spaces (Severin, 1987; Severin and Erskian, 1981). Some individuals make vertical burrows to escape to the sediment surface while others make near-horizontal burrows in the surface $1 \, cm$ of sediment (Severin *et al.*, 1982). Tubular forms such as *Bathysiphon* and *Siphonammina* can move at up to $1 \, cm \, h^{-1}$ and they are responsible for bioturbating the sediment (Geslin *et al.*, 2004). In *Iridia*, the organism can leave the test and move around independently (Cushman, 1922). Where foraminifera have moved across a surface they commonly leave a trail of plasma membrane fragments which may be mistaken for mucus (Travis and Bowser, 1991). It is not known whether the same grip-and-tug process operates for infaunal taxa as traction by pseudopodia may move surrounding detrital particles rather than the foraminifera. However, infaunal species certainly move and leave microscopic burrows, especially in fine-grained sediment (speed generally $<100 \, \mu m \, min^{-1}$, Kitazato, 1988; 4 to $8 \, \mu m \, min^{-1}$, Wetmore, 1988; <2 to $>5 \, \mu m \, min^{-1}$ for bathyal forms, Gross, 2002).

Geotaxis. The response of an organism to gravity (Myers, 1943; Manley and Shaw, 1997). Although foraminifera commonly climb onto raised substrates, the cause may not be in relation to gravity but another parameter such as food or light.

Phototaxis. The response of an organism to light. Positive: *Cibicides* and *Elphidium*; negative: *Ammonia, Astrononion, Bolivina, Buliminella, Fissurina* and *Uvigerinella* (Kitazato, 1981). In some cases the intensity of light controls whether the response is positive or negative as it is concerned with obtaining suitable illumination for the symbionts (Zmiri *et al.*, 1974; Lee *et al.*, 1980). Some epifaunal foraminifera hide in small marine plants in shallow waters in tropical environments where the light is intense; others gather plant fragments and detritus to provide shade. *Amphisorus* swept off plants in the Red Sea crawl back to their hosts within a few days and this has been interpreted as a response to light regulation, specifically to reduce exposure to bright light (Reiss and Hottinger, 1984). Positive phototaxis in *Elphidium crispum* causes it to climb up onto algae or other raised substrates prior to reproduction and this may aid dispersal of gametes or juveniles (Myers, 1943; Manley and Shaw, 1997).

Squatting. Use of a test made by a previous occupant as a shelter. A soft-shelled foraminiferan was placed in a tank together with an empty *Quinqueloculina* test and within one hour had occupied the test as its new home (Moodley, 1990a). Some soft-shelled benthic foraminifera occupy planktonic tests; they are easily recognised from their dark stercomata (Hemleben and Kitazato, 1995).

3.2.10 Food and feeding

Primary production is carried out by marine plants, mainly microscopic algae (phytoplankton and microphytobenthos), which produce organic matter through photosynthesis with sunlight as the energy source. In the oceans, primary production takes place in the phytoplankton in the surface euphotic zone. In marginal marine settings, especially temperate intertidal zones, most primary production is from benthic algae (microscopic unicellular; macroscopic multicellular seaweeds). In the intertidal zone, the microphytobenthos (composed mainly of diatoms, euglinid flagellates and cyanobacteria) forms transient biofilms that move through the sediment in response to tidal movement and exposure to light. On tidal marshes and in some shallow lagoons, higher plants play a key role. Secondary production is based on the consumption of primary producers by other organisms including herbivores, detritivores, carnivores and decomposers (bacteria, fungi, etc.). Because of the current interest in food as a fundamental limiting resource controlling benthic foraminiferal distribution patterns it is essential to know the feeding preferences of different species. Also, foraminifera are not only commonly abundant but are also the most abundant organisms in some marine environments (e.g., deep sea) so they may be important in recycling organic carbon (Gooday et al., 1992a).

The reticulopodia play a key role in both gathering food and in digestion. They scour surfaces gathering potential food and are even capable of pulling up pieces of biofilm, which coats most sediment substrates (Jepps, 1942; Bernhard and Bowser, 1992). Individual *Amphisorus hemprichii* normally have separate random feeding territories and if the pseudopodia of two individuals touch one another they take avoidance action (Lee et al., 1980). It is known that many epifaunal benthic foraminifera also pick up organic and inorganic detritus (Murray, 1963; Lee, 1980) to cover the test and reticulopodia or to form a bolus around the aperture. This may resemble feeding but it cannot be assumed that everything picked up is for this purpose. Gathering food around the aperture and digestion may occur at quite different times (Faber and Lee, 1991a). In feeding experiments using polystyrene microspheres, it was found that although they were picked up by the reticulopodia of allogromiids they were not ingested, whereas bacteria were rapidly taken into food vacuoles (Bowser et al., 1985). The authors suggest that the process of taking bacterial food into a food vacuole is triggered by specific molecules present in the outer envelopes of the prey. Presumably the same might apply to other potential food resources. Feeding is affected by the age of the food organism (various algae), the concentration of food and the stage in the life cycle of the foraminifera (Lee et al., 1970). Consumption is directly proportional to the concentration of food near the pseudopodial web. Digestion may take place in the reticulopodia, in the cytoplasmic mass around the aperture and within the cell-body cytoplasm (Jepps, 1942; Lee et al., 1991c). However, in those cases where the aperture is small or just a row of pores (as in *Elphidium crispum*) no evidence of food particles is seen in the cell-body cytoplasm so digestion must take place entirely in the reticulopodia (Jepps, 1942). In larger foraminifera with symbionts, acid phosphatase activity indicates that digestion takes place in food clumps around the individuals and in the final chamber; whereas in *Discorbis*, which lacks symbionts, it also occurs in the outer chambers (Faber and Lee, 1991b). *Notodendrodes* without food vacuoles in the body cytoplasm may take up dissolved organic material (DeLaca, 1982; DeLaca et al., 1981; 2002). Some species form a secondary test around themselves that Cedhagen (1996) considers to be a camouflage strategy although other functions, including feeding, are also possible. Experiments using ^{13}C-labelled food show that uptake is rapid and the foraminifera play an important role in carbon recycling (Moodley et al., 2000). However, foraminifera may have low net production efficiencies and spend a lot of energy on movement, much of which is associated with feeding (Hannah et al., 1994). On the Vøring Plateau,

Norwegian Sea, *Cribrostomoides subglobosus* accounted for <0.2% of sediment community respiration but after feeding this increased to 1.8% (Linke *et al.*, 1995). Nevertheless, the metabolic rate of unicellular organisms is lower than that of multicellular organisms of equal mass.

Many different feeding strategies have evolved (see reviews by Lee, 1980; Lipps, 1983; Goldstein, 1999). Until recently, determination of feeding mode was mainly circumstantial; if the organism gathers a particular food item it was assumed to feed on it. Confirmation that an item is being eaten can come from TEM studies of the cytoplasm, from feeding experiments and from bio-marker analysis (which shows what has been eaten). Without such checks, the interpretations are speculative. The feeding strategies of agglutinated for-aminifera listed by Jones and Charnock (1985) were acknowledged by them to be a mixture of observation and assumption. Caution is required in using their categories. For most foraminiferal species there is no reliable information on choice of food.

Herbivory and bactivory. Examples: *Ammonia, Astrammina, Bolivina, Discorinopsis, Elphidium, Glabratella, Notodendrodes, Quinqueloculina, Rosalina, Rota-liella, Spiroloculina*. Herbivory is restricted to the euphotic zone. Active herbi-vores gather algae and bacteria with their reticulopodia. Passive herbivores are sessile epifauna that gather food from around the site of attachment. Epifau-nal, sessile or clinging *Rosalina globularis* is active when food is scarce and passive when it is abundant (Sliter, 1965). Most littoral forms feed on pennate diatoms and small chlorophytes (Arnold, 1974; Lee, 1980; Anderson *et al.*, 1991) and bacteria also seem to be an essential requirement of their diet (Muller and Lee, 1969; Muller, 1975). But whereas allogromiids feed on bacterial films, neither calcareous nor agglutinated forms were found to do so in experiments (Bernhard and Bowser, 1992). Glycosaminoglycans are used by bacteria and fungi as a source of nutrients. Therefore foraminifera secreting this substance (*Textularia bocki* and *Quinqueloculina ungeriana*) may do so in order to promote the growth of bacteria which they can then use as food (Gerlach, 1978; Langer and Gehring, 1993). Algal blooms are patchy in occurrence and this may be a prime cause of spatial patchiness in the abundance of for-aminifera as they respond to the stimulus of a large supply of food (Buzas, 1965, 1969; Lee *et al.*, 1969; 1977; Matera and Lee, 1972). Experiments on *Ammonia 'beccarii'* show that because they consume large amounts of bacteria and algae, they may completely deplete the sediments of these resources. This affects the distribution and abundance of a harpacticoid copepod suggesting foraminifera – copepod amensalism (Chandler, 1989). Although specific algae may stimulate population growth, an excess of food sometimes depresses it

(Lee *et al.*, 1966). Some taxa with symbionts also need to feed (e.g., *Peneroplis*, Faber and Lee, 1991a).

Passive suspension feeding. Examples: *Astrorhiza, Fontbotia, Miliolinella, Planulina, Rupertia, Saccorhiza.* Foraminifera spread their reticulopodia in the water column but they cannot draw water through the net. They rely on water currents to bring food to them and as currents increase in velocity with height above the sea bed, they need to be raised up above the substrate. They are epifaunal and sessile on hard substrates or rooted in soft sediment with the test held erect and the aperture positioned well above the substrate (Lipps, 1983; Schönfeld, 2002). They are known from shelf and bathyal environments.

Detritivores. Examples: in sediment – *Globobulimina, Hyperammina, Reophax, Uvigerina*; phytodetritus – *Adercotryma, Alabaminella, Cribrostomoides, Epistominella, Fursenkoina, Globocassidulina, Oridorsalis, Tinogullmia.* Forms that colonise phytodetritus also live in small numbers in the surface sediment. Foraminifera feeding on organic detritus may be selective or unselective. They are especially abundant in the surface layer of sediment and although this mode of feeding can take place in almost any environment, it must be the main method in areas of the sea floor below the euphotic zone. In seasonally influenced environments there is episodic input of phytodetritus and in the deep sea it forms a layer above the sediment. This provides a rich food source for opportunistic species that move out of the surface sediment to colonise it (reviewed by Gooday, 2002). Degraded organic detritus and its associated bacteria are important for infaunal taxa (Heeger, 1990; Goldstein and Corliss, 1994). In the subsurface sediment the food resources are more stable and less affected by seasonality (Rudnick, 1989).

Carnivory. Examples: *Allogromia, Amphistegina, Astrammina, Astrorhiza, Bathysiphon, Elphidium, Floresina, Glabratella, Myxotheca, Nemogullmia, Patellina, Peneroplis, Pilulina, Spiculosiphon.* The reticulopodia can be spread as a trap to catch small animals with their adhesive pseudopodia (Jepps, 1942; Bowser *et al.*, 1992). The adhesive secretion is not thought to be toxic (Langer and Bell, 1995). The prey may die of exhaustion once captured and digestion is normally in the reticulopodia outside the test (Buchanan and Hedley, 1960; Banner, 1978). It was speculated that higher speeds of movement might be characteristic of carnivorous foraminifera (Gross, 2000). *Floresina amphiphaga* attacks *Amphistegina* (Hallock and Talge, 1994) and possibly *Globorotalia menardii* (Nielsen *et al.*, 2002).

Dissolved organic material. Examples: *Astrammina, Notodendrodes.* In theory a cell with a large surface area should be able to take up dissolved organic material from its ambient environment (Lipps, 1983).

Omnivory. Examples: *Astrammina, Astrorhiza.* Both Lee (1980) and Lipps (1983) consider that many benthic foraminifera are opportunistic and feed on what is available, whether animal or plant. However, that remains to be proved as experiments show that certain taxa are highly selective in their choice of food (Ward *et al.*, 2003).

Parasitism. Among the forms said to be ectoparasitic are *Fissurina marginata* (as *Entosolenia*, on *Discorbis*, Le Calvez, 1947), *Fissurina submarginata* (on *Rosalina*, Collen and Newell, 1999), *Hyrrokkin sarcophaga* (on bivalves and a deep-water coral, Cedhagen, 1994; Freiwald and Schönfeld, 1996), *Planorbulinopsis parasita* (an endoparasite of *Alveolinella quoyi*, Banner, 1971), and *Cibicides refulgens* (on the extrapallial fluid of scallops, Alexander and Delaca, 1987). The elongate neck of *Lagena* has been considered as a possible adaptation for parasitic feeding (Haynes, 1981).

Resource partitioning. This is where species feed on different parts of the available food. Competition between species is minimised by resource partitioning at least among herbivorous forms (Lee *et al.*, 1977). This may explain how high diversity is maintained. It follows that if resource partitioning is widespread, the flux of organic carbon (C_{org}) is a very crude, and possibly misleading, measure of 'food' supply.

3.2.11 Endosymbiosis

This is the association between two different species of organism that is of benefit to both partners. Since algal symbionts require light but are dependent on the host habitat they have evolved photoacclimation, i.e., adaptations to the specific environment of their host. The chemical environment of an endosymbiont and the inorganic carbon acquired through the cell of the host are different from those of a free-living alga. The interface between host and symbiont is critical to the metabolic relationships yet is the least understood aspect of symbiosis. Translocation of photosynthetically fixed carbon is the main benefit brought to the host and the latter controls the environment surrounding the symbiont. The host needs to supply inorganic carbon (as CO_2) from the external environment and in corals this is done from the bicarbonate pool. Rubisco is the primary CO_2 fixation enzyme. Translocation of carbohydrate to the host is often as glucose but may include lipids and some amino acids. There are extensive reviews of foraminiferal–algal endosymbiosis by Lee and Anderson (1991b) and Hallock (1999) giving a great deal of detail beyond that presented here.

Types of symbiont. Autotrophic endosymbiotic algae are hosted in the cell of larger foraminifera, a 'group' considered to have evolved to benefit from this strategy (Hallock, 1985). A range of algae serve as symbionts in different groups of foraminifera and this suggests that the process has arisen independently in each group. For the larger foraminifera, in the Miliolina the symbionts are red and green algae, dinoflagellates and diatoms whereas in the Rotaliina only diatoms are found. Although Leutenegger (1983) considers that the symbionts are host specific, in the case of diatoms the species taken up by the same foraminiferal species vary both temporally and spatially (Lee and Anderson, 1991b; Lee *et al.*, 1995). The possible benefits of endosymbiosis to foraminifera are threefold: gaining energy from photosynthetic primary production of their symbionts, enhancement of test calcification and uptake by the algae of waste metabolites from the host. As all shallow-water larger foraminifera contain symbionts it may be assumed that this was the case in the past. In some smaller foraminifera chloroplasts sequestered from algae are husbanded in the cell-body cytoplasm. Although this may not be true symbiosis, the benefit to the host is thought to be gain of energy from photosynthesis.

Transmission. Symbionts can be transmitted from parent to offspring during asexual reproduction (Röttger, 1974, 1981) but this is not possible during sexual reproduction because there is merely fusion of gametes. In larger foraminifera asexual reproduction is more frequent than sexual reproduction (Röttger and Schmaljohann, 1976). Symbionts can be taken up from the environment in the same way that chloroplasts are sequestered, i.e., they are gathered by the reticulopodia. Lee and co-workers (Lee *et al.*, 1991b; Faber and Lee, 1991b) suggest that those algae that manage to avoid the initial stage of digestion (which may take place near the aperture) can establish themselves as endosymbionts in the chambers. The chambers subdivide streams of protoplasm and allow microhabitats to be developed. Chambering in foraminiferal tests may be a pre-adaptation to endosymbiosis. In soritids the dinoflagellates taken up as symbionts may be derived from zooxanthellae expelled by corals (Lee *et al.*, 1997). In the case of diatoms, there is a signal–receptor system involving molecules on the surfaces of the diatoms and the reticulopods. This controls whether the foraminifer recognises the diatoms as potential symbionts. These molecules are termed the common symbiotic surface antigen (CSSA). It is suggested that the CSSA functions like a theatre ticket telling the host where to place the ingested particle (Chai and Lee, 2000), which is well away from areas where digestion takes place (Leuteneger, 1984; Lee et al., 1991c). The CSSA also serves to protect the symbionts from being digested once they are in the cell-body cytoplasm.

Role in nutrition. The gain in energy from photosynthesis (photo-synthate: sugars and glycerol) is in addition to that gained from feeding (protein) and this gives endosymbiont-bearing organisms a considerable energy advantage over those lacking symbionts (Hallock, 1981a). However, the gain is higher in some groups than others and the host needs to control the availability of fixed nitrogen to the symbionts to obtain greatest benefit. Photosynthesis is limited by the availability of sunlight and CO_2; when symbionts receive plenty of sunlight but are nutrient limited they produce sugars and glycerol. If there is abundant fixed nitrogen, the symbionts keep these sugars for their own growth and reproduction (to the disadvantage of the host), but if the fixed nitrogen is controlled this energy source is given to the host, enabling it to utilise the proteins ingested from its food to promote growth and reproduction. There is a range of heterotrophic feeding by the host from active to apparently negligible. Active feeders include *Amphisorus, Archaias, Peneroplis* and *Sorites*. Feeding experiments show that *Amphisorus hemprichii* uses food as a source of carbon and energy (Ter Kuile et al., 1987; Ter Kuile and Erez, 1991) and this applies to other porcelaneous forms (Lee and Bock, 1976; Faber and Lee, 1991a). *Amphistegina lobifera* uses feeding mainly as a source of nutrients (Ter Kuile et al., 1987). The more advanced rotaliids with canal systems (nummulitids and calcarinids) feed on their endosymbionts and have not been observed to feed on anything else (Röttger et al., 1984; Röttger and Kruger, 1990). However, they may also feed on bacteria, as shown experimentally for *Heterostegina depressa* (Lee et al., 1980). The organic elastic sheaths produced by these forms may serve as a substrate for farming bacteria as a further source of food (Hallock, 1999). Endosymbiotic larger foraminifera need trace substances provided by bacteria. Both *Amphisorus hemprichii* and *Amphistegina lobifera* died when kept in the dark even though they were fed.

Role in test calcification. The role in test calcification is controversial. One view is that, in the presence of calcium, symbionts split bicarbonate and remove CO_2 for photosynthesis leading to the formation of $CaCO_3$ and water; however no evidence was found for this in the calcification of larger foraminifera (Ter Kuile, 1991). An alternative is the poison-removal theory: that at the site of calcification, the organism removes those ions that inhibit calcification (ammonium, phosphate, magnesium). This might apply in porcelaneous foraminifera where calcification occurs in vesicles (Hemleben et al., 1986). In perforate tests, the presence of an organic lining seems to be of great importance for calcification. Experimental growth of Amphistegina shows that their reaction to sunlight intensity is the same as that expected for microalgae: photosynthesis and growth takes place between a lower threshold where there is insufficient light and an upper threshold where there is too much light (Hallock, 1981b). This relationship between growth and optimum light conditions makes larger foraminifera completely dependent on their endosymbionts for growth (Röttger and Berger, 1972; Röttger, 1976; Röttger et al., 1980; Ter Kuile and Erez, 1984; 1991; Hallock et al., 1986a; Lee and Anderson, 1991b). Forms kept in the dark grow very slowly (Duguay, 1983). However, in the Mediterranean, the lower limit of Amphistegina lessonii is controlled not by light but by a thermocline at 60–70 m water depth (whereas other light-dependent organisms exist down to around 135 m). Hollaus and Hottinger (1997) suggest that the lower temperature below the thermocline has an adverse effect on the growth and reproduction of the symbionts and this, in turn, limits the growth and range of Amphistegina lessonii. This might be because the biogeographic limit of the species is close by.

Bleaching. The loss of algal symbionts/pigment is termed bleaching and may be induced by elevated water temperature. It leads to deterioration of cellular and intracellular membranes, digestion of endosymbionts, disintegration of organelles, and vacuolation, granulation and breakdown of cytoplasm in larger foraminifera (Hallock and Talge, 1993; Talge and Hallock, 1995). Bleaching on a global scale became obvious in 1982–83 and by 1991–92 had affected Amphistegina spp. (Williams et al., 1997). In the Florida Keys photosynthetically active radiation (PAR) values exceed the levels determined from laboratory experiments to be harmful. Although the cause of the damage has not been determined, it may be related to solar radiation and particularly to the increase in UV-B radiation due to ozone depletion (Hallock et al., 1992; 1995; Williams et al., 1997; Hallock 2000a). Subsequent experiments show that Amphistegina gibbosa has a limited optimum irradiance under laboratory conditions (1–2 W m^{-2} delivered by PAR proton flux densities (PFDs) of 6–8 μM m^{-2} s^{-1}, representing a daily dose rate of 6–8$\times 10^4$ J m^{-2} d^{-1}). Lower PAR PFD reduces growth rates while higher PAR PFD induces bleaching (Williams and Hallock, 2004). At equal PAR PFD, blue radiation (20% higher usable energy) causes significantly faster growth than white radiation. Under natural conditions, Amphistegina gibbosa is found over a depth range where visible and UV irradiance is up to 100 times the optimum. The main behavioural mechanism by which the species control their exposure to irradiance is phototaxis. *Amphistegina* is positively phototactic and negatively geotactic. Forms from deeper water have a less cryptic mode of life, moving about and climbing to gain greater irradiance. In shallow waters they become more cryptic. Another experiment has shown that at irradiance levels significantly greater than the optimum, a thicker test wall is secreted (Hallock *et al.*, 1986b). Also, at least ten species of diatom endosymbionts are present in *Amphistegina gibbosa* (Lee *et al.*, 1995) and these may have their own individual light requirements allowing some flexibility in the response to irradiation. All these strategies mean that *Amphistegina gibbosa* is highly adaptable (Hallock, 2000a, b).

Chloroplast husbandry. Among the smaller foraminifera, members of the families Nonionidae, Rotaliellidae and Elphidiidae sequester chloroplasts from the algae that they partially consume. The chloroplasts are husbanded in the cell-body cytoplasm (Lopez, 1979; Lee, 1983; Lee *et al.*, 1988; Lee and Lee, 1990; Cedhagen, 1991; Bernhard and Bowser, 1999). The chloroplasts do not grow in the foraminiferan and therefore they are not subject to temperature control hence foraminifera with husbanded chloroplasts can live down to the maximum depth of light penetration (Hollaus and Hottinger, 1997) and they are also known from greater depths (Grzymski *et al.*, 2002). The chloroplasts are from diatoms (Lopez, 1979; Knight and Mantoura, 1985; Lee *et al.*, 1988). This type of activity seems to be common in temperate intertidal habitats although it has also been recorded in warm water *Cycloclypeus* and in sublittoral *Nonionellina labradorica* down to 300 m and *Stainforthia fusiformis* down to 115 m. The benefit of husbandry to the foraminifera is presumed to be gain of energy from photosynthesis but the precise mechanism remains to be investigated and its functional value in the aphotic zone remains to be determined (Bernhard and Bowser, 1999). In *Nonionella stella*, which lives much deeper than the euphotic zone, it is thought that the chloroplasts assist the assimilation of inorganic nitrogen (Grzymski *et al.*, 2002).

Endosymbiotic bacteria. *Buliminella tenuata*, *Globocassidulina* cf. *biora* and *Nonionella stella* living in oxygen-depleted environments have possible endo-symbiotic bacteria (Bernhard and Reimers, 1991; Bernhard, 1993; 1996; Bernhard *et al.*, 2000). They also retain chloroplasts in their cell-body cytoplasm even though they live well below the euphotic zone. Bernhard (1996) speculates that the bacteria might either reduce sulphur compounds or be magnetotactic and that the foraminifera might receive some as yet unknown benefit from the chloro-plasts. Gram negative, non-photosynthesising bacteria morphologically similar to marine nitrifying bacteria are present in deep-sea *Spiculodendron corallicolun*.

These are free-living bacterial endosymbionts with as many as 2.2×10^9 in a single foraminiferan (Richardson and Rützler, 1999).

Ecological consequences. Endosymbiosis is a key process allowing mixotrophic nutrition, i.e., to both feed (heterotrophic) and photosynthesise (autotrophic; Hallock, 1981a). This gives great ecological advantage because it reduces the energy loss (typically 90%) between the links in the food web (Hallock, 1999). It is an adaptation for survival and growth in severely oligotrophic tropical/subtropical shallow-water environments. The nature of the symbionts controls the choice of habitat (Hallock and Peebles, 1993) and depth distribution of larger foraminifera: chlorophyceans 0–15 m, rhodophyceans 0–70 m, diatoms 0–130 m (Leutenegger, 1984). In addition, porcelaneous walls are less transparent than hyaline walls (Hottinger, 2000). The benefit to the host of chloroplast husbandry has yet to be quantified.

3.2.12 Commensalism

Where two species co-exist in a way that one benefits and the other is unaffected, the process is termed commensalism. There are numerous instances of foraminifera being sessile on other animals, some of which are themselves sessile (hydroids and bryozoans: Dobson and Haynes, 1973; brachiopods: Zumwalt and Delaca, 1980; sponges: Voigt and Bromley, 1974; tunicates: Nyholm, 1961) while others are active (various crustaceans and molluscs: Bock and Moore, 1968; Delaca and Lipps, 1972; Haward and Haynes, 1976; Zumwalt and Delaca, 1980; Moore, 1985). The advantages are that the foraminifera are supported above the sediment surface (important for passive suspension feeding). In some cases the attachment site is close to the inhalent inlet or exhalent outlet of the host and this may enhance the food supply.

3.2.13 Predation on foraminifera

Foraminifera are a potential food resource for other organisms especially as they are major contributors to meiofaunal biomass in many environments. Predation is important in structuring marine communities and in controlling the numbers of benthic organisms (Lipps, 1988; Buzas et al., 1989). Predation may be selective or unselective (as in deposit feeding). Some foraminifera eat other foraminifera (Christiansen, 1964; Hallock and Talge, 1994) but the predators are mainly from other groups: gastropods (Hickman and Lipps, 1983; Herbert, 1991; Chester, 1993; Berry, 1994), scaphopods (Bilyard, 1974; Langer et al., 1995), nematodes (Sliter, 1971), holothurians (Mateu, 1968), echinoids (Reidenauer, 1989), prawns (Rainer, 1992; Wassenberg and Hill, 1993), fish (Palmer, 1988; Lipps, 1988) and unknown (Kaminski and Wetzel,

2004b). In the Forth estuary, Scotland, the gastropod *Retusa obtusa* consumed $\sim$4800 foraminifera m^{-2} d^{-1} in September and in the course of one year of growth, one *Retusa* consumed $\sim$2750 foraminifera, many of which were *Haynesina germanica* (Berry, 1994). An unknown predator selectively feeds on *Aschemonella* (Kaminski and Wetzel, 2004b). Fish are particularly important croppers of foraminifera (Lipps, 1988) and responsible for regulating the standing crop of benthic foraminifera in a Florida estuary (Buzas, 1978; 1982; Buzas and Carle, 1979).

Predation involves consumption and disturbance. In flume experiments juvenile fish consumed $\sim$45% of the live foraminifera and disturbance during feeding was responsible for 30–50% of foraminiferal mortality (Palmer, 1988). During feeding, sediment is drawn over the gills and discharged into the water and in addition, movement of the tail disturbs the sediment surface and may cause rearward transport of foraminiferal tests. Some predators drill holes in the foraminiferal tests and although these are documented the organisms responsible remain to be identified (Nielsen, 1999). Incidental consumers, mainly deposit feeders, include flatworms, polychaetes, chitons, gastropods, nudibranchs, bivalves, crustaceans, asteroids, ophiuroids, holothurians, echinoids, crinoids, tunicates and fish (Lipps, 1983). Because they feed on algae and other biota on tropical reefs, herbivorous and omnivorous fish incidentally eat large numbers of foraminifera. Consequently, many reef foraminifera are more common in cryptic habitats that the fish cannot reach. Other defences include growing to a size too large to be eaten and cementation to a substrate (Lipps, 1988).

3.2.14 Disease

Virus-like particles have been found in *Elphidium clavatum* from the Baltic Sea. Foraminifera may be potential vectors of virus diseases in other organisms (Heeger, 1988). Viruses are one of the prime factors leading to bacterial mortality in coastal waters, equivalent to grazing by protists (Fuhrman and Noble, 1995).

3.2.15 Test function

The evolutionary and ecological success of foraminifera depends in part on the functional significance of the test. However, some agglutinated foraminifera can live as naked organisms outside the test (*Astrorhiza limicola*: Schultz, 1915; Buchanan and Hedley, 1960; Christiansen, 1971; Cedhagen, 1988; *Iridia*: Cushman, 1922; *Astrammina rara*: Bowser and Delaca, 1985). Such forms may be in between growth stages for when they are supplied with detrital material, *Astrammina* soon makes a new test. Indeed, Cedhagen and

Tendal (1989) suggest that juveniles of this species smaller than 1.5 mm in size might live without a test. Individuals of *Astrammina rara* from which the test was experimentally removed soon formed a new one (Bowser *et al.*, 1995).

There are morphological features of the test that may be of functional importance (see reviews by Hallock *et al.*, 1991 and Hottinger, 2000) but experimental evidence supporting functional interpretations of specific morphological features is still very sparse. Six possible functions for the test as a whole have so far been proposed although, as discussed below, the results are inconclusive (Marszalek *et al.*, 1969; Murray, 1991).

To provide shelter. The test gives protection against unfavourable environmental conditions and, in addition, some species close their test openings for several hours by sealing them with debris. The organic lining of calcareous tests is the ultimate defence against low pH as Bradshaw (1961) showed that *Ammonia 'beccarii'* survived pH 2.0 for $11\frac{1}{4}$ h even though dissolution of the test took place. *Spirillina vivipara*, which lacks an organic lining, did not survive. Marszalek *et al.* speculated that chambered tests, such as that of *Quinqueloculina*, would give protection against sudden osmotic changes but this was shown by experiment not always to be true (Murray, 1968a). There may be some protection against certain wavelengths of light in shallow waters (Haynes, 1965; Banner and Williams, 1973).

To serve as a receptacle for excreted matter. There are two aspects to excretion. It is known that some foraminifera store the waste products (stercomata or xanthosomes) in their test (Tendal, 1979). However, others consider that the test itself may be a consequence of excretion (e.g., the organic lining: Banner *et al.*, 1973; or removal of toxic Ca^{2+}, Brasier, 1986).

To aid reproduction. There is no direct evidence that the test is of particular use during reproduction (except that some parent tests are partially dissolved to supply material for the offspring).

To control buoyancy. It is likely that the density of the soft parts is similar to that of seawater. In high-energy environments tests are commonly heavy and robust and, since they are made of material much denser than the seawater they displace, they counteract any buoyancy of the soft parts. Also, in many environments individuals gather detrital particles in their reticulopodia while feeding, as camouflage and possibly as additional counter buoyancy aids.

To offer protection from predators. There is no direct evidence of this. Some species are more conspicuous because of their form or colour and are therefore preferentially selected by predators.

To assist growth of the cell. Foraminiferal cells are relatively large and some exceed 3 cm (Figure 3.1). The test serves as a container that not only houses the cell but also provides additional space that may be used for growth or for storage, e.g., of stercomata (Mullineaux, 1987). In active individuals the test is incompletely filled with cytoplasm but when the reticulopodia are withdrawn into the test it may become filled. In the large form, *Alveolinella quoyi*, on average only 43% of the test is filled (Severin and Lipps, 1989) and deep-sea forms have highly vacuolated cytoplasm filling only part of the test. In shallow-water environments rich in food, the test is normally filled with dense cytoplasm.

The six possible functions of the test are not mutually exclusive. It is clear that some of the features of the test described by taxonomists as 'ornament' are definitely functional. Tubercles and structures such as teeth in the aperture serve to break up aggregates of food and detritus, ribs channel the extra-thalamous cytoplasm (Banner and Culver, 1978; Kitazato, 1990; Bernhard and Bowser, 1999) and spines support pseudopodia and help stabilise the test on soft substrates. Some rotaliids have a canal system that replaces primary and secondary apertures and allows communication between the chambers and the test surface for the extrusion of extrathalamous cytoplasm and reticulo-podia, for removing waste products and for release of juveniles during repro-duction (Röttger *et al.*, 1984).

3.2.16 Selection of test grains

Although many foraminifera gather inorganic and organic particles in a seemingly random way (as during feeding), agglutinated foraminifera have the ability to select grains of different shapes and sizes to make the three-dimensional form of the test and some even select just one type of grain such as a single species of coccolith, sponge spicules or flakes of mica (Heron-Allen, 1915). Some take grains from the tubes of worms to which they are attached (Langer and Bagi, 1994). The grain-size distribution of the agglutinated walls of two taxa (*Hormosina mortenseni* and *Cyclammina cancellata*) follows a fractal (statistically self-similar) pattern (Allen *et al.*, 1998). This is a very efficient mechanism for test construction that limits the need for biologically produced cement. Modelling shows that both fractal and log normal grain-size dis-tributions may be present in different parts of the test and this demonstrates that the reticulopodia are able to sort grains (Tuckwell *et al.*, 1999).

3.2.17 Test strength

Test breakage can result from predation and from high energy (currents, waves) in the environment of life. It is a matter of observation that tests in high-energy environments are thicker walled and more robust than those living in low-energy environments. The crushing strength of tests has been measured experimentally and confirms this relationship. However, test strength correlates better with sediment grain size than with water turbulence because delicate tests are associated with algae in the wave-swept zone. Test architecture is more important than the amount of material used in construction; thicker walls and a biconvex test shape increase the cross-sectional area over which the force is distributed (Wetmore, 1987). Larger foraminifera are subject to forces great enough to cause breakage and tests showing repair are found in the sediment near reefs. *Archaias angulatus* collected from the more exposed outer reef are stronger than those from sheltered habitats (Wetmore and Plotnick, 1992). Endolithic algae dissolve out narrow tunnels that may lead to weakened calcareous test walls (Zebrowski, 1937). There is no evidence that any single wall type is stronger than another.

3.2.18 Summary of key points

- The incessant movement of reticulopodia is fundamental to their roles in gathering food and its digestion.
- The rates of movement are commonly less than $50\,\mu m\ min^{-1}$, individuals respond positively or negatively to gravity and light but otherwise movement is random. Even supposedly attached forms are sometimes mobile, and forms can be epifaunal or infaunal according to prevailing conditions. Therefore, the terms infaunal and epifaunal are not restrictive.
- There are many feeding strategies utilised by benthic foraminifera; some taxa may employ more than one, and there is evidence of resource partitioning within some feeding strategies to reduce interspecific competition.
- Larger foraminifera are entirely dependent on their symbionts for growth. Chloroplast and bacterial husbandry operates at water depths down to several hundred metres, i.e., in the aphotic zone.
- The possible functions of the test include protection against adverse conditions, a receptacle for waste products, to control buoyancy, and to aid reproduction or growth of the cell. There is no evidence that the test offers protection against predation.

- Some test structures described as ornament (ribs, pustules) have a functional role in controlling the flow of cytoplasm and sieving out selected particles.

3.2.19 Some of the problems awaiting solution

- In view of the importance attached to food, much more information is needed about what foraminifera eat and the role of resource partitioning. This will require carefully conducted experiments.
- Understanding the physiological responses to varying environmental conditions must hold the key to understanding the niches of individual species and the interpretation of patterns of distribution.
- There is a need to quantify the annual production of foraminifera, especially in regard to carbonate production and the role they play in the carbon cycle including benthic–pelagic coupling.
- The physiological roles of chloroplast and bacterial husbandry need to be determined, especially in foraminifera living in dysoxic environments.

3.3   Basic ecology

3.3.1   Introduction

In this chapter, some basic ecological principles relevant to foraminiferal ecology are outlined prior to the discussion of specific environments (Chapters 4–8). Ecology is the study of the causes of patterns of distribution and abundance of organisms. It concerns the interaction between individuals of the same and different species, and interaction with the chemical and physical environment. A major feature of ecology is its complexity. Research on marine ecology of benthic macrofauna has been primarily field-data driven and has revealed the great variability in community patterns. However, it continues to be difficult to identify the most important processes regulating the communities (Lenihan and Micheli, 2001). In a review of community ecology and the possibility that there are general laws in ecology, Lawton (1999) favoured greater concentration on macroecology and less on microecology. Patterns are defined by Lawton (1999) as regularities in what we observe in nature; that is, they are 'widely observable tendencies.' Lawton notes that patterns are contingent, that is, 'if A and B hold, then X will happen, but if C and D hold, then Y will be the outcome.' 'The same pattern can be generated by different rules (pattern does not imply process), and that the underlying rules (for example, a mathematical model) can generate different patterns, depending upon the contingent details.' On a global (macroecological) scale, terrestrial environments are distributed in relation to two environmental variables: mean annual temperature and mean annual precipitation (Whittaker, 1975). On a local (microecological) scale, no two environments are ever exactly the same or have identical species assemblages. Thus it is almost impossible to make generalisations about the contingent details controlling their patterns. At a local scale there are too many contingencies to comprehend and they prevent the recognition of general patterns. The same applies to studies of benthic foraminiferal ecology.

3.3.2   Critical thresholds and the niche

The ways in which environmental factors control the distributions of organisms remain poorly understood. For limiting factors, there is a relationship with successful existence (including the ability to feed and reproduce). The critical thresholds are the lower and upper limits for survival (outside this range the organism will die) and within these limits there are critical thresholds where the factor begins to cause serious stress (Figure 3.3A). In a simplistic way, it is often assumed that under optimum conditions, the organism will achieve its maximum abundance. In reality, it is likely that once all factors are within the range of the critical thresholds so that the species can flourish, its abundance is likely to be variable because not all factors will be at their optimum at the same time; also the main constraints on numbers are likely to be food supply, competition and stochastic events such as predation.

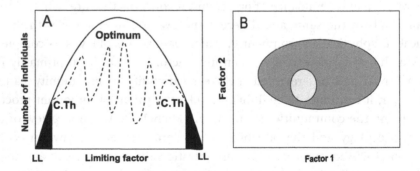

Figure 3.3. The niche. A. Abundance–factor relationships: the continuous line shows the theoretical relationship if abundance is related to the optimum for a given factor; the broken line shows the more likely abundance pattern within the two critical thresholds (C.Th); LL indicates limits of life; outside these limits the species cannot survive. B. The larger shape represents the fundamental niche and the smaller inset the realised niche occupied by the species.

A distinction is made between habitat and niche. The place where an organism lives is its habitat. The term niche is used for the habitat with ecological parameters within the range for the successful existence of a species. Thus, a niche is not a place but a set of circumstances particular to one species. Each species has its own niche influenced by a large number of abiotic and biotic factors. For survival, the numerical value of all these factors must lie within the upper and lower critical-threshold tolerance limits particular to that species (Lee, 1974; Pielou, 1974). Ecologists also recognise that a distinction must be made between the fundamental niche, the ecospace where a species could potentially exist and the realised niche, where the species really does exist and which is always a smaller part of the ecospace. This is shown in a very simplistic way for two variables in Figure 3.3B. The sum of the realised niches of species in different geographic areas probably occupies a greater part of the fundamental niche than in any individual area.

A species does not have to be perfectly adapted to its habitat; it must be sufficiently adapted to survive and be able to do so better than any competitor (Hallock *et al.*, 1991). So far no niche has been satisfactorily defined for any benthic foraminiferal species. As pointed out by Lee (1974, p. 219) 'It staggers the imagination to consider just the field work involved . . .'. Under natural conditions some factors co-vary and operate together and this may change thresholds. Experimental studies artificially isolate individual factors in order to determine critical thresholds and these commonly show that the critical limits for a given factor are more extreme than those suggested through field studies (Murray, 2001). Early attempts by Muller at defining the niches of four species are given by Lee (1974). At that time neither food supply nor oxygen were considered important controlling factors although both Phleger (1960a, p. 189) and Murray (1973, p. 230) had already pointed out the importance of food. This illustrates another difficulty in defining the niche: with time, previously ignored factors are brought into consideration or found to be important. In distributional studies, the species with the higher abundances are invariably considered to be important. Yet the key to determining thresholds is where species die out, so this means looking at low abundances. Perhaps we should be asking the following question. *Why is a given species absent from one area or environment while it is present in an adjacent area or environment?*

The separation of niches of related species may be quite subtle. For instance, *Amphistegina lessonii* (having thin walls) is inhibited by higher light intensities at depths <3 m and is less competitive than *Amphistegina lobifera*, whereas both appear to compete equally at depths of 3–15 m where the light level is lower (Hallock, 1981b).

The niche controls species distributions on both regional and local scales. For example, at the scale of the individual there may be small but significant differences (around the critical thresholds) in the values of factors associated with patch-scale spatial or temporal variability in the environment. Also, foraminifera have the ability to withstand (for some days or even longer) the effects of some factors that have exceeded the expected critical thresholds (e.g., anoxia; see review by Bernhard and Sen Gupta, 1999). Many foraminifera are opportunistic and great survivors; they exist in small numbers for long periods even when conditions are far from optimal, but when conditions change in their favour they rapidly increase their numbers. The so-called 'phytodetritus species' (Gooday, 1988, 1993) and *Stainforthia fusiformis* (Alve, 1994, 2003) are good examples of this.

The importance of biotic factors in limiting niches should not be neglected although they are often difficult to quantify. For example, through experimental studies the critical thresholds of temperature, salinity, oxygen, and sediment grain size were determined for the fundamental niche of the meiofaunal polychaete *Protodriloides symbioticus*; the realised niche (based on field observations) was just a small part of the fundamental niche. Not all apparently suitable substrates were colonised, and it was found that only those containing a particular species of bacterium were attractive (due to a tactile chemical response). Furthermore, *Protodriloides symbioticus* does not colonise sands occupied by the gastrotrich *Turbanella hyalina* because it produces a chemical that is offensive to the polychaetes (Gray, 1981). The distribution of epifaunal foraminifera such as *Cibicides lobatulus* and *Rosalina globularis* is limited by the availability of suitable firm substrates that project above the sea floor (Schönfeld, 2002), and they do not live on the associated sandy sediment (although their dead tests accumulate there). In these cases the response to the factors (offensive chemical; suitable substrates) is essentially binary rather than gradational.

The niche concept is fundamental to understanding faunal distribution patterns and proxy relationships (Murray, 2001). At any one time the factor(s) close to the threshold of tolerance for any given species are those that will limit its local distribution. It follows that for each species living in continually varying environments it is probable that different factors or a combination of factors are limiting distributions both temporally and spatially. This in turn explains why in one variable environment there is a strong correlation between certain foraminifera and one particular factor while in another there is not, and it also accounts for the lack of a consistent regional pattern of correlation between individual species and any single factor (Murray, 1991). At the present state of knowledge, we can define broad patterns of distribution in

relation to major environmental controls. However, as is the case with macrofaunal ecological studies, it is often difficult to identify the most important
processes regulating the communities (Lawton, 1999; Lenihan and Micheli,
2001).

3.3.3   Environmental factors

In order even to begin to understand the niche of each species there
are several questions to be addressed. How quickly does it respond to a change
in an abiotic environmental factor? What are the values of each factor at the
critical thresholds? What are the optimum values?

With the exception of a rapid, catastrophic change, the response time is
likely to be slow because most species seem tolerant of a range of environmental variability – otherwise they would frequently become locally extinct.
Species do not necessarily have critical thresholds for all factors and the concept of optimum values may be an ideal rather than reality. Therefore, there is
major problem to know which factors to measure. For example, current
thinking is that food is an important controlling factor. In order to test this
hypothesis properly, rather than just measuring sediment C_{org} or fluxes of C_{org}
to the sea floor, measurements need to be made of food type, quantity, quality,
variability of supply and acceptability to a given foraminiferal species. This will
undoubtedly prove to be a more daunting task than enumerating the foraminifera.

In the majority of studies very few factors are measured at the time of
collection of the samples. This may not matter if the response of the organism
is dependent on the range of variability rather than an instantaneous record.
For some environmental parameters estimates of short-term or long-term
ranges are obtained from the literature (e.g., flux of organic material), while
for others non-quantitative estimates are made (e.g., low or high energy).
The problem is that estimates from the literature may not be an entirely
reliable guide to the conditions actually experienced by each species prior to
the survey.

Where direct measurements are made either interval or ratio scales are
used. An interval scale does not start at zero (e.g., temperature). A ratio scale
starts at zero and can be discrete (standing crop of foraminifera) or continuous
(test size, all chemical or physical measurements). Often descriptive terms
rather than numbers are used. To a large extent, the use of terms (even when
they are quantified) is an indication of uncertainty or a way of expressing
environmental variability. It is convenient to be able to group environments
having some similarity but not exactly the same absolute values for a variable,
e.g., brackish. However, there are disadvantages if the terms are vague or if

different users have different ranges for them, e.g., nutrient terms, especially oligotrophic and eutrophic. For an ecological parameter that varies continuously over a range, describing its level as 'high' or 'low' does not give any real indication of the absolute value. In some cases, terms have been introduced for segments of the range, with or without quantification. The choice of the limits of segments may be arbitrary (as for tides) or related to critical limits of ecological significance (as for salinity and dissolved oxygen, the latter only in the lower part of the range). Because some environments vary both diurnally and seasonally, it may not be realistic to assign one term to describe any one variable. For example, a lagoon may be brackish during the rainy season but may become hypersaline during the dry season. Some of the more commonly measured variables are listed below and this is followed by further comments on selected variables (controls on anoxia, primary production, sediment macrofaunal communities and disturbance).

Abiotic factors

Salinity. This may be brackish or hyposaline (0–33), euhaline or normal marine (33–37), hypersaline (>37). Organisms that are confined to water of normal salinity are said to be stenohaline. Those tolerant of brackish or hypersaline water are euryhaline. The range tolerated by foraminifera is from fresh water to strongly hypersaline (0 to at least 70).

Temperature. This may be arctic or polar, temperate, tropical or warm. Organisms that are confined to water of restricted temperature are said to be stenothermal. Those tolerant of a wide range of temperatures are eurythermal. The upper temperature limit for marine organisms is around 45 °C.

Dissolved oxygen. The concentration depends on the temperature and salinity of the water. In seawater the range is from zero to around $8.5 \, ml \, l^{-1}$; $379 \, \mu M$. Parts of the range are anoxic (zero dissolved oxygen), dysoxic (0.2–$1.0 \, ml \, l^{-1}$; 9–$45 \, \mu M$) and oxic ($>1.0 \, ml \, l^{-1}$; $>45 \, \mu M$) (after Allison et al., 1995). The critical range for foraminifera is from anoxia to dysoxia. Once an environment is oxic, there is no stress from oxygen. The terms applied to faunas are anaerobic, dysaerobic and aerobic, in anoxic, dysoxic and oxic environments, respectively. To convert $ml \, l^{-1}$ to μM, divide by 0.0224. To convert μM to $ml \, l^{-1}$, multiply by 0.0224.

Nutrients. These refer to the chemicals necessary to promote production through photosynthesis by plants. Terms are used to describe nutrient availability – oligotrophic (low), mesotrophic (moderate) and eutrophic (high). In the scheme proposed by Nixon (1995) the definitions are related to annual productivity. Oligotrophic: environments having a low concentration of nutrients (in the euphotic zone) or low food supply (below the euphotic zone); production $<100\,g\,C_{org}\,m^{-2}\,y^{-1}$. Mesotrophic: moderate abundance of nutrients; intermediate between oligotrophic and eutrophic; production 100–$300\,g$ $C_{org}\,m^{-2}\,y^{-1}$. Eutrophic: environments in which the nutrient flux is sufficiently high that it supports phytoplankton densities that limit light penetration; production 301–$500\,g\,C_{org}\,m^{-2}\,y^{-1}$.

Tides. These can be microtidal (range $<2\,m$), mesotidal (2–$4\,m$) or macrotidal ($>4\,m$).

Substrate. This can be either sediment or a hard surface.

Biotic factors

Competition. This can be between individuals of the same species (intraspecific) or between different species (interspecific). It is very difficult to measure and is ignored in most foraminiferal studies.

Space. This is not readily measured but may be important.

Food supply. As discussed above, foraminifera eat a wide range of food and some have symbionts or husband chloroplasts. Although some species have more than one feeding strategy, there is resource partitioning (i.e., feeding on different parts of the available food).

Both abiotic and biotic factors

Disturbance. (after Sousa, 2001). This is damage, displacement or mortality caused by physical-, chemical- or incidentally by biotic-activity. Disturbances range in duration from discrete events that last for a fraction of the average lifespan of the affected species to much longer periods affecting several generations. The causes of disturbance are: physical (storm waves and currents, sediment slumps, volcanic ash fall, extended subaerial exposure, very high or very low temperature, activities of man particularly trawling, dredging); chemical (freshwater, hypersalinity, anoxia, pollution); biological (bioturbation, sediment movement by foragers, accumulation of living or dead plant material – seasonal algal mats, deposits of algae, seagrass or marsh plants).

Controls on anoxia. Organic matter settles on the sea floor becoming part of the sediment. Organic matter is decomposed by bacteria using oxygen but when oxygen is used up they utilise other oxidised compounds (e.g., nitrates, sulphates, phosphates) and produce ammonia, hydrogen sulphide, ferrous ions and other toxic reduced compounds (Fenchel and Reidl, 1970). Where microbial oxygen consumption exceeds oxygen renewal, anoxia and sulphides form in the sediment. The change from the presence of dissolved oxygen to anoxia in sediments is a transition zone called the redox potential discontinuity (RPD) which is normally *within* the yellow-brown surface layer rather than at the colour change from yellow-brown to grey or black. Macrofaunal burrows often extend below the RPD and introduce haloes of oxygen around themselves and these provide microhabitats for benthic foraminifera.

3.3.4   Primary production

Photosynthesis, by green plants using solar energy, converts inorganic carbon to organic carbon (primary production). Marginal marine environments receive nutrients from freshwater inflow and from the decomposition of organic material in the sediment. Water movement through wave and tidal circulation keeps mixing the nutrients in the water column and promotes enhanced primary production. Phytoplankton production in marginal marine environments is typically three to five times that of the open ocean. In addition, there is benthic primary production by seagrass and marsh plants and by a range of micro- and macroalgae (diatoms, chlorophyceans, seaweeds). There is seasonality in primary production especially in higher latitudes and there is commonly a spring bloom of phytoplankton and benthic algae. Controls include the amount and intensity of sunlight, temperature and nitrogen. Benthic algae rely on nutrient fluxes from the sediment and photosynthesise over a wider range of light intensities than plankton (Lenihan and Micheli, 2001). Even when other criteria are met, nitrogen availability is an important limiting factor in marine-plant productivity. In areas where there is excess input of nitrogen through pollution, there may be eutrophication leading to the development of free-living seaweeds such as *Ulva* that are left stranded on the intertidal sediment when the tide ebbs. This excludes light from the sediment surface and may lead to dysoxic or anoxic conditions. It is a common feature during the summer months in estuaries in southern England. This process may be responsible for the decline of seagrass beds in many temperate estuaries and lagoons. Apart from primary production, other potential sources of food for benthic foraminifera are bacteria and organic debris from decomposing plants. Intertidal sediments are covered with a biofilm of bacteria and microalgae (microphytobenthos) that helps to stabilise the surface through the production of biopolymers that reduce the effects of erosion from waves and currents. Transient biofilms in the intertidal zone form the basis of the food chain and mediate the flux of nutrients across the sediment–water interface. In open marine environments, the supply of food is largely from plankton primary production and the settling of marine snow or phytodetritus.

3.3.5   Sediment macrofaunal communities

The nature of the sediment partly controls the distribution of the benthos but organisms also play an important role in modifying the substrate in which they live. Some stabilise the sediment because they are anchored, and provide protection through acting as baffles, thus reducing hydrodynamic stress at the sediment surface (e.g., seagrasses, tube worms). Others destabilise the sediment through movement over the surface (gastropods, flat fish) or the formation of burrows (bioturbators such as molluscs, worms and crustaceans). They disturb the sediment, causing mixing and resuspension and they allow oxygen to penetrate to greater depths than would otherwise be the case. Even some larger foraminifera have the same effect (Geslin *et al.*, 2004). Deposit-feeding animals package fine-grained sediment into sand-sized faecal pellets, thus reducing the likelihood of erosion through current and wave attack. The modification of environments by habitat-forming species (e.g., biogenic reefs, seagrass, halophytic plants) is a powerful structuring force.

3.3.6   Disturbance

Disturbance may release limiting resources for use by survivors and for new colonisers. In an extreme case it may destroy the habitat (e.g., volcanic ash fall). The direct effect of disturbance is to reduce abundance and biomass, but its net effect may be to increase the local abundance of those species that would otherwise be rare or absent. This is achieved through freeing existing resources or generating new ones. Although species diversity may be reduced by disturbance, during recovery it increases, even if only temporarily. If there is asynchronous disturbance in a broad area it may cause a mosaic of patches and lead to an overall higher species diversity than would otherwise be the case (Sousa, 2001).

3.3.7   Populations, assemblages, associations, communities

There is an hierarchical structure in the occurrence of organisms. Each species comprises a population of individuals co-existing at the same time. Assemblages are groups of species found together in the same sample. Associations are groups of species (often of a single taxonomic group) that commonly occur together (sometimes called recurrent associations). However, in foraminiferal studies the term assemblage is also often used in the sense of association. In a broader context, there are communities of organisms spanning a wider range of taxonomic groups. Whereas there is no implication of interaction between species in assemblages, for associations and communities there is interaction between species in time and space. Groups of communities comprise ecosystems. The ecological role of species is to serve as a reservoir of genetic information. Local populations are ephemeral and local extinction is compensated for by recruitment from nearby populations. Thus, within an ecosystem there is constant replacement of local populations (Eldredge, 1990).

Individuals of each species compete with one another for resources such as space and food (intraspecific competition) and so do species (interspecific competition). The definition of a community is intended to imply that the organisms show interdependence. However, there is a range of ecological opinion on this. At one extreme it is considered that because of inter-dependence, communities are larger than the sum of the parts (holistic approach). At the other extreme it is considered that each species (and individual) behaves independently so that the community concept is meaningless (reductionist approach; Clements and Newman, 2002). Probably the truth is in between; while interactions between some species may be stronger than for others, the fact that there are predictable patterns of species associations demonstrates that there is some organisation or structure in communities. Nevertheless, the associations of benthic foraminifera defined by Murray (1991) normally have only the first few rank-order species in common. This suggests that many species are behaving fairly independently so here the term assemblage is used in preference.

3.3.8   Summary of key points

The niche concept is fundamental to understanding faunal distribution patterns. At any one time the factor(s) close to the threshold of tolerance for any given species are those that will limit its local distribution. It follows that for each species living in continually varying environments it is probable that different factors or a combination of factors are limiting distributions both temporally and spatially.

4

Marginal marine environments

Wherever possible, citations of references and data tables are given in the captions to figures to save repetition and to avoid interrupting the flow of the text. There are more comprehensive versions of figures 4.3–4.7, 4.11 and 4.13 on the linked website (indicated by '+ web').

4.1 Introduction

Marginal marine environments (lagoons, estuaries, fjords and deltas) are those that form the boundary between the land and the sea. Sandy beaches facing continental shelves are considered in Chapter 5. Because of the post-glacial rise of sea level of 120–130 m over the past 12 000 years (Kennett, 1982), marginal marine environments are geologically very young. They show varying degrees of freshwater input and evaporation loss and therefore have salinity gradients. These gradients may be horizontal and/or vertical depending on river flow and the tidal regime, ranging from brackish to hypersaline depending on climate. Where tides are present, there is an intertidal zone; sometimes this is bordered on the landward side by a vegetated area of halophytes (marsh in temperate regions, mangrove or mangal in tropical regions). In non-tidal areas there are fluctuations of water level due to atmospheric pressure and wind forcing. All these ecosystems are sensitive to changes (e.g., of sea level, temperature, freshwater runoff, salinity, pollution). Because of their accessibility and the low cost of research, they have been investigated more than any other environment.

The two most important physical factors affecting coasts are waves and tides because both impart energy into the environment. Wave-dominated coasts lying obliquely to the crest of the incident waves develop coastal barriers comprising coastal dune, beach and shoreface that are linked by sediment transport feedback and that may separate sheltered lagoons or estuaries from the open sea. Tides vary in magnitude from microtidal (<2 m) to macrotidal (>4 m) and the main environments dominated by them are estuaries; the highest macrotidal ranges are found mainly in funnel-shaped estuaries and embayments. Tidal currents are generated by the rise and fall of sea level and the maximum velocities occur at the mid points between high and low water or vice versa. Minimum velocities occur at slack water (at high or low tide) and sediment deposition occurs mainly at these times.

4.2   Marsh and mangal/mangrove

Vegetated intertidal sediments are termed marshes in temperate zones and mangals or mangrove swamps in tropical climates. They exist because of the foundation species (salt-tolerant angiosperm halophytes) that build and maintain them. Mangroves are restricted to the tropical zone between $32°$ N and $38°$ S; they grow where the monthly minimum air temperature is $20°$C and their poleward limit is the winter seawater $20°$C isotherm. Marshes form in low-energy areas sheltered from wave attack and floating ice. Extensive marshes are a feature of accretionary shorelines such as deltas, estuaries and lagoons, but are present only in localised sheltered areas along coastlines that are erosional or steep (as in fjords). Marshes are highly dynamic environments that accrete and prograde but are also subject to localised erosion from waves, currents and floating ice. The highest marshes are flooded less regularly than low marshes and consequently the high marsh sediments are often more oxygenated because the water has more time to drain away. The pattern of sediment salinity varies with marsh elevation (but not necessarily in any orderly way), with the presence or absence of creeks and non-vegetated areas, and also with latitude. The sediment pore-water salinity is influenced by tidal input and flushing, and by freshwater input from adjacent land areas and rainfall. Although subsurface anoxia excludes some plants, others tolerate it and introduce oxygen into the sediment around their roots. The loss of water through plant transpiration can lower the water table and also introduce oxygen into the sediment. Thus, there is habitat modification by plants and animals and positive feedback between these and the stability, salinity and oxygen content of the pore water. The marshes of North America are organic-rich compared to those in Europe (Edwards, 2001) and they extend into the lower intertidal zone whereas in Europe they are essentially confined to the uppermost intertidal zone. In non-tidal areas marshes may be exposed to the air or submerged by a veneer of water often for periods of days rather than hours. Several environmental parameters vary with respect to elevation in the intertidal zone. In tidal areas the most important is exposure to the atmosphere and this is greatest in the position of the highest spring tides and smallest at the position of lowest spring tides. Exposure can result in sediment drying and a more extreme range of surface pore-water temperature and salinity; higher light intensity; and the effects of rain or snow.

4.2.1   Marsh/mangal foraminifera

There is no fundamental difference in the low-diversity foraminiferal assemblages of marshes and mangals. The characteristic feature is the abundance of agglutinated taxa (Figure 4.1). In pre-1978 references by various authors, the forms now distinguished as *Jadammina macrescens* and *Balticammina pseudomacrescens* were not separated and were commonly grouped under *Jadammina* or *Trochammina macrescens*. Species may be infaunal or epifaunal, the latter mainly free living but sometimes clinging to algal filaments. They are a mixture of detritivores and herbivores. *Siphotrochammina lobata* and *Trochammina inflata* are epiphytic on algae in Brazilian mangrove swamps (Eichler *et al.*, 1995). *Trochammina inflata* forms a rigid cyst of detrital material in which asexual reproduction takes place. Within the cyst it concentrates fine detrital particles that will be used to form the wall of the juveniles. Within 24 hours of forming a cyst, the juveniles are dispersed (Angell, 1990). Living *Jadammina macrescens* and *Balticammina pseudomacrescens* occur with random orientation on filamentous algae whereas *Tiphotrocha comprimata* is more firmly attached by its umbilical side. *Jadammina macrescens* is most abundant on the decaying leaves of *Carex* (Alve and Murray, 1999). *Miliammina fusca* sometimes occur aperture downwards on dead leaves. This author has never observed foraminifera on the stems of the living halophytes. With the exception of *Miliammina fusca*, the agglutinated species listed above are confined to marsh/mangal. However, although calcareous species from adjacent tidal flats and subtidal areas may extend onto low to mid marshes and sometimes occur in high abundance, none is confined to marsh/mangal: *Ammonia* group, *Elphidium* spp. and *Haynesina germanica* (Figure 4.2).

Marsh foraminifera have been recorded living in areas not connected to the sea. For instance, in northern Germany there are inland marshes where salt-rich waters come to the surface from underlying evaporite deposits and these have a fauna solely of *Jadammina macrescens* (as *Jadammina polystoma*, Haake, 1982). In Canada *Polysaccammina ipohalina* and *Balticammina pseudomacrescens* (as *Jadammina macrescens*) have been recorded living in salt springs (Patterson *et al.*, 1990) and a new species has been recorded from Lake Winnipegosis

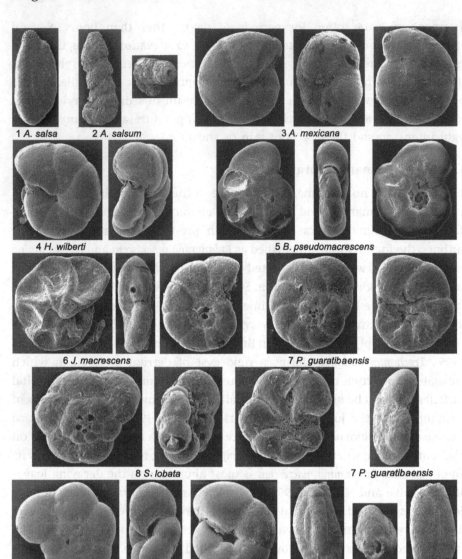

1 *A. salsa*   2 *A. salsum*   3 *A. mexicana*

4 *H. wilberti*   5 *B. pseudomacrescens*

6 *J. macrescens*   7 *P. guaratibaensis*

8 *S. lobata*   7 *P. guaratibaensis*

9 *T. inflata*   10 *M. fusca*

Figure 4.1. Scanning electron micrographs of marsh agglutinated foraminifera (longest dimension, μm). 1. *Ammoastuta salsa* (400). 2. *Ammotium salsum* (620, 200). 3. *Arenoparrella mexicana* (315, 350, 220). 4. *Haplophragmoides wilberti* (395, 300). 5. *Balticammina pseudomacrescens* (370, 340, 500). 6. *Jadammina macrescens* (250, 260, 400). 7. *Paratrochammina guaratibaensis* (400, 210, 230). 8. *Siphotrochammina lobata* (385, 290, 275). 9. *Trochammina inflata* (460, 430, 460). 10. *Miliammina fusca* (350, 220, 425).

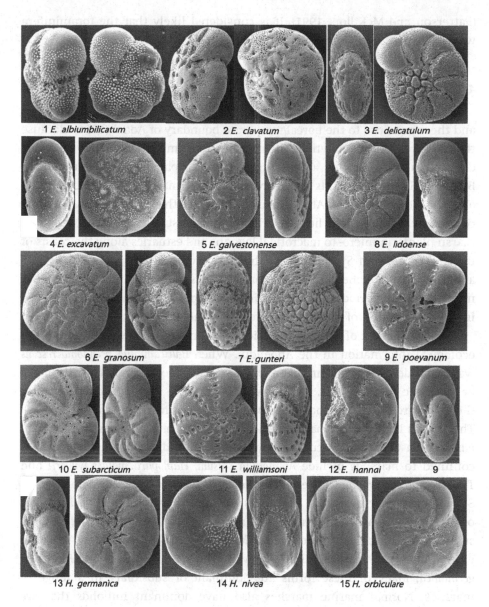

Figure 4.2. Scanning electron micrographs of lagoon foraminifera (longest dimension, μm). 1. *Elphidium albiumbilicatum* (120, 150). 2. *Elphidium clavatum* (400, 420). 3. *Elphidium delicatulum* (285, 285). 4. *Elphidium excavatum* (300, 310). 5. *Elphidium galvestonense* (325, 325). 6. *Elphidium granosum* (420, 275). 7. *Elphidium gunteri* (200, 450). 8. *Elphidium lidoense* (390, 440). 9. *Elphidium poeyanum* (290, 350). 10. *Elphidium subarcticum* (440, 510). 11. *Elphidium williamsoni* (330, 410). 12. *Elphidiella hannai* (240). 13. *Haynesina germanica* (420, 470). 14. *Haynesina nivea* (180, 200). 15. *Haynesina orbiculare* (450, 450).

(Patterson and McKillop, 1991). It is considered likely that the foraminifera were transported inland on the feet of migrating sea birds.

4.2.2   European marshes

The most northerly marshes studied are those of southern Scandinavia and these are close to the boreal/temperate boundary of Adam (1990). Because the area is essentially non-tidal, the marshes are compressed, do not show any clear plant zonation, and are developed as small patches in sheltered areas. Between the marsh plants there are sometimes mats of filamentous green algae and cyanobacteria. At the landward margin there is commonly several centimetres of tree leaf litter overlying the moist sediments between the marsh plants. In micro- to macrotidal lagoons and estuaries along the coasts of Britain, France and Spain, the marshes are extensive and often cut by meandering channels. Because of evaporation these basically brackish marshes may experience periods of normal salinity or hypersalinity and this is reflected in the composition of their foraminiferal assemblages.

The distribution of the main species is summarised in Figure 4.3 and the ecological information in the Appendix. When *Balticammina pseudomacrescens* was first described, Brönnimann *et al.* (1989a) considered that it might be endemic to the Baltic. It is now known to be widespread but only at the landward side of the marsh (around the highest high water in tidal areas). The typical high- to mid-marsh species are *Jadammina macrescens* and *Trochammina inflata* and these occur as dominant or subsidiary taxa. Other species confined to marshes include *Ammotium salsum*, *Haplophragmoides wilberti* and *Tiphotrocha comprimata*. Of these, only *Haplophragmoides wilberti* is dominant (in Scandinavian marshes) while the others are subsidiary or minor. Species found on the low seaward side of marshes and on adjacent intertidal and shallow subtidal areas include agglutinated *Miliammina fusca* and calcareous *Ammonia* group, *Aubignyna hamblensis*, *Elphidium williamsoni*, *Fissurina lucida* and *Haynesina germanica*. Many of these forms are dominant or subsidiary in a range of marshes. Normal marine marshes also have dominant miliolids that are usually small smooth-walled *Quinqueloculina* (probably incorrectly identified as *Quinqueloculina seminulum*). There are regional differences in the distribution of some taxa: *Elphidium albiumbilicatum* is a northern species that has so far been recorded only from Norwegian marshes (although it occurs elsewhere in other environments). *Ammonia* group and *Haynesina germanica* are southern forms; the former extends as far north as the Oslo Fjord but is absent from the Baltic. *Haynesina germanica* is absent from the Baltic and only rarely occurs on Scandinavian marshes. In the latter, *Miliammina fusca* and *Elphidium williamsoni* occur

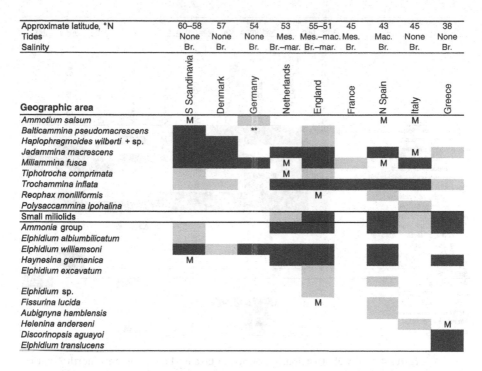

Figure 4.3. + web. Distribution of marsh taxa in Europe. Black = dominant, grey = subsidiary, M = minor. ** Later recorded by Brönnimann *et al.*, 1989a. Tides: mes. = meso; mac. = macro. Salinity: br. = brackish, mar. = marine. Sources of data: S Scandinavia: Alve and Murray, 1999 (WA-1); northern Germany: Lutze, 1968 (WA-1); The Netherlands: Phleger, 1970 (WA-1); England: Phleger, 1970 (WA-1), Horton *et al.*, 1999 (WA-2), Horton *et al.*, 1999 (WA-2), Swallow, 2000 (WA-3); France: Le Campion, 1970 (WA-4); northern Spain: Cearreta *et al.*, 2002b (WA-5); Italy: Petrucci *et al.*, 1983 (WA-6); Greece: Scott *et al.*, 1979 (WA-6).

on sediments with a mud/silt content less than ~66% whereas the *Ammonia* group and *Haynesina germanica* are common in sediments with >80% mud/silt. A time-series study of a low-marsh intertidal pool in northern Spain shows that the principal forms, the *Ammonia* group and *Haynesina germanica*, alternate in dominance in relation to their reproductive activity (Cearreta, 1988). In the Mediterranean the Venice marshes are dominated by *Miliammina fusca* or *Trochammina inflata* with subsidiary miliolids, the *Ammonia* group and *Helenina anderseni*. The Greek marshes, which also include non-vegetated mudflats, are dominated by the *Ammonia* group, *Discorinopsis aguayoi* or *Haynesina germanica* with subsidiary *Jadammina macrescens*, *Trochammina inflata* or *Quinqueloculina seminulum*. The warm water taxa *Discorinopsis aguayoi* and *Helenina anderseni* also make their only appearance in Europe here.

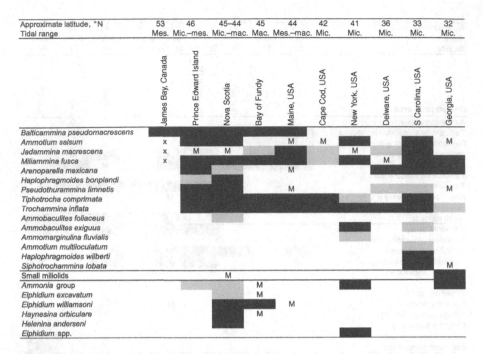

Figure 4.4. + web. Distribution of marsh taxa (0–1 cm) in eastern north America. Black = dominant, grey = subsidiary, M = minor, x = sample < 50 individuals. Tides: mic. = micro, mes. = meso, mac. = macro. Sources of data: Canada: Scott and Martini, 1982 (WA-7), Scott *et al.*, 1981 (WA-8), Scott and Medioli, 1980b (WA-9, WA-10, WA-11), Smith *et al.*, 1984 (WA-12), Patterson *et al.*, 2004 (WA-13); USA: Gehrels, 1994, Gehrels and van der Plassche, 1999 (WA-14), Saffert and Thomas, 1998 (WA-15), Steineck and Bergstein, 1979 (WA-16), Hippensteel *et al.*, 2000 (WA-17), Collins *et al.*, 1995 (WA-18), Goldstein and Harben, 1993 (WA-19), Goldstein *et al.*, 1995 (WA-20).

4.2.3   Atlantic North American marshes

Marshes are widely developed from James Bay, Canada, to Georgia, USA (Figure 4.4). They range from microtidal to macrotidal and climatically from subarctic in the north to warm temperate in the south. In the analysis of the results presented here, the live abundances have been calculated to exclude thecamoebians so in some cases the percentage occurrences differ from those in the source papers. The typical marsh species are *Ammotium salsum*, *Arenoparrella mexicana*, *Haplophragmoides* spp., *Jadammina macrescens*, *Tiphotrocha comprimata* and *Trochammina inflata*. Another common species is *Miliammina fusca* but this also lives on bare intertidal flats. Most of these species occur throughout the area from James Bay to Georgia. *Balticammina pseudomacrescens* is most tolerant of low salinities and low temperatures; it dominates marshes from James Bay to Maine and is known to occur in Massachusetts (de Rijk and Troelstra, 1999). *Pseudothurammina limnetis* is dominant around Nova Scotia and subsidiary to minor elsewhere. Species of *Ammobaculites, Ammomarginulina* and *Polysaccammina ipohalina* are minor components and irregularly distributed. *Siphotrochammina lobata* is confined to the warm temperate marshes of South Carolina and Georgia.

In many cases, for example in the publications of Scott and his co-authors, foraminiferal distributions in the 0–1 cm-sediment samples have been related to marsh elevation, which is a proxy for the frequency of tidal immersion. In any one sampling transect going seawards from the shoreline, salinity is generally inversely related to elevation (Scott and Medioli, 1980b; Scott *et al.*, 1981) although the magnitude of the salinity range will vary according to the influence of the open sea. Thus the inner part of an enclosed bay will be more brackish while the mouth will be more marine. Whereas the elevation of the marsh is constant, the salinity regime may change from one season to another. The salinity measurements taken at the time of sampling are instantaneous and do not give an indication of temporal variability. Thus, if frequency of exposure/immersion is an ecological control on species distributions, it is reasonable to measure elevation as a proxy. *Balticammina pseudomacrescens*, commonly regarded as an indicator of high marsh, generally shows a positive correlation with elevation (Chezzetcook, Nova Scotia, correlation coefficient $r = 0.69$–0.90) but in Wallace Basin, Nova Scotia, the correlation is negative ($r = -0.72$). *Trochammina inflata* also shows both positive and negative correlations (Prince Edward Island, $r = -0.4$; Nova Scotia, 0.54). This means that with respect to elevation these species are behaving differently in different parts of the marsh. *Miliammina fusca* shows mainly negative correlations as would be expected for an essentially tidal flat species. Multivariate analysis of the dataset for Chezzetcook inlet, Nova Scotia (Scott and Medioli, 1980b; n 147, WA-10) also shows the complexity of the relationship between foraminiferal abundance and marsh elevation. Q-mode cluster analysis of the dataset (without transformation) for transects I–V divides the assemblages into two main groups: those with elevations from -34 to 89 cm MSL (mean sea level) and those from 69 to 109 cm MSL. Principal component analysis roughly divides the assemblages according to elevation: those with a positive score on PC1, elevation -34 to 89 cm MSL; those with a negative score on PC1, 33–109 cm MSL. However, PC1 accounts for only 14.9% of the variation and the cumulative percentage variation for factors 1–5 is only 54.6%. R-mode cluster analysis groups *Elphidium excavatum clavatum* and *Elphidium excavatum selseyensis* at 83% similarity but otherwise all species pairs or groups have <33% similarity. From these results, it appears that there is a great deal of 'noise' that makes any linkage between living-species distributions and elevations more difficult to see.

The standing crop in the 0–1 cm samples is extremely variable not only throughout the Atlantic seaboard but also within individual marshes. In every marsh there are some barren samples and the maximum values range from a few tens to 2464 individuals 10 cm^{-3}. Some studies have shown the occurrence of rose Bengal-stained marsh foraminifera down to 60 cm although the numbers decline with increasing depth. On the Great Marshes, Massachusetts, foraminifera live infaunally down to 20–30 cm: high marsh, 19% living in the top 2.5 cm and 75% in the top 10 cm; intermediate-level marsh, 33 and 54%; and low marsh, 16 and 49%, respectively. On Kelsey marsh, Connecticut, the figures are 55 and 85% for high marsh, 22 and 71% for intermediate marsh and 25 and 63% for low marsh (Saffert and Thomas, 1998). Three sites from high, mid and low marsh in Delaware Bay were sampled every three months over three years, from the surface down to 60 cm (Hippensteel *et al.*, 2000, 2002; Martin *et al.*, 2002). The surface 0–1 cm is dominated by *Arenoparrella mexicana* on the high marsh and by *Trochammina inflata* on the mid and low marsh. The subsidiary species at all times is *Jadammina macrescens*, also with *Arenoparella mexicana* on the low marsh and on the mid marsh in year three. The peak of subsurface abundance is 1–40 cm in each case and for individual species it is mainly 3–10 cm except in the low marsh where three species peak at 30 cm in years one to two (*Arenoparella mexicana, Miliammina fusca, Trochammina inflata*). All species show temporal changes in abundance and subsurface distribution, with reproduction and abundance being greatest in the spring. In Virginia, normal-marine to hypersaline marshes are dominated mainly by the *Ammonia* group with subsidiary *Elphidium excavatum* and by *Ammobaculites exiguus, Textularia earlandi* or *Trochammina inflata* in more brackish conditions (Woo *et al.*, 1997). Four cores from marshes on Sapelo Island, Georgia, were sampled to 30 cm (Goldstein and Harben, 1993). The surface 0–1 cm assemblage of a marsh adjacent to a creek is dominated by *Triloculina oblonga* (small miliolids) with subsidiary *Ammonia* group. By 8 cm the live calcareous forms have disappeared and the dominant species is *Arenoparrella mexicana*. Two cores from *Spartina* marsh are dominated by *Arenoparrella mexicana* both at the surface and down to 30 cm. The transitional marsh is dominated throughout all depths by the *Ammonia* group. Thus, the true standing crop is not measured by the 0–1 cm sample. The presence of deep live foraminifera is most probably due to macrofaunal burrows of crabs and worms which allow water to circulate and bring oxygen into sediments that may otherwise be anoxic. This is clearly shown by the mottled appearance and subsurface patterns of pH and Eh in Georgia marshes (Goldstein *et al.*, 1995). The foraminifera may live in the oxic haloes around the burrows. It seems unlikely that the foraminifera are actively transported downwards through bioturbation by macrofauna as not all species are equally affected (Hippensteel *et al.*, 2000, 2002).

4.2.4   Gulf of Mexico and Atlantic South American marshes/mangals

All the studied marshes and mangals are microtidal. The mangals of southwestern Florida are developed on islands in a network of channels. Salinities are brackish (down to 5) in the summer but normal marine (32–36) during the winter. The sediments are calcareous quartz sands and silts ($CaCO_3$ 20–30%). In Mississippi and Texas there are marshes with *Spartina*, *Salicornia* and sparse mangroves. On the Mississippi delta, marsh salinities range from 21–40 with low values especially after rain. Along the Texas coast, marshes fringe lagoons and deltas. These examples span brackish–normal marine and possibly hypersaline regimes due to the high summer air temperatures. In Texas, Phleger sampled ponds and adjacent channels or lagoons as well as the vegetated marsh. The species distributions (Figure 4.5) show some differences in dominance of agglutinated species between areas: *Ammoastuta inepta*, *Balticammina pseudomacrescens*, *Miliammina fusca* and *Trochammina inflata* in Mississippi delta marshes; *Ammotium salsum*, *Arenoparrella mexicana*, *Jadammina macrescens*, *Miliammina fusca*, *Tiphotrocha comprimata* and *Trochammina inflata* in Texas marshes (N.B. *Jadammina macrescens* here may include *Balticammina pseudomacrescens*). Miliolids are also dominant and hyaline calcareous taxa are important in Texas both on the marsh and in the adjacent shallow lagoon or channels, which form their more typical habitat: *Ammonia* group, *Elphidium* spp., *Palmerinella palmerae*. Freshwater marshes on the Mississippi delta contain a few living *Haplophragmoides manilaensis* and *Balticammina pseudomacrescens* especially towards the channel, suggesting some marine influence. The Florida mangal has overall dominant *Ammonia* group and other taxa occasionally dominant include *Ammotium salsum*, miliolids and *Elphidium matagordanum*. The maximum standing crops are much higher in Texas marshes (2450 individuals $10\,cm^{-3}$) than on the Mississippi delta (959 individuals $10\,cm^{-3}$). In Florida they reach 1489 individuals $10\,cm^{-3}$.

In a Colombian mangal, Boltovskoy and Hincapié de Martínez (1983) distinguish an indigenous fauna comprising *Ammotium salsum*, *Ammobaculites exiguus* and *Arenoparrella mexicana* from marine calcareous forms that have invaded from the shelf. The mangal of Tobago, West Indies, described as 'stagnant', has a living assemblage of *Siphotrochammina lobata* and *Trochammina laevigata* (Radford, 1976a). South American marshes have been sampled from southern Brazil to Argentina (Figure 4.5). Assemblages from a channel, ponds

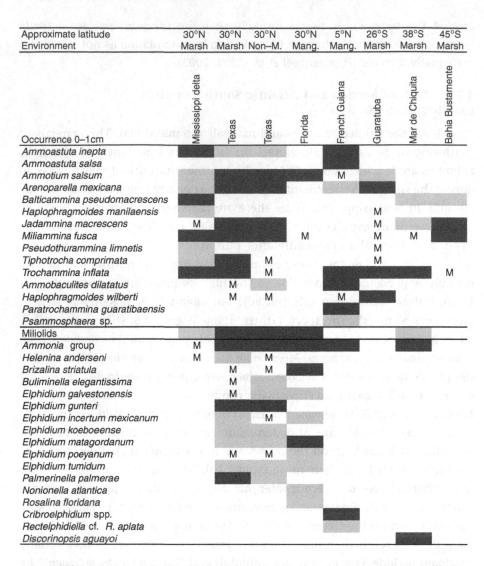

Figure 4.5. + web. Distribution of marsh and mangrove (mangal) taxa, Gulf of Mexico to South America. Non-M. = non-marsh, mang. = mangal. Black = dominant, grey = subsidiary, M = minor. Sources of data: Mississippi delta, USA: Scott *et al.*, 1991 (WA-21); Texas, USA: Phleger, 1965a, 1966 (WA-22); Florida, USA: Phleger, 1965b (WA-23); French Guiana: Debenay *et al.*, 2002, 2004 (WA-24); Brazil and Argentina: Scott *et al.*, 1990 (WA-25); Barbosa *et al.*, 2005 (WA-26).

and between mangrove plants in French Guiana are dominated by *Paratrochammina guaratibaensis*, *Ammoastuta salsa*, *Ammobaculites dilatatus*, *Trochammina inflata*, *Cribroelphidium* spp. or the *Ammonia* group. The canopy of mangroves and the litter produced from it protects the sediment and fauna from the sun. This may favour the *Ammonia* group. Guaratuba, southern Brazil, is the southernmost limit of mangroves and is dominated by *Arenoparrella mexicana*, *Haplophragmoides wilberti* and *Trochammina inflata*. The marsh assemblages of Bahia Bustamente, Argentina, are entirely agglutinated and dominated by *Jadammina macrescens*, *Miliammina fusca* with subsidiary *Balticammina pseudomacrescrens*. Those from Mar de Chiquita, Argentina (salinity 18–40), are dominated by *Trochammina inflata*, *Jadammina macrescens*, *Ammonia* group or *Discorinopsis aguayoi* with subsidiary *Elphidium* spp. The maximum standing crop here is very high in the surface 1cm (6593 individuals 10 cm^{-3}).

4.2.5   Pacific American marshes

Marshes from British Columbia, Canada, to Oregon, USA, are brackish but those in California, USA, and Baja California, Mexico, are normal marine to hypersaline (salinity 33–50). *Jadammina macrescens* and *Miliammina fusca* are dominant in all these marshes and *Trochammina inflata* in all except Baja California (Figure 4.6). The normal marine and hypersaline marshes also have dominant miliolids and hyaline forms such as *Elphidium* spp. Inner-shelf taxa such as *Glabratella ornatissima*, *Rosalina columbiensis* and *Spirillina vivipara* are also present due to the favourable salinities. Rather unusually, *Glomospira* sp. is a dominant form in Baja California. The maximum standing crop values are very high (4610–7250 individuals 10 cm^{-3}).

Subsurface distributions of living forms on British Columbian marshes were sampled down to 30 cm (Ozarko *et al.*, 1997, WA-28). Living forms extend down to the limit of sampling but 50% of the fauna lives from 0 to 6–7 cm and generally 90% lives from 0 to 16–22 cm. There are differences in infaunal distributions between high and low marsh. *Jadammina macrescens* occurs from 0–11 cm in the low marsh and 0–20 cm in the high marsh. The comparable figures for *Trochammina inflata* are 0–20 cm and 0–25 cm, respectively. *Miliammina fusca* occurs primarily from 0–10 cm with highest abundances from 0–3 cm. *Haplophragmoides wilberti* has its maximum occurrence between 3 and 7 cm. The authors conclude that to obtain a representative sample of the live assemblage, it is necessary to examine the top 10 cm of sediment. Similar studies carried out by Patterson *et al.* (1999) yielded few samples with an adequate living assemblage.

4.2.6   Other marsh/mangal faunas

Marshes are rare on Japan and the only study is that of the microtidal freshwater–brackish marsh of Hokkaido, Japan. This has dominance of *Balticammina pseudomacrescens* or *Miliammina fusca* (rarely *Ammobaculites exiguus*) with subsidiary *Haplophragmoides manilaensis* or *Trochammina inflata* (Scott *et al.*,

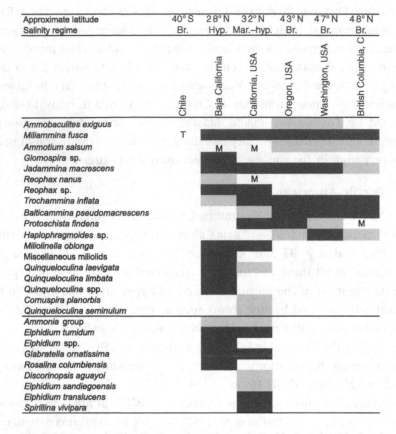

Figure 4.6. + web. Distribution of marsh taxa in Pacific America. Black = dominant, grey = subsidiary, M = minor, T = total. Salinity: br. = brackish, mar. = marine, hyp. = hypersaline. Sources of data: British Columbia, Canada: Phleger, 1967 (WA-27), Ozarko *et al.*, 1997 (WA-28); Washington, Oregon, USA: Phleger, 1967 (WA-27); California, USA: Phleger, 1967 (WA-27), Scott, 1976a, b (WA-29); Baja California, Mexico: Phleger, 1965c (WA-30), 1967 (WA-27), Phleger and Ewing, 1962 (WA-31).

1995; WA-32). The maximum standing crop is 10908 individuals 10 cm^{-3} and many values are >1000.

Living assemblages from around North Island, New Zealand are dominated by *Trochammina inflata* on high marsh and *Ammotium salsum*, *Ammonia* group or *Elphidium williamsoni* on low marsh. *Haplophragmoides* sp. and *Miliammina fusca* occur on both low and high marsh (Phleger, 1970; WA-33). At macrotidal Kaipara and Miranda, Auckland, there is a range from shell ridge, mangal and marsh to adjacent mudflat with S > 30 (Hayward *et al.*, 1999, WA-34). The shell ridge, marsh and mangal have dominant *Jadammina macrescens* and *Trochammina inflata* except in one Kaipara mangal where there is *Miliammina obliqua* and *Elphidium excavatum*. The mudflat has *Ammonia* group sometimes with *Cornuspira involvens* and *Quinqueloculina seminulum* may be introduced in by tidal currents. Micro–mesotidal brackish Pauatahanui inlet, Wellington, has dominant *Haplophragmoides wilberti* except on the high marsh where *Jadammina macrescens* and *Trochammina inflata* take over. Subsidiary forms include *Haynesina germanica* and *Helenina anderseni*.

A normal marine mangrove microtidal creek system in Indonesia (Horton *et al.*, 2005 (WA-35) has dominant *Quinqueloculina* spp. and *Ammonia* group extending from mudflats onto the mid marsh. The high marsh has *Quinqueloculina* spp., *Jadammina macrescens*, *Trochammina inflata* and *Miliammina fusca*.

Little is known of marshes and mangals in the Indian Ocean with just a single study in the Red Sea (Halicz *et al.*, 1984). No living forms are found on the pneumatophores of *Avicennia* but on *Laurencia* and *Halodule* there are *Sorites*, miliolids, *Spirolina*, *Peneroplis* and *Pararotalia*. The sediment assemblages are primarily miliolids, *Spirolina* and *Peneroplis*. Most of the samples are from pools rather than the marsh surface and this is essentially a lagoon rather than a marsh assemblage. At the northernmost limit of mangal development, in the Red Sea, the fauna is essentially that seen elsewhere in lagoons (Reiss and Hottinger, 1984; see also Chapter 6).

4.2.7 Summary of marsh/mangal faunas

Marshes are separated from mangals on the basis of their floras. Marshes occur from high to low latitudes while mangals are restricted to low latitudes. There is no fundamental difference between marsh and mangal foraminiferal faunas and that is why they are considered together here. The relationship between foraminifera and determination of sea level is discussed in Section 10.5.10. The ecological controls on marsh foraminifera seem to be duration of tidal submergence, shade from vegetation or leaf litter, effects of drying, salinity and pH of pore waters. Infaunal occurrences may be related in part to the influence exerted on sediment geochemistry by plant root systems. Although typical marsh agglutinated foraminifera do not live epiphytically, the presence of plants seems to be essential as they live only rarely in adjacent non-vegetated areas. The distinction between brackish, normal marine and hypersaline examples is not clear-cut because many marshes range through these conditions from one season to another. There is no obvious relationship between marsh faunas and the magnitude of the tidal range. Even division into high-, mid- and low-marsh assemblages are somewhat arbitrary. For instance, meso–macrotidal high-marsh taxa such as *Balticammina pseudomacrescens*, *Jadammina macrescens* and *Trochammina inflata* also occur in non-tidal marshes where no high, mid or low zones can be recognised (as in Scandinavia).

Perhaps the most remarkable feature of marshes is that so many of the species are worldwide in their distribution. Unlike continental shelves, marshes are discontinuously developed and in some cases they are remote from all others. For instance, there are few marshes in Japan yet the only described example (Scott *et al.*, 1995) has the typical marsh assemblage found elsewhere. These disjunct distribution patterns lend support to the concept of propagules as a means of dispersal over large distances (see Alve and Goldstein, 2002, 2003) although transport on the feet of migrating birds is another possibility over short distances (Patterson *et al.*, 1990). Nevertheless, not all marsh species are worldwide in their distribution. For example, *Ammoastuta inepta* and *Tiphotrocha comprimata* have not yet been recorded from the marshes of Pacific North America.

As pointed out by Zaninetti (1979), Boltovskoy (1984) and Scott *et al.* (1990, 1996), the northern and southern hemispheres show some faunal similarities and patterns of species distributions. Not all latitudinal zones have been sampled to the same extent. Information is sparse for southern hemisphere occurrences on all continents and there is only a small amount of information on mangals. Marshes of the far north >50 °N (St James Bay, Canada) and far south >50 °S (Tierro del Fuego, Argentina) of the Americas are similar in having extremely low diversity and low abundance assemblages. *Balticammina pseudomacrescens* is the principal form in the north. Rare individuals of *Balticammina pseudomacrescens*, *Trochammina advena* and *Trochammina inflata* occur in the south. By contrast, the same latitudinal belt in Europe is warmed by the North Atlantic drift and has a much more diverse fauna including calcareous species. Here a similar fauna stretches from 36–60 °N and resembles that seen only from 36–50 °N and S elsewhere.

Other widely distributed forms include *Ammotium salsum*, *Arenoparrella mexicana*, *Jadammina macrescens* and *Tiphotrocha comprimata*. *Miliammina fusca* is common on low marshes and it also, more typically, lives in adjacent lagoons or estuaries. Only a few species have restricted distributions: *Siphotrochammina lobata* is confined to warm, low latitude, marshes and mangals; *Paratrochammina guaratibaensis* is confined to tropical South America. Calcareous forms that otherwise live in lagoons invade the lower parts of marshes where conditions are favourable. These include the *Ammonia* group, *Elphidium* spp., miliolids (smooth-walled forms mainly of *Quinqueloculina* but sometimes *Triloculina*), *Discorinopsis aguayoi*, *Helenina anderseni* and *Palmerinella palmerae*. On the Pacific coast of the Americas, *Ammotium salsum* is present only in the northern brackish marshes and *Arenoparrella mexicana* is generally absent although Phleger and Bradshaw (Phleger and Bradshaw, 1966; Phleger, 1967) record rare occurrences in southern California, USA. Other absentees include

Ammoastuta inepta and *Tiphotrocha comprimata*. The standing crops on brackish marshes are highly variable over short distances and show great patchiness; this is not surprising considering the patchy distribution of the flora. Exceptionally high values of standing crop >10 000 individuals 10 cm^{-3} are present in Japan.

4.3   Intertidal–subtidal brackish lagoons and estuaries

Major processes controlling the morphology of lagoons, estuaries and deltas are fluvial input, wave action and tidal regime. Intertidal areas are subject to physical forcing (seasonal, tidal, weather) which cause considerable physical, chemical and biological variability. In addition, there is forcing due to the alternation of emersion and immersion and these induce thermal and radiance changes. Therefore, variation of primary production and respiration during a single tidal cycle could be almost the same as long-term variability such as seasonal or inter-annual processes. This variability has a potentially profound effect on the foraminifera even though their life cycles are measured in months rather than hours. Intertidal areas are visited by wading birds when the tide is rising or falling. They cause physical disturbance of the surface layer of sediment through locomotion and feeding, and chemical disturbance (addition of nutrients) through deposition of faeces. Both may contribute to the patchiness of the meiofauna and benthic algae. A feature of the intertidal zone is the development of transient biofilms composed of diatoms, euglenid flagellates and cyanobacteria. The biofilms move up and down in the sediment in response to tidal and light influences. They provide a source of food for benthic foraminifera and they help to stabilise the sediment surface making it more resistant to erosion.

4.3.1   Europe and Africa

Estuaries between the Arctic and tropics range from entirely brackish (the Arctic to southern England) to brackish – normal marine (southern England to Spain) to hypersaline (salt pans in the Mediterranean, Senegal; see Figure 4.7 for sources of data). Notwithstanding this, some taxa are common to many of these varied environments, particularly *Ammonia* group and *Haynesina germanica* (Figure 4.2). The identification of *Eggerella advena* off Russia awaits confirmation as this species is otherwise not known from Europe. *Eggerelloides scaber* may include *Eggerelloides medius* in some studies.

Europe   The arctic river Ob in northern Russia is large and it generates a salinity gradient that extends hundreds of kilometres into the Kara Sea. A

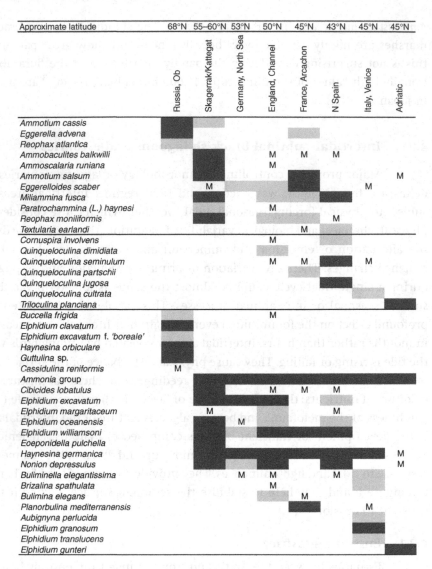

Figure 4.7. + web. Distribution of taxa in northwestern European estuaries and lagoons. Black = dominant, grey = subsidiary, M = minor. Sources of data: Ob, Russia: Korsun, 1999 (WA-36); Skagerrak–Kattegat: Alve and Murray, 1999 (WA-37); North Sea: Haake, 1962, van Voorthuysen, 1960 (WA-38), Wang, 1983 (WA-39); England: Murray, 1965a (WA-40), 1983 (WA-41), 1968b (WA-42), Alve and Murray, 1994 (WA-43), Murray and Alve, 2000a (WA-44, WA-45), Ellison, 1984 (WA-46); France: Le Campion, 1970 (WA-47); Spain: Cearreta, 1988 (WA-48), 1989 (WA-449); Italy: Donnici et al., 1997 (WA-50), Hohenegger et al., 1989 (WA-51).

transect of samples along the length of the estuary into the Kara Sea spans a depth range of 13–31 m (Korsun, 1999). The halocline at ~20 m separates surface water of salinity 20–25 from more saline deeper water of salinity 32. Above the halocline assemblages are dominated by brackish species such as *Haynesina orbiculare*, *Guttulina* sp., *Elphidium clavatum* and *Elphidium excavatum* forma *boreale*. Below the halocline the dominant forms are mainly agglutinated: *Ammotium cassis*, *Eggerella advena*, *Reophax* sp. together with subsidiary calcareous species *Buccella frigida*, *Elphidium clavatum* and *Elphidium excavatum* forma *boreale*.

Intertidal to shallow-subtidal (6 m) temperate brackish assemblages from the Skagerrak, Oslo Fjord and Kattegat include common *Miliammina fusca*, *Ammonia* group, *Elphidium excavatum*, *Elphidium williamsoni* and *Haynesina germanica*, any of which may be dominant with the others subsidiary. Additional local subsidiary species include *Ammobaculites balkwilli*, *Ammoscalaria runiana*, *Ammotium cassis*, *Reophax moniliformis* and, occasionally, *Ammotium salsum*. Locally, there are assemblages dominated by *Ammobaculites balkwilli*, *Ammoscalaria runiana*, *Reophax moniliformis*, *Elphidium oceanensis* or *Eoeponidella pulchella* (Figure 4.8). In near marine conditions there are assemblages dominated by *Cibicides lobatulus* or *Nonion depressulus* (Alve and Murray, 1999). None of these species is confined to the intertidal zone but several are restricted to the subtidal zone (*Ammotium cassis*, *Eggerelloides scaber*, *Ophthalmina kilianensis*). Bottsand lagoon in the Baltic is a rather extreme environment (salinity 12–18, frozen in winter to 28–30 °C in summer) and the >100μm living assemblages are dominated either by *Elphidium williamsoni* (= *Cribrononion articulatum* of Lutze, 1968) or *Miliammina fusca* with few other species present. The standing crop varies from 1–343 individuals 10 cm^{-2}.

The macrotidal Elbe estuary, Germany, has mainly an *Elphidium excavatum* assemblage with subsidiary *Ammonia* group, *Haynesina germanica* and *Nonion depressulus*. In the estuary mouth there is an *Ammonia* group assemblage with subsidiary *Elphidium excavatum*, *Elphidium williamsoni* and *Haynesina germanica*. On the shelf immediately in front of the estuary there is a return to an *Elphidium excavatum* assemblage with *Ammonia* group. Although there is no relationship between test size and sediment grain size for living individuals, the dead forms are sorted to match the sediment grain size (Wang, 1983). In the mesotidal southern North Sea the brackish tidal flats of Germany have assemblages dominated by calcareous taxa. The main species is *Haynesina germanica* and Haake (1962) noted a correlation between its abundance and that of the diatom *Gyrosigma baltica*. *Elphidium excavatum* is widely distributed especially on sandy substrates. Other species commonly present included *Ammonia* group, *Elphidium oceanensis* and *Elphidium williamsoni*. The same species are present in

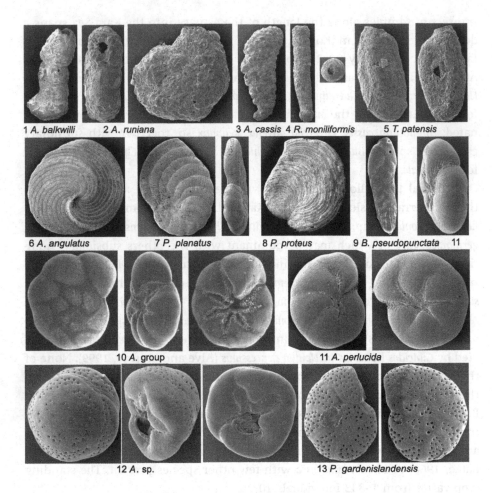

Figure 4.8. Scanning electron micrographs of lagoon foraminifera (longest dimension, μm). 1. *Ammobaculites balkwilli* (330). 2. *Ammoscalaria runiana* (340, 340). 3. *Ammotium cassis* (2400). 4. *Reophax moniliformis* (700, 130). 5. *Trilocularena patensis* (225, 225). 6. *Archaias angulatus* (1700). 7. *Peneroplis planatus* (1170, 1000). 8. *Peneroplis proteus* (2750). 9. *Bolivinellina pseudopunctata* (420). 10. *Ammonia* group (all 300). 11. *Aubignyna perlucida* (300, 300, 330). 12. *Asterigerinata* sp. (all 150). 13. *Palmerinella gardenislandensis* (150, 180).

the Dollart–Ems estuary (van Voorthuysen, 1960; Reinöhl-Kompa, 1985) and the southern North Sea intertidal zone (Langer *et al.*, 1989) and all of them are dominant in different assemblages.

The estuaries of southern England range from microtidal (Christchurch Harbour, Murray, 1968b) through mesotidal (Exe, Murray, 1983; Hamble, Alve and Murray, 1994; Murray and Alve, 2000a) to macrotidal (Tamar, Murray, 1965a; Ellison, 1984). The annual salinity range in Christchurch harbour is much more extreme than that of the other estuaries because it is small and shallow and is fed by two rivers. During the winter salinities are very low due to higher river discharge but in the summer seawater penetrates the lower part of the estuary. This estuary also experiences diurnal salinity variation related to tidal exchange. In the other estuaries the salinity falls below 30 only occasionally and then never below 22 and the diurnal variation is modest. *Haynesina germanica* is dominant in all the estuaries and is probably the most opportunistic of the brackish species. *Paratrochammina (Lepidoparatrochammina) haynesi, Ammonia* group, *Elphidium williamsoni* and *Nonion depressulus* are dominant in one or two estuaries though they are subsidiary in others. The occurrence of these taxa reflects subtle differences between the estuaries. R-mode analysis of Christchurch harbour data (Howarth and Murray, 1969) shows that *Ammonia* group is the most euryhaline, while the other common taxa (*Miliammina fusca, Elphidium oceanensis, Elphidium williamsoni* and *Haynesina germanica*) show no correlation with any of the measured factors. *Eggerelloides scaber, Quinqueloculina dimidiata, Bulimina elegans* and *Elphidium margaritaceum* are shelf species that are able to colonise the seaward parts of estuaries.

The Bassin d'Arcachon on the Atlantic coast of France ranges from normal salinity to brackish. Le Campion (1970) recognised a number of subenvironments. Subtidal marine sands with *Venus* are dominated by *Planorbulina mediterranensis* or *Elphidium lidoense* with subsidiary *Eggerella advena, Eggerelloides scaber, Nouria polymorphinoides* or *Triloculina rotunda*. Muddy sands with *Abra* are dominated by *Eggerelloides scaber* or *Planorbulina mediterranensis* with *Quinqueloculina partschii* or *Nonionoides boueanum*. Brackish intertidal muds with a vegetation of *Zostera* are dominated either by *Ammotium salsum, Eggerelloides scaber, Triloculina inflata, Ammonia* group, *Elphidium williamsoni* or *Haynesina germanica* with subsidiary *Miliammina fusca* or *Reophax moniliformis*. Brackish *Scrobicularia* mud is dominated by *Ammotium salsum* with subsidiary *Miliammina fusca, Reophax moniliformis* or *Trochammina inflata*.

Two areas in northern Spain have been sampled seasonally over two years. Santoña estuary and San Vicente de la Barquera lagoon are shallow and mesotidal; there are extensive tidal flats at low water. The inner part of Santoña is brackish (salinity 0–25 during the winter, 9–40 during the spring and summer) but the main part of the estuary is normal marine to slightly hypersaline (salinity 31–42) as is San Vicente de la Barquera (salinity 33–43). The principal species throughout both areas and all seasons are the *Ammonia* group and *Haynesina germanica* with *Quinqueloculina seminulum* becoming one of the dominant species in the summer in Santoña and in spring and summer in San Vicente de la Barquera (Cearreta, 1988; 1989). *Miliammina fusca* is locally dominant in the Santoña inner estuary and *Elphidium williamsoni* and *Nonion depressulus* in San Vicente de la Barquera. In salt works on the Mediterranean coast of Spain, foraminifera survive until salinities exceed 60–65. Sandy substrates support assemblages of *Trochammina inflata* and *Jadammina macrescens* or *Ammonia* group, *Haynesina germanica*, *Nonion depressulus* (*Haynesina depressula* of Zaninetti, 1984) but only at salinities < 50. *Jadammina macrescens* is controlled more by substrate than salinity. Areas with algae support small miliolid assemblages with *Miliolinella*? sp. surviving up to 50 while *Triloculina* sp. survives up to salinity 65.

Africa. The elongated Casamance estuary in Senegal is hypersaline in its inner part. The two most abundant living forms are the *Ammonia* group and *Ammotium salsum*. The former is most abundant towards the centre of the estuary and the abundance is lowest during the rainy season (July, August) and reaches a peak in January–February. It is able to exist where the salinity is 35–50 for at least three months. *Ammotium salsum* is most abundant in the inner part of the estuary where salinities sometimes exceed 60 (Debenay and Pagès, 1987).

Time-series studies. These have been carried out in several areas. Bottsand Lagoon showed a peak in standing crop in November 1965 but this was not repeated in 1966. *Elphidium williamsoni* (given as *Cribrononion articulatum* by Lutze, 1968) was dominant in 1965 but in 1966 *Miliammina fusca* was mainly dominant. *Elphidium excavatum* showed peaks of abundance in March and October in a lagoon in Wales (Haman, 1969). *Haynesina germanica* was dominant for 11 months out of 12 in the Plym estuary (Castignetti, 1996) and its peak standing crop was from May to August. The seaward part of Christchurch Harbour is partly colonised by inner-shelf taxa during the autumn when salinities are at their highest (Murray, 1968b). In the outer Exe estuary, the standing crop and the dominant (inner-shelf) species, *Nonion depressulus*, showed peaks in the spring and autumn of two years but not of the third (Murray, 1983). *Nonion depressulus* was the dominant species throughout the 32-month study period.

In the Hamble estuary, two intertidal stations were sampled monthly over a 27-month period in one of the most detailed time-series studies so far undertaken. The sediment was sampled at 2.5 mm intervals in the top 1 cm and 1 cm intervals down to 4 cm (giving 7 samples per replicate; two replicates each month from each of the two stations; Murray and Alve, 2000a; Alve and Murray, 2001). Patchiness occurred on a scale of a few centimetres and this affected the temporal record of the standing crop. The dominant and subsidiary species were most abundant in the upper 0.25 cm of sediment

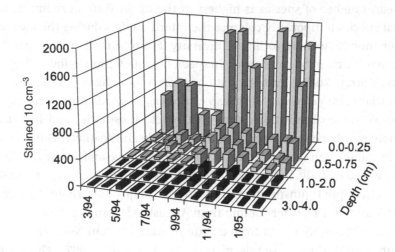

Figure 4.9. Time/depth distribution of standing crop at an intertidal location in the Hamble estuary, England (from Alve and Murray, 2001).

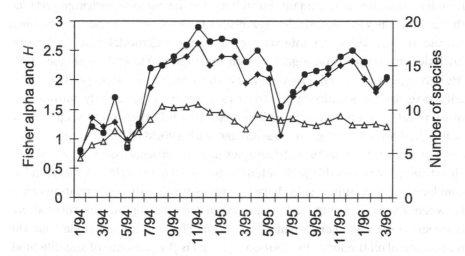

Figure 4.10. Variation in species diversity (Fisher alpha, circles; *H*, triangles; and number of species, diamonds) in the surface 0–1 cm of sediment, Hamble estuary, England (from Murray and Alve, 2000a).

throughout the study period (Figure 4.9) showing that seasonality is not a controlling factor for the vertical distribution of individuals in this area. There is no clear seasonal pattern in the standing crop of the whole assemblage or the dominant species but species diversity (Figure 4.10) and biomass show some cyclicity. The total number of living species encountered over the 27 months at the two stations is 35 but no sample contains more than 22 species.

The mean number of species is highest in the upper 0.25 cm sediment layer. Subtidal species temporarily colonise the intertidal flats during the favourable summer months leading to higher diversity. From SHE analysis it was determined that the cumulative increase in species through time follows a log series pattern (Murray, 2003a). At both stations there was a change in the dominant species (*Haynesina germanica, Elphidium excavatum, Ammonia* group) during each year. Thus, these observations show that brackish estuaries and lagoons are extremely variable environments both temporally and spatially. This should be borne in mind when making comparisons between live and dead assemblages.

Venice lagoon, Italy, is microtidal, has an annual salinity range of ~28–32, a peak in chlorophyll (unspecified) from July to August and there is more suspended matter in the water column from August to December than for the rest of the year. Time-series data from a suite of stations from November 1992 to September 1994 show a simple picture (Donicci *et al.*, 1997). The *Ammonia* group is dominant at most stations and during most months, followed by *Haynesina germanica* over numerous months, and there is just one month when *Elphidium granosum* is dominant. Subsidiary species include *Aubignyna perlucida* throughout the year, *Buliminella elegantissima* in October 1993 and *Edentostomina cultrata* in July 1994. An intertidal pool in the microtidal Gulf of Trieste, Adriatic, was sampled on a grid pattern in August 1985 and September 1987 (Hohenegger *et al.*, 1989). The pool has a substrate of fine sand partly covered with seagrass, the salinity ranges from 18 in spring to 33 in early autumn and most points are continuously submerged. In 1985, with one exception, all points had an *Ammonia* group assemblage with subsidiary *Triloculina planciana* and *Elphidium gunteri*. In 1987, 10 samples had an *Ammonia* assemblage and 8 an *Elphidium gunteri* assemblage. In addition to *Triloculina planciana*, *Quinqueloculina dimidiata* was a common subsidiary species. Strong positive correlations exist between *Ammonia* group and *Elphidium gunteri* and significant correlations between these and *Quinqueloculina dimidiata*. The authors conclude that the main control on the local distribution patterns is the presence of suitable food. *Ammonia* group and *Elphidium gunteri* show a strong preference for 'blue-green algae' especially *Oscillatoria*, a positive relationship with the diatoms *Pinnularia* and *Navicula* but a negative relationship with the diatom *Gyrosigma*. The pattern is similar for *Quinqueloculina dimidiata* except that its preferred 'blue-green alga' is *Anabaena*. *Triloculina planciana* shows a positive correlation with seagrass. This study reveals some of the complexities of local controls on species microdistribution patterns. See Section 8.6 for further discussion.

Summary. The distribution of the main species is summarised in Figure 4.7; ecological information on species ecology is given in the Appendix.

In the Arctic (Russia), typical brackish forms are *Ammotium cassis, Eggerella advena*?, *Reophax atlantica, Elphidium clavatum* and *Haynesina orbiculare*. In temperate brackish conditions (Scandinavia–English Channel), typical forms are *Ammotium salsum, Miliammina fusca*, the *Ammonia* group, *Elphidium excavatum, Elphidium williamsoni* and *Haynesina germanica*. The inner-shelf species *Eggerelloides scaber* and *Nonion depressulus* occur in the mouths of estuaries. In warm temperate areas (France, Spain), salinities may become hypersaline during the summer. Then miliolids such as smooth *Quinqueloculina* and *Triloculina* become more abundant. Some species are always subtidal (*Ammotium cassis*). Others are mainly intertidal but may extend down to a few metres (*Ammotium salsum, Elphidium williamsoni, Haynesina germanica*).

Along the seaboard of Europe, there is overlap in the distribution of northern and southern species. For instance, the northern species *Elphidium albiumbilicatum* reaches its southern limit in the Kattegat and in Loch Etive, Scotland. The southern species *Haynesina germanica* dies out in the Oslo Fjord and does not extend into the Baltic. Other brackish species that occur in the Kategatt but do not extend into the Baltic include *Ammobaculites balkwilli, Goesella waddensis, Haplophragmoides wilberti, Paratrochammina (Lepidoparatrochammina) haynesi, Reophax moniliformis* and *Elphidium oceanensis* (Alve and Murray, 1999). The reason may be partly due to the lower salinities in the Baltic. Also, because the Baltic surface brackish layer has such a low salinity (<10–15) and extends to tens of metres, forms such as *Haynesina germanica* that feed on diatoms and husband chloroplasts may find the light level too low for them and their food in the less brackish deeper waters.

The standing crop is very variable showing that distributions are patchy. Above the halocline in the Ob estuary it is very low (2–14 individuals 10 cm^{-3}) while below there is a seaward increase from 44 to 543 individuals 10 cm^{-3}. High maximum values 10 cm^{-3} are reported from estuaries in southern England and northern Spain: 2288 Hamble, 1828 Santoña, 3404 San Vicente de la Barquera.

4.3.2 Atlantic North America

These estuaries extend from latitude 50 to 27 °N and cross climatic belts from warm summer, cold winter in Canada, through hot summer, cold winter from Cape Cod to Cape Hatteras, to humid temperate hot summer, warm winter down to Florida. All are brackish to a certain extent and Indian River, Florida, is partly hypersaline. Most have muddy substrates because they are low-energy environments. The Canadian examples are meso- to macrotidal with an extensive intertidal zone but all the USA examples are microtidal. Although there is some oxygen depletion in Long Island Sound it is not to a level that affects foraminifera. As elsewhere, foraminifera live infaunally down to several centimetres (Akers, 1971; Buzas, 1965; 1974; 1977; Brooks, 1967; Schafer, 1971; Matera and Lee, 1972) but the results discussed here concern the top 1 cm of sediment. The sources of data and overall distribution are given in Figure 4.11.

Canada. There are *Miliammina fusca* assemblages in the upper-most Miramichi estuary and inner Chezzetcook inlet. There is also a single *Ammotium salsum* assemblage in inner Chezzetcook. Assemblages dominated by *Ammotium cassis* or by *Haynesina orbiculare* occur in the transition from inner to outer Miramichi Bay and LaHave estuary while *Elphidium clavatum* with *Haynes-ina orbiculare* occurs in outer Miramichi Bay. The absence of the *Ammotium cassis* assemblage from Chezzetcook inlet is attributed by Scott *et al.* (1980) to the lack of a turbidity maximum zone. The *Haynesina orbiculare* assemblage with subsidiary *Elphidium selseyensis* and *Ammonia* group or *Elphidium williamsoni* is by far the most common in Chezzetcook inlet occurring in the lower-estuarine and open-bay divisions. In LaHave estuary, where salinities are 25–40, the *Haynesina orbiculare* assemblage has subsidiary *Elphidium excavatum* in the inner part and *Ammotium cassis* in the outer part. This is the northernmost occurrence of assemblages dominated by the *Ammonia* group. Scott *et al.* (1980) suggest that in northern latitudes the *Ammonia* group is restricted to intertidal areas because these are the environments that achieve sufficiently high tempera-tures during the summer months for reproduction to take place.

Northern USA. From Buzzards Bay to Long Island Sound, there are *Ammonia* group, *Elphidium clavatum* and *Elphidium incertum* assemblages. *Haynesina germanica* is present throughout these examples but is only common in Long Island Sound. Here, in intertidal pools, the assemblages are primarily dominated by the *Ammonia* group (1966–1967 but not 1968), *Elphidium incertum* (mainly 1969) or *Haynesina germanica*. Standing crop is greatest on *Enteromorpha* in early summer and then on other algae, but by the end of the summer there are few foraminifera living on *Enteromorpha*. *Haynesina germanica* is most abun-dant in early summer but by late July–August *Elphidium incertum* is most abun-dant (in 1968). *Haynesina germanica* is rare on sediment but that is certainly not true elsewhere. The main sediment species is *Elphidium incertum*. More living foraminifera occur in the sediment than epiphytically on algae. For these rea-sons, Matera and Lee (1972) consider that the epiphytic assemblages are not continuous with those in the sediment. These results show that in this very variable environment, there is major faunal change not only within a single season, but also no repetition from one year to the next (e.g., *Ammonia* group). Other assemblages occurring locally include *Stainforthia fusiformis* in the centre

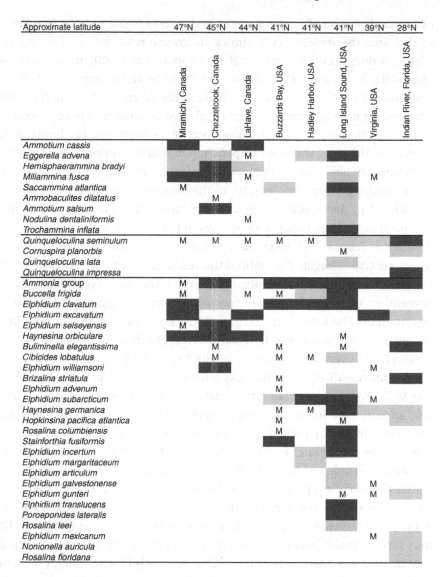

Figure 4.11. + web. Distribution of taxa in Atlantic N American estuaries and lagoons. Black = dominant, grey = subsidiary, M = minor. Sources of data: Canada: Scott *et al.*, 1977 (WA-52), Scott *et al.*, 1980, (WA-53), Allen and Roda, 1977 (WA-54); USA: Murray, 1968a (WA-55), Buzas, 1968a (WA-56), 1965 (WA-57), 1969 (WA-58), Lee *et al.*, 1969, Matera and Lee, 1972 (WA-59), Akpati, 1975 (AA-60), Ellison, 1972 (WA-61), Culver *et al.*, 1996, Woo *et al.*, 1997 (WA-62), Buzas and Severin, 1982 (WA-63).

of Buzzards Bay, amd *Eggerella advena* and *Buccella frigida* in Long Island Sound. In this area the standing crop shows an inverse relationship with depth; the highest values occur at depths < 20 m. Thomas *et al.* (2000), using total assemblage data, have suggested that since the 1960s there has been faunal change in Long Island Sound with a decline in *Eggerella advena* and an increase in the *Ammonia* group. This remains to be confirmed by studies of living forms.

Chesapeake Bay, Virginia, has low diversity assemblages. In Choptank River, Maryland, there are three main species: *Ammobaculites exiguus, Ammonia* group, *Elphidium clavatum*. Buzas (1969) took four replicate samples each month from three stations along the length of the estuary. The innermost station is dominated by *Ammobaculites exiguus* every month in all replicates. The middle station is dominated by all three species in the January replicates, by *Elphidium clavatum* in February, *Elphidium clavatum* and *Ammobaculites exiguus* in March and by *Ammobaculites exiguus* throughout the rest of the year. The outermost station is dominated by *Elphidium clavatum* throughout the year except during July and September when *Ammobaculites exiguus* and *Ammonia* group are respectively co-dominant. The temperature regime is the same for all stations and salinity is higher by ~2 at the outer station compared to the inner but the annual range is small (12–17 overall). Correlation of the three species with a set of variables (temperature, salinity, dissolved oxygen and chlorophyll *a, b, c*) is significant at the 95% level but none of the variables is significant individually. A similar situation is found in James River where the boundary between an *Ammobaculites crassus* assemblage in the inner estuary and an *Elphidium clavatum* assemblage in the outer estuary coincides with the average position of the 14 isohaline (Nichols and Norton, 1969). In Rappahannock River an *Ammobaculites crassus* assemblage is present where the salinity is low (12–15) and the substrate fine grained with abundant organic matter and faecal pellets (Ellison, 1972). *Ammobaculites crassus* of Ellison (1972) and *Ammobaculites exigua* of Buzas (1969) may be the same (Ellison and Murray, 1987). Data on the *Ammonia* group show that in 1998 it predominates where salinities are 9–12 with a slight decrease as salinity rises to 20. It occurs in both oxic and seasonal dysoxic conditions (oxygen 0.47–1.72 ml l^{-1}) and is also considered to be more abundant today than in the 1960s (Karlsen *et al.*, 2000). Near the entrance to Chesapeake Bay, a mainly intertidal lagoon in Virginia has only a small presence of *Ammobaculites exiguus* (except on marshes) with most assemblages composed of the *Ammonia* group and *Elphidium excavatum* (Culver *et al.*, 1996; Woo *et al.*, 1997).

Southern USA.   Indian River on the Atlantic coast of Florida is a shallow, elongated lagoonal estuary (195 km long) separated from the ocean by a barrier island. The average depth is 1.5 m and the greatest depths of about 3.5 m occur in dredged channels. Three connections with the open sea are maintained artificially. It has been studied in detail over many years (Buzas, 1977; 1978; 1982; 1989; Buzas and Severin, 1982; 1993; Buzas and Hayek, 2000; Buzas *et al.*, 2002). Throughout most of the lagoon annual temperature range is ~14–32 °C but in the northern part the range is higher at 11–35 °C. Likewise, salinities are mainly 20–38 but in the north they range from 18–46 so become hypersaline. By far the most abundant forms are the *Ammonia* group and they occur with a variety of subsidiary species including *Quinqueloculina impressa, Quinqueloculina seminulum, Brizalina striatula, Buliminella elegantissima, Elphidium excavatum* and *Elphidium mexicanum*. In the northern, more saline part, there are *Quinqueloculina impressa* and *Quinqueloculina seminulum* assemblages. There are also rare tropical forms such as *Peneroplis pertusus* and *Sorites marginalis* that are at the northern limit of their distribution. Canonical analysis of the fifteen most abundant species was carried out by Buzas and Severin (1982). Canonical variate 1 contrasts the inlets plus locality Haulover with the rest of the area and shows them to be distinct in that some of the minor species are more/less abundant in the former than the latter. However, in most cases the 15 most abundant species make up 95% of the standing crop. Living foraminifera are present down to 6–7 cm below the sediment surface and there is no stratification of species (Buzas, 1977). Experiments using cages on the sediment surface to exclude macrofaunal predators indicate that predation is responsible for keeping down the standing crop and that a wide variety of organisms consume foraminifera, many of them incidentally while deposit feeding (Buzas, 1978; 1982; Buzas and Carle, 1979). Experiments to assess the effect of quartz sand versus calcareous sand as a substrate on standing crop reveal no differences (Buzas, 1989). During 1978, fortnightly sampling of two different substrates, bare quartz sand and quartz sand with a dense cover of seagrass, showed there is no significant difference in standing crop between three of the main taxa: *Quinqueloculina* (mainly *impressa* and *seminulum*), *Elphidium* (mainly *gunteri* and *mexicanum*) and the *Ammonia* group. Comparison of the foraminiferal results with ten environmental variables (which are all highly correlated) shows that for *Quinqueloculina* at least seven variables are required to explain the principal component results. The *Ammonia* group appears to be least influenced by any of the variables measured. During the main survey in 1975–76 (Buzas and Severin, 1982) the standing crop of the top 2 cm of sediment was 8–472 individuals 10 cm^{-3}. Sampling continued at one locality for 11 times spread over a 20-year period (Buzas and Hayek, 2000). Seasonal, interannual and decadal variation is evident: 1994 is the year with the lowest values for four of the five main taxa; for all species values in the 1980s are higher than those of the 1970s or 1990s. However, overall there is no apparent long-term change in standing crop (results are expressed in terms of statistics rather than primary data). Further monthly replicate sampling of three stations over a five-year period (1992–1996; Buzas *et al.*, 2002) led to the concept of pulsating patches of standing crop. These temporal and spatial differences in the patches are caused by asynchronous reproduction of the main species. Thus there is short-term variability, but no long-term increase or decrease of any species is evident. These various studies carried out by Buzas and his co-workers have considerably advanced our understanding of the processes operating in marginal marine environments.

Summary.   The distribution of species (Figure 4.11) shows a progressive change from north to south. There are areas of faunal change: a cool water zone extends from the Gulf of St Lawrence to between the southern border of Canada and Cape Cod; there is another boundary at Cape Hatteras separating a temperate region from a warmer one to the south. Along the Canadian seaboard there are assemblages dominated by *Ammotium cassis*, *Hemisphaerammina bradyi*, *Miliammina fusca*, *Elphidium excavatum* and *Haynesina orbiculare* and the first two of these species do not occur further south even as minor components of assemblages. The southern limit of this group is between the Canadian border and Cape Cod. *Elphidium clavatum* dominates assemblages between Canada and Long Island Sound which is its southern limit (on the data available). The *Ammonia* group are dominant along the entire seaboard except in the northernmost estuary (Miramichi, Canada) although the group has been recorded as a minor component in nearby Northumberland Strait (Schafer, 1967). *Buccella frigida* is dominant in Long Island Sound but not elsewhere although it occurs as far north as Canada. *Elphidium incertum*, *Elphidium subarcticum*, *Elphidium translucens* and *Haynesina germanica* dominate assemblages in Long Island Sound. At the southern extreme, Indian River, Florida, has a very different fauna from all the other estuaries with only the *Ammonia* group in common.

The faunas of the Atlantic seaboards of Europe and North America are remarkably similar although this is not always immediately obvious because of different usage of species names. Figure 4.12 lists those taxa that are dominant in one or both areas. Ten are dominant in both but the rest are dominant in only one; in a few cases a species is present in subsidiary or minor amounts in the other area. However, several species have not been recorded from estuaries and lagoons in one or other area. This difference is almost certainly real for agglutinated forms. *Eggerelloides scaber* has not been recorded from the Atlantic seaboard of North America. There, the equivalent form is *Eggerella advena*. There are a few records of *Eggerella advena* in Europe but whether they have been correctly identified requires confirmation. Species of *Elphidium* are variable in morphology and consequently difficult to identify so the differences between

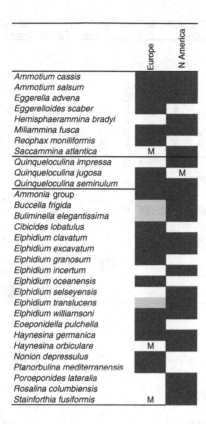

Figure 4.12. Comparison of North Atlantic dominant-species distributions.
Black = dominant, grey = subsidiary, M = minor.

the two areas may be apparent rather than real. This problem may be addressed
in the future via molecular genetics. There are major differences in the latitudinal positions of the northern limits of certain taxa. Both *Ammonia* group and
Haynesina germanica have their northern limit in Europe in southern Scandinavia
(latitude 59–60° N) whereas off North America the limits are in Nova Scotia
(latitude 42 and 45° N respectively). This reflects the influence of the North
Atlantic Drift on the European climate.

4.3.3   Gulf of Mexico and South America

In the Gulf of Mexico, the shallow waters of lagoons may reach 38 °C
during summer. Salinities may range from brackish to hypersaline according
to season. There are two basic sedimentary types of lagoon: carbonate and
clastic. The carbonate environments are discussed in Chapter 6 (Florida, USA;
Mexico; Jamaica; Barbuda; Tobago). Sources of data for clastic estuaries and
lagoons are given in Figure 4.13. Alvarado lagoon, Mexico is brackish. In the

Salinity	Mar.	Mar.	Mar.	Br.–hyp.	hyp.	Br.	Mar.	Mar.	Mar.	Mar.	Mar.	Mar.
Sediment	Carbonate		Clastic				Carbonate			Clastic	Carbonate	
	Mollasses Reef, Florida, USA	Buttonwood, Florida, USA	Tampa, Florida, USA	Texas, USA	Laguna Madre, Mexico	Alvarado, Mexico	Campeche, Mexico	Belize	Jamaica	Puerto Rico	Barbuda	Tobago
Miliammina fusca	M			M	M		M					
Textularia agglutinans	M							M	M	M		
Ammobaculites dilatatus		M		M			M					
Martinottiella occidentalis												
Valvulina oviedoiana								M		M		M
Ammotium salsum							M					M
Deuterammina cf. rotaliformis												
Trochammina cf. multiloculata												
Ammoscalaria pseudospiralis												M
Ammoastuta inepta												
Bigenerina irregularis							M					M
Textularia candeiana									M		M	
Eratidus foliaceus												
Reophax comprima												
Reophax scorpiurus												
Archaias angulatus	M							M	M			
Hauerina bradyi	M							M			M	
Hauerina ornatissima	M							M				
Miliolinella circularis	M											M
Peneroplis carinatus	M											
Peneroplis proteus	M							M	M			
Quinqueloculina bidentata	M							M				M
Quinqueloculina bosciana								M				
Quinqueloculina bradyana								M				M
Quinqueloculina funafutiensis	M		M			M						
Quinqueloculina laevigata	M											M
Quinqueloculina lamarckiana	M	M					M	M		M	M	M
Quinqueloculina poeyana	M						M	M		M		
Quinqueloculina polygona	M	M					M	M			M	M
Schlumbergerina alveoliniformis	M											M
Triloculina bassensis	M											
Triloculina bermudezi	M											
Triloculina linneiana	M							M			M	M
Triloculina oblonga	M	M										M
Triloculina rotunda		M						M				
Triloculina trigonula	M	M	M				M	M		M	M	M
Quinqueloculina sabulosa												
Quinqueloculina seminulum		M	M	M		M	M				M	M
Quinqueloculina subpoeyana		M					M			M		M
Triloculinella obliquinoda		M										
Triloculina cf. brevidentata												
Edentostomina cultrata												
Quinqueloculina rhodiensis					M		M			M		
Quinqueloculina wiesneri							M					
Cornuspira planorbis							M		M			
Amphisorus hemprichii											M	M
Articulina sulcata												
Triloculina circularis												
Cornuspira involvens											M	M
Miliolinella subrotunda												
Cornuspiramia antillarum												
Parrina bradyi												
Quinqueloculina quadrilateralis												
Triloculina eburnea												M
Quinqueloculina sclerotica												
Sigmoilina distorta												

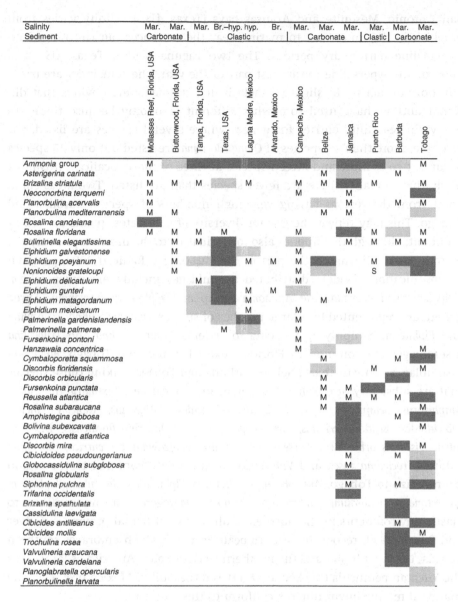

Figure 4.13. + web. Distribution of taxa in Gulf of Mexico lagoons.
Black = dominant, grey = subsidiary, M = minor. Sources of data: Florida: Lynts,
1971 (WA-64), Wright and Hay, 1971 (WA-65), Walton, 1964 (WA-66); Texas:
Phleger, 1956 (WA-67), 1960b (WA-68), Phleger and Lankford, 1957 (WA-67), Poag,
1976 (WA-69); Mexico: Ayala-Castañares and Segura, 1968 (WA-70), Phleger and
Lankford, 1978 (WA-71), Ayala-Castañares, 1963 (WA-72); Belize: Cebulski, 1961,
1969 (WA-73); Jamaica: Buzas et al., 1977 (WA-74); Puerto Rico: Seiglie, 1975
(WA-75); Barbuda: Brasier, 1975a, b (WA-76, WA-77); Tobago: Radford, 1976a,
b (WA-78).

San Antonio, Mesquite and Aransas bays (Texas, USA) salinity varies from brackish in the inner part to marine near the entrances but can range into hypersaline during dry periods. The two Laguna Madre, Texas, USA and Mexico, are hypersaline for at least part of the year. The remainder are basically normal marine to slightly brackish. It is immediately obvious that different authors have used somewhat different taxonomy because there are many single-locality entries for genera where several species are listed. For example, more than 40 species of *Quinqueloculina* are listed but only 25 species occur at two or more and fifteen species at three or more localities. It is most likely that in reality there are fewer species than are listed. Two areas stand apart from the rest in having very high numbers of species: Barbuda and Tobago. This may reflect the greater diversity of substrates (plants as well as sediment; although that applies also to Florida), or to the more tropical aspect, or to greater splitting on the part of the authors (S., Radford 2004, personal communication). Widely distributed assemblages include *Ammotium salsum*, *Quinqueloculina poeyana*, the *Ammonia* group and *Elphidium poeyanum*. Three genera are represented by a large number of species: *Quinqueloculina*, *Triloculina* and *Elphidium*. Simplifying the data to genera, *Quinqueloculina* and *Triloculina* assemblages are common in Florida, Texas, Barbuda and Tobago. *Elphidium* assemblages occur in Texas, Belize, Barbuda and Tobago. Taking those species that are subsidiary or dominant at one or more localities, *Textularia candeiana*, *Amphisorus hemprichii*, *Parrina bradyi*, *Cibicides antilleanus*, *Cibicides mollis*, *Cibicidoides pseudoungerianus*, *Cymbaloporetta* spp., *Discorbis floridensis*, *Discorbis mira*, *Discorbis orbicularis*, *Fursenkoina puncata*, *Hanzawaia concentrica*, *Siphonina pulchra*, *Trochulina rosea* and *Valvulineria candeiana* are confined to the region from Belize to Tobago. *Ammobaculites dilatatus*, *Elphidium delicatulum*, *Elphidium galvestonense*, *Elphidium gunteri* and *Elphidium matagordanum* are confined to clastic environments in the northern Gulf. For shelf faunal provinces, Culver and Buzas (1999) recognised a Caribbean Province which embraces southern Florida, the West Indies and the northern coast of South America as far west as the Yucatan peninsula (see Section 8.11) and the Gulf of Mexico province. The marginal marine environments conform to this pattern.

Alvarado is the only lagoon with a *Miliammina fusca* assemblage. Poag considers that the *Elphidium delicatulum* assemblage is characteristic of low salinity and low temperature in Texas lagoons. The lagoon off Belize (with clastic sediments) is separated from the open sea by a barrier reef. There are two main assemblages: undifferentiated Miliolidae with subsidiary *Amphisorus hemprichii*, *Articulina sulcata*, *Peneroplis carinata* or *Elphidium poeyanum* present mainly on the reef; *Elphidium poeyanum* with subsidiary Miliolidae, *Ammonia* group and *Planorbulina acervalis* in the lagoon. Species found only on the reef include

Valvulina oviedoiana, Articulina mucronata, Archaias angulatus, Asterigerina carinata, Borelis pulchra, Cymbaloporetta squammosa and *Discorbis orbicularis.* The highest standing crops (1000–2800 individuals $10\,cm^{-3}$) are found leeward of mangal or coral sand cays.

Guayanilla Bay on Puerto Rico was subject to pollution from chemicals and hot water discharged from a power plant in the 1970s. The sediments are mainly fine grained with up to 34% $CaCO_3$ and 4–6% organic carbon. Salinities are normal (35–36) and temperatures at the time of sampling 29–36 °C. A warm-water lagoon with depths of <4.5 m is present in the inner part of the bay. In this lagoon the assemblages are dominated by the *Ammonia* group with subsidiary *Quinqueloculina* spp.; locally there are assemblages dominated by *Ammotium salsum.* In the main bay, at depths of 8–25 m there is an *Ammonia* group assemblage with subsidiary *Fursenkoina punctata* and locally there are assemblages dominated by *Fursenkoina punctata.*

There are few studies of living foraminifera in South American lagoons and estuaries. Strongly hypersaline Araruama lagoon, Brazil, has a high proportion of deformed tests which Debenay *et al.* (2001b) attribute to the extreme salinity conditions. Microtidal estuaries situated on the coastal plain south of São Paulo, Brazil, have either dominant *Ammonia* group with *Elphidium* spp. or vice versa or *Miliammina fusca* with these taxa as subsidiaries (Duleba and Debenay, 2003, WA-79). There is no difference in the faunas between clear and dark (high humic-acid content) water estuaries. Although Duleba and Debenay (2003) recognise seasonal differences based on total assemblages these are not evident from the live data because the same stations did not yield adequate assemblages at both seasons. Standing crops are generally very low as relatively few samples yield >50 live individuals $10\,cm^{-3}$. Patos lagoon has dominance of either *Miliammina fusca* or *Haynesina germanica* (Closs and de Madeiros, 1965, Closs and Madeira, 1966 (WA-80)). Live assemblages from brackish to normal marine Laguna de Mar Chiquita, Argentina, have standing crops of 0–172 individuals $10\,cm^{-3}$ and two abundant taxa, *Ammonia* group and *Elphidium discoidale*, either of which may be dominant (Lena and L'Hoste, 1975, WA-81).

4.3.4 Pacific Ocean

Overall, there are few studies and those are mainly on North America: brackish Samish Bay, Washington (Jones and Ross, 1979), California (normal marine Tomales Bay, McCormick *et al.*, 1994, WA-82; normal marine Agua Hedionda and brackish to hypersaline Los Penasquitos, Scott *et al.*, 1976, WA-83, WA-84), Baja California (normal marine to hypersaline Ojo de Liebre and Guerrero Negro, Phleger and Ewing, 1962, WA-31; Phleger, 1965c, WA-30).

In the brackish lagoon the most abundant species are *Trochammina pacifica* and *Miliammina fusca*. The *Ammonia* group has its maximum abundance in August while that of *Elphidium* spp. is June–July. It is only during the summer months that these calcareous species form a significant part of the assemblages. In the normal marine lagoons, a *Glabratella ornatissima* assemblage is common in summer especially in the more turbulent sandy areas near the mouth. Other higher-energy assemblages are those of *Rotorbinella campanulata* and *Elphidium lene*. In less turbulent areas with finer sediment there are *Ammonia* group, *Elphidium excavatum*, *Elphidiella hannai* and *Buccella tenerrima*, and assemblages of basically inner-shelf taxa such as *Brizalina acuminata*, *Brizalina vaughani*, *Bulimina denudata*, *Buliminella elegantissima*, *Fursenkoina pontoni* and *Hopkinsina pacifica*. The hypersaline lagoons have *Ammonia* group, *Elphidium* spp., *Rosalina columbiensis*, *Trochulina lomaensis-versiformis* and *Quinqueloculina* spp. In Laguna Ojo de Liebre there is high organic production (47.2 mg C m^{-3} d^{-1}). Some of the standing crop values are high (maximum 1652 individuals 10 cm^{-3}). Phleger (1976a) considers that nutrients are introduced by inflowing ocean water. By contrast, lagoons in southern Gulf of California have somewhat lower standing crops with many values less than 100 and maximum values around 485 individuals 10 cm^{-3} (Phleger and Ayala-Castañares, 1969).

One common species living in brackish waters of Japan is *Trochammina hadai* (Matsushita and Kitazato, 1990; Kitazato, 1994). Hamana Lake has stratified water masses during the summer when the bottom water becomes strongly dysoxic or anoxic. Then the sediment surface becomes covered by mats of sulphur bacteria (*Beggiatoa*). During the winter, the waters are mixed from top to bottom, oxygenated, diatom growth is resumed and bioturbating macrofauna return. *Trochammina hadai* has its greatest abundance in mid winter (maxima, 1987: 400; 1988: 800 individuals 10 cm^{-2}) and is scarcely present during the summer anoxia. It lives primarily in the surface flocculent sediment layer. The *Ammonia* group live infaunally from the surface to deeper levels (7–9 cm) and they are able to move up and down through the sediment (Kitazato, 1988). Their numbers do not decrease during the winter so they are inferred to have the ability to survive anoxia for some months. *Trochammina hadai* has a reproductive cycle with microspheric forms in spring and summer and megalospheric forms in the autumn. During asexual reproduction, the parent tests are destroyed whereas during sexual reproduction they are undamaged. In Matsushima Bay the dominant species is *Trochammina* cf. *japonica* with subsidiary *Trochammina hadai*, *Ammonia* group and *Elphidium subarcticum* (Matoba, 1970, WA-85). In Inchon, Gyunggi Bay and Asan Bay, Korea, the main living forms are *Ammonia* group, *Elphidium etigoense*, *Elphidium subincertum* and *Nonion nicobarense* (Chang and Lee, 1984, WA-86).

Living forms from the brackish coastal areas of the East China Sea and Huanghai (Yellow) Sea are divided into supralittoral, littoral and estuarine (Wang et al., 1985). The area is mesotidal and experiences strong tidal currents. The supratidal area includes pools and marshes flooded only on spring tides (salinity 1–17, rarely >35). The North Jiangsu littoral zone is 20–30 km wide with muddy sediments and salinity 26–30; that of Zhejiang is only a few kilometres wide, with muddy sediment and salinity 24–33. The channels of the meso–macrotidal estuaries are scoured by strong tidal currents and are barren of foraminifera. However, foraminifera live in the silty intertidal estuarine sediments. Two assemblages are recognised: supralittoral, dominated by the *Ammonia* group with subsidiary *Miliammina fusca*; littoral, dominated by *Ammonia* with subsidiary *Elphidium* sp. Species diversity is lowest in the supralittoral zone (Fisher alpha < 2), intermediate in the estuarine and North Jiangsu littoral-zones (< 4) and highest in the Zheijiang littoral zone (3–6). The majority of assemblages are composed almost entirely of hyaline tests with six supralittoral assemblages having agglutinated forms and four North Jiangsu littoral-zone assemblages having a porcelaneous component. Three living assemblages from a Taiwan tidal flat have dominant *Ammonia* group or *Elphidium gerthi* with subsidiary *Elphidium incertum* while the assemblage from the river bank has a low-diversity marsh assemblage with *Jadammina macrescens*, *Haplophragmoides wilberti* and *Trochammina inflata* (Haake, 1980a, WA-87).

4.3.5   Indian Ocean

On the east coast of India, brackish Bendi lagoon (salinity 26–39.5; temperature 30–33 °C) has low-diversity assemblages mainly of *Ammonia* group, *Haynesina* sp., *Miliammina fusca*, *Quinqueloculina seminulum*, *Quinqueloculina stalkeri*, *Triloculina brevidentata* and *Trochammina advena*. Optimum conditions prevail in the winter with lower temperatures and salinities compared with unfavourable conditions in the summer (Naidu and Subba Rao, 1988). This is true also for the Vembanad estuary (Antony, 1980). Fisher diversity is low in the brackish Pennar estuary (maximum value 3; Reddy and Jagadiswara Rao, 1983). On the west coast, Cochin lagoon has *Ammonia* group assemblages in the brackish areas and one *Textularia earlandi* assemblage with subsidiary *Ammonia* group (Seibold and Seibold, 1981, WA-88).

St. Lucia lagoon in South Africa has four rivers draining into it but salinities are commonly hypersaline even during the rainy season during some years while in other years salinities are brackish. At the time of sampling salinities did not exceed 50. The living assemblages are strongly dominated by the *Ammonia* group (mainly >80%) and the only other common forms are miliolids in the lagoon and *Bolivina* spp. in the 20 km long channel connecting it to the open sea (Phleger, 1976b, WA-89). Most of the standing crops are 50–150 individuals 10 cm^{-3} for the May 1973 samples but some have much higher values (200–4000) in May and November 1972. Nutrients are brought in from the sea (because there is a net inflow of seawater due to evaporation) and they also come from nitrogen fixation by cyanobacteria in the adjacent marshes and rivers. The extreme dominance of *Ammonia*, very low species diversity, and variable standing crops are measures of the extreme conditions in this lagoon.

4.4   Deltas

Deltas are progradations of clastic sediment onto a shelf sea and because rates of deposition are high, the environments are geologically very young (decadal–centennial). The major physical processes controlling the form of deltas are river discharge, waves and tides. It could be argued that from a biological point of view, a delta is merely an assembly of environments found elsewhere – river estuary, lagoon, and inner shelf. It is because of their sedimentary and geological importance that they are treated as a unit. However, as shown below, the foraminiferal faunas found on deltas are the same as those found in the individual environments elsewhere. The Ebro, Spain, and Mississippi, USA, have been studied in most detail. Other records are sparse. Small numbers of live *Ammonia* group and *Arenoparrella mexicana* occur on a delta in Papua New Guinea (Haig and Burgin, 1982). From seasonal studies of the Rhône delta, France (Vangerow, 1972; 1974; 1977) it is known that *Miliammina fusca*, *Ammonia* group and *Elphidium* sp. are tolerant of varying brackish conditions. However, the distribution of the *Ammonia* group is patchy on a small scale. Taxa that avoid areas of freshwater are *Trochammina* sp. and miliolids.

Ebro delta.   This is on the Mediterranean coast of Spain in a non-tidal setting and is influenced primarily by fluvial and wave processes. Sampling by Scrutton in 1967–1968 was briefly reported by Maldonado and Murray (1975) and the dataset has now been made available by Scrutton (personal communication 2004, WA-90). Although the Rio Ebro drains 15% of the area of Spain (85 835 km^2), much of the water is extracted for irrigation and the amount of water and sediment reaching the sea even in the 1960s had been much reduced from its natural state. The salinity of the open sea is 38 and the outflow of freshwater does not influence depths greater than 10 m even close to the river mouth; the freshwater floats as a surface layer ~3 m thick on the denser seawater. However, silty mud sediment settles out under the area of discharge (deltaic marine) with sands along the shore. Wave attack on the flanks of the delta causes erosion of the sea floor to form a delta front platform;

with transport of fine sediment offshore into water >20–25 m deep and longshore transport of sands to form spits to the north and south of the delta. The spits enclose shallow lagoons with variable salinities in the shallower parts (16–38) but 38 in the deeper areas. There are patches of seagrass on the delta front platform and in the lagoons. Adjacent to the lagoons there are almost freshwater lakes (salinity <3); when the lakes are periodically flooded by the sea the salinity rises to 16–30. The sediments are rich in organic matter and depleted in oxygen below the surface.

Because there is little difference in salinity throughout most of the delta, the fauna is much the same in the marine subenvironments. *Ammonia* group, *Asterigerinata* sp., *Bolivinellina pseudopunctata*, *Bulimina aculeata*, *Elphidium lidoense*, *Nonion* spp., *Nonionella opima* and *Valvulineria complanata* are abundant in the deltaic marine, delta front platform and prodelta slope environments. On the sandy substrates of the delta front platform miliolids are also present and there are also two occurrences of common *Leptohalysis scottii*. The lagoons have a miliolid, *Ammonia* group, bolivinid, *Hopkinsina pacifica/atlantica*, *Nonionella opima* assemblage. The organic-rich sediments in the deeper parts of the lagoon are low in oxygen and this may account for the abundance of bolivinids and *Nonionella*. The shallow lakes have an assemblage of *Haynesina germanica* but where the water is brackish the *Ammonia* group may be present. Overall, the faunas of the delta are essentially those to be expected on a normal-marine–slightly hypersaline inner continental shelf.

Mississippi delta. The shape is controlled by fluvial processes. The marsh assemblages have already been discussed above. Phleger (1955, WA-92) and Lankford (1959, WA-93) studied the eastern margin of the delta and recognised a series of intergrading environments. The lagoon ('Sound' of Phleger) lies between the delta, Breton and Osier islands, and the normal shoreline. The delta platform is made up of the interdistributary bays, i.e. the bays between the leveed river channels. The fluvial marine environment occupies the river channels, with depth <10 m and salinity dependent on river flow with freshwater during floods and a salt wedge with salinity up to 32 during dry periods as in a stratified estuary. The prodelta slope or deltaic marine is the progradational face of the delta. The true shelf is more or less beyond the direct influence of the delta. Phleger comments on the extreme variability of the oceanography of the area due to tides (microtidal), winds and variation in freshwater discharge. The data for the two papers have been organised into the environments defined by Lankford. Although some of the dominant species found by the two authors were the same, some were different. This may be due to different times of sampling. In the lagoon, the *Ammonia* group is dominant above all others. Other forms dominant in one or a few samples are *Ammobaculites* sp. 'young', *Ammoscalaria pseudospiralis*, *Ammotium salsum*, *Elphidium* spp., *Fursenkoina pontoni* and *Nonionella opima*. Miliolids are also common in some samples. In the fluvial marine setting, *Palmerinella gardenislandensis* is the key form (Figure 4.8). Lankford notes that it is associated with river water even when it occurs in interdistributary bays. Apart from the species listed for fluvial marine, the *Ammonia* group is a common subsidiary together with *Ammotium salsum*, *Miliammina fusca*, and *Elphidium* spp. On the delta platform *Ammonia* group is the main form but other dominant species include *Ammobaculites* sp. 'young', *Ammotium salsum*, *Miliammina fusca*, *Elphidium* spp. and *Nonionella opima*. The prodelta slope and shelf are mainly dominated by *Nonionella opima* or *Brizalina lowmani* but other species sometimes dominant including *Ammonia* group, *Buliminella* cf. *bassendorfensis*, *Epistominella vitrea*, *Fursenkoina pontoni* and *Elphidium* spp. The size of individuals is generally small and Lankford suggests that this reflects favourable conditions with frequent reproduction.

Comparisons.   The two deltas have some points in common and some major differences. In both, the lagoons have *Ammonia* group, miliolids and *Nonionella opima* but *Elphidium* spp. are not important on the Ebro Delta which is less affected by freshwater input. The delta front platform and prodelta slope have *Ammonia* group and *Nonionella opima* in common but otherwise the assemblages are quite different. There are fewer agglutinated forms and more porcelaneous forms on the Ebro Delta and the species diversities are much higher than those of the Mississippi. A major difference is that the standing crops on the Ebro Delta (maximum 240 individuals 10 cm^{-2}, average values 10–100) are very low compared with those of the Mississippi (maximum 8240, average values 80–500). The latter is a large river draining an enormous catchment and it brings large quantities of nutrients to the sea; these clearly benefit the foraminifera through increased food supply. There are no species confined to deltas because the environments are essentially the same as those found elsewhere in lagoons, estuaries and on inner shelves.

4.5   Fjords

Fjords are a special type of estuary found in high latitudes and some are the deepest estuaries in the world. They have been, or are at present, being excavated by ice from land. They have steep ice-scoured sides, may have a U-shaped profile, and are typically long and narrow. Some have a glacier occupying their inner part but many lack ice and receive drainage from one or more rivers. Unlike shallow estuaries and lagoons, deep fjords may be geologically much older, pre-dating the last glacial event. Shallow fjords and those separated from the sea by a shallow sill are geologically young (<12 000 years). Because most fjords are relatively deep and there is freshwater inflow from the head, there is normally two-layer circulation: a seawards-flowing surface brackish layer separated from the deeper more-saline water by a halocline. Fjords commonly have one or more transverse sills, the outermost of which separates the fjord from the open sea, while other sills divide the depositional area into separate basins. If the outermost sill is shallower than the depth of the halocline, there may be restricted entry of shelf water into the basin and this may lead to ephemeral or permanent oxygen depletion of the bottom waters and even to anoxia. However, the majority of silled fjords have oxic bottom waters. In silled fjords, there is limited supply of sediment from the adjacent shelf and most fjord sediments are derived from the glaciers or rivers flowing into them. However, plankton may be introduced and may locally bloom to contribute organic material to the sediment. Because of rapid sedimentation with few or no breaks (at least away from steep slopes), fjords preserve a detailed record of changing environmental conditions. They are especially sensitive to small-scale changes in the cyclic exchange of water (which may involve phases of anoxia, dysoxia and oxia), climatic changes (especially in relation to the North Atlantic oscillation), relative changes of sea level (due to the interplay of rising sea level and isostatic uplift following the last glacial) and pollution.

4.5.1   European fjords

Wherever there is either summer melting of sea or glacier ice or inflowing meltwater, there is a surface brackish layer of water which is relatively warm compared with the underlying deep water, which is typically more saline and cold throughout the year. In some areas there is a transitional water layer between the two. In this review, an attempt has been made to separate the faunas of the two water masses. Surface-water assemblages are least varied in the Arctic and more varied in temperate southern Scandinavia. Also there are major differences in the dominant species between all three areas (Figure 4.14). By contrast the deep-water assemblages are more widely distributed especially *Adercotryma glomeratum*, *Cassidulina reniforme*, *Elphidium clavatum* and *Nonionellina labradorica*.

Arctic fjords. The surface water is an extreme environment. In Scoresby Sund, east Greenland, foraminifera are absent at depths of a few metres then down to 13 m there are assemblages (>100 μm) dominated by *Haynesina nivea*, *Elphidiella hannai* (Figure 4.2) or *Cornuspira planorbis*. An *Islandiella islandica*

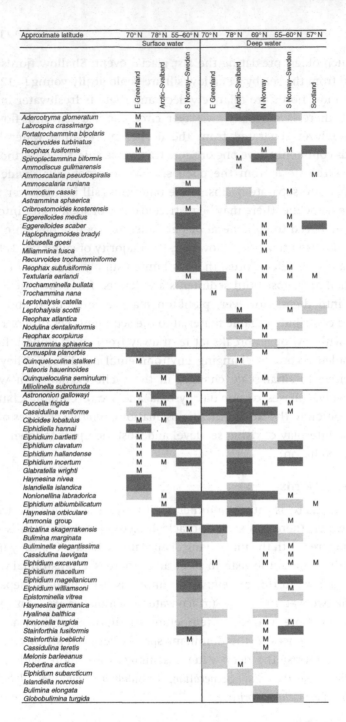

Figure 4.14. Distribution in fjords. The surface water includes the transitional layer. Black = dominant, grey = subsidiary, M = minor. Sources of data: East Greenland: Jennings and Helgadottir, 1994 (WA-93), Madsen and Knudsen, 1994 (WA-94); Arctic–Svalbard: Korsun and Hald, 1998 (WA-95), 2000 (WA-96), Hald and

assemblage with *Cassidulina reniforme* extends from 20–30 m and a *Spiroplectammina biformis* assemblage with *Portatrochammina bipolaris* and *Recurvoides turbinatus* from 23–26 m. At 36 m there is a *Portatrochammina bipolaris* assemblage with *Rhabdammina* sp. In Freemansundet, Svalbard, there are five main assemblages: (1) close to the glacier, *Astrononion gallowayi* with *Pateoris hauerinoides* and *Elphidium hallandense*; (2) in the middle and outer parts, *Elphidium clavatum* with *Astrononion gallowayi, Cibicides lobatulus, Elphidium incertum* and *Elphidium hallandense*. (3) *Cassidulina reniforme* with *Elphidium clavatum*; (4) *Haynesina orbiculare* with *Elphidium clavatum* and *Cibicides lobatulus*; and (5) *Cornuspira involvens*.

A deep-water assemblage on the east coast of Greenland from about 600 m is dominated by *Cassidulina reniforme* with subsidiary *Nonionellina labradorica* while another from ∼500 m has a *Textularia earlandi* assemblage with subsidiary *Cassidulina teretis* and *Cribrostomoides* sp. Korsun and Hald (1998) consider that off Novaya Zemlaya *Cassidulina reniforme* and *Nonionellina labradorica* appear where the impact of the glacier decreases and nutrients increase. Although dissolution of calcareous tests takes place away from the glaciers in the proximal clays, high sedimentation rates inhibit dissolution and lead to the preservation of calcareous tests. Hald and Korsun (1997) reported on the total assemblages > 100 μm from several fjords on Svalbard but only the living data are presented here. *Cibicides lobatulus*, a sessile epifaunal species, shows good correlation with high-energy gravel sediments. *Cassidulina reniforme* is associated with cold bottom water in the inner fjord together with *Elphidium clavatum* near the ends of glaciers. *Adercotryma glomeratum* and *Nonionellina labradorica* correlate with Atlantic water in the outer fjord. There is no marked evidence of dissolution.

Tempelfjorden, Svalbard (Korsun and Hald, 2000) has high turbidity of the surface water during the summer (when the clay-silt sediment accumulates) and ice cover during the winter. *Cassidulina reniforme* is dominant throughout the year: at 40 m depth close to the glacier termination (surface water); 67 and 99 m depth away from the glacier. The only subsidiary species is *Elphidium clavatum*. *Cassidulina reniforme* shows no obvious pattern of reproduction but *Elphidium clavatum* and *Nonionellina labradorica* reproduce in the period March–May, the latter perhaps in response to increased availability of fresh phytoplankton detritus. There was a decline in the summer abundance of *Elphidium*

Caption for **Figure 4.14** (*cont.*)

Korsun, 1997 (WA-97), Hansen and Knudsen, 1995 (WA-98); northern Norway: Austin and Sejrup, 1994 (WA-99), Corner *et al.*, 1996 (WA-99), Husum and Hald, 2004, (WA-100); southern Norway–Sweden: Alve and Nagy, 1990 (WA-101), Alve and Nagy, 1986 (WA-102), Alve, 1995a (WA-103), Gustafsson and Nordberg, 1999 (WA-104), 2000 (WA-105), 2001 (WA-106); Scotland: Murray *et al.*, 2003 (WA-107).

clavatum and *Cassidulina reniforme* between 1995 and 1996. This is attributed to a surface cooling in 1995 but not in 1996.

Temperate fjords. These are present along the coasts of Norway and southern Sweden. Tanafjord on the north coast of Norway has a surface brackish layer 2–5 m thick in summer; at depths 0 > 20 m the salinity is permanently 33–34 and the water is well oxygenated. On a delta growing into the fjord living foraminifera are absent or rare down to 23 m and rare to abundant deeper than this (Corner *et al.*, 1996). Only the deep water of Malangen Fjord was sampled (Husum and Hald, 2004). Fensfjord and Lurefjord in southwestern Norway have a surface brackish layer that extends down to 20–30 m and this was the water mass that was sampled by Austin and Sejrup (1994). Central Oslo Fjord has a range in depth down to >200 m and is separated from the Skagerrak by a sill at 150 m. The water is layered with a seasonal brackish layer down to around 26 m, a transitional layer down to around 90 m and deep water (salinity >33) below this. None of them is oxygen deficient. Gullmar Fjord and Havstensfjord, Sweden, also have a brackish surface-water layer. Four of the investigated fjords have stagnant basin waters leading to periods of low oxygen or anoxia: Frierfjord, Norway, due to pollution from 1870 (Alve, 2000); Gullmar Fjord, Sweden, from the 1970s (Gustafsson and Nordberg, 2001); Drammensfjord, a branch of Oslo Fjord, experienced a period of anoxia in the 1980s (Alve, 1995a); and Koljö Fjord, Sweden, in the 1990s (Gustafsson and Nordberg, 1999). Loch Etive, Scotland, has a halocline at ~20 m with episodic replenishment of deep water which becomes depleted in oxygen but never to a level adversely affecting the foraminifera. The surface water and transitional water masses in these temperate fjords have a wide range of assemblages dominated by many different species and with numerous subsidiary species including agglutinated, hyaline and even some porcelaneous forms (Figure 4.14). In the deeper-water masses, there are agglutinated assemblages in northern Norway and Scotland with *Adercotryma glomeratum* and *Spiroplectammina biformis* as the dominant forms on sandy sediments in the former and *Eggerelloides scaber* and *Reophax fusiformis* in the latter. With the exception of Scotland, hyaline forms are commonly dominant too: *Brizalina skagerrakensis* (outer fjord), *Bulimina marginata* (inner fjord), *Cassidulina reniforme, Globobulimina turgida* and *Nonionellina labradorica* in the north, where even live individuals show evidence of dissolution in Tanafjord, and *Elphidium* spp. In those fjords where there is increased stress due to low oxygen conditions, *Stainforthia fusiformis* is highly dominant.

Time series. These provide a valuable insight into how the fauna changes through time in response to changing environmental conditions. In

Drammensfjord, southern Norway, foraminifera recolonised the basin after a prolonged period of anoxia (>5 years) at water depths >30–35 m (Alve, 1995a). Re-oxygenation started slowly in 1983 but by 1984 there were still no living foraminifera at depths >30 m and even above this, in the transition layer, the faunas were sparse. In 1988, the living assemblages were sampled in February, May, August and October. In the transitional layer (salinity ~30, variable temperatures), an *Ammodiscus gullmarensis* assemblage was present in February, August and October. Other assemblages include *Eggerelloides scaber, Elphidium albiumbilicatum* and *Elphidium excavatum*. The basin water >30 m has a salinity of 30 and a small range in temperature (6–8 °C). An *Ammodiscus gullmarensis* assemblage is present in organic-rich sediment close to anoxic bottom water except in October, a *Stainforthia fusiformis* assemblage except in May, and an *Adercotryma glomeratum* assemblage only in August. *Stainforthia fusiformis* is the most successful coloniser. Between 1984 and 1988, standing crops in the transitional layer increased significantly (>125 μm assemblages, from <14 to a maximum of 144 individuals 10 cm^{-3}).

In southern Sweden time series studies have been carried out in Koljö, Havstens and Gullmar fjords. Koljö Fjord is silled with strong water stratification (pycnocline at 15–20 m, a 5–10 m thick transitional layer below this) and stagnant bottom water, which periodically becomes dysoxic or anoxic due to natural causes (Gustafsson and Nordberg, 1999). The cores were sampled to 7 cm but the majority of forms live in the surface 0–1 cm layer and these are the results discussed here. At 12 and 18 m in the surface-water conditions were oxic throughout. At 12 m the assemblages are generally dominated by *Ammotium cassis* with *Elphidium incertum*. At 18 m in the transitional layer *Elphidium incertum* is dominant throughout with subsidiary *Ammotium cassis* and *Elphidium clavatum*. The situation is similar in the deep water at 24, 28 and 43 m except that *Elphidium clavatum* and *Elphidium magellanicum* are frequently subsidiary and occasionally dominant. Only at 43 m are the standing crops adversely affected by low oxygen. Elsewhere, the lowest standing crops occur during periods of greater oxygenation but this may be related to the seasonality of the food supply. Standing crop is generally low with no values reaching 100 and many below 50 individuals 10 cm^{-3}. The arrival of phytodetritus from the spring phytoplankton bloom stimulated population growth in *Elphidium clavatum* and *Elphidium incertum* and they reached maturity within one month.

At 12 m in the surface brackish layer of Havstens Fjord the dominant species throughout most of the time series is *Ammotium cassis* but during four months *Elphidium clavatum* is dominant or the two species are co-dominant. Subsidiary species include *Astrammina sphaerica, Eggerelloides scaber, Ammonia* group and

Elphidium incertum. The standing crop ranges from 3–67 individuals 10 cm^{-3} with most values between 10 and 28; the peak is in June. At 20 m, in the transitional layer (salinity 25–31), the standing crop is much higher (54–213 individuals 10 cm^{-3}) with peak values in November. The dominant species is generally *Quinqueloculina seminulum* or *Stainforthia fusiformis* but *Bulimina marginata* was dominant during March. Subsidiary species include *Eggerelloides scaber*, *Reophax subfusiformis*, *Ammonia* group, *Buliminella elegantissima*, *Epistominella vitrea* and *Nonionella turgida* but generally these reach >10% abundance for only one or two months. At 30 m, in the more stable deep water (salinity 31–33), the standing crop is yet higher (126–701, with approximately half the values >400 individuals 10 cm^{-3}). *Stainforthia fusiformis* is dominant for all but one month when it is co-dominant with the second-most-abundant form, *Elphidium clavatum*. Subsidiary species include *Eggerelloides scaber*, *Bulimina marginata*, *Elphidium magellanicum* and *Epistominella vitrea*. During August to November 1993, the dissolved oxygen in the bottom water was <1 ml l^{-1} and Gustafsson and Nordberg consider that the pore water would have been anoxic. At 40 m, conditions are even more stable (salinity 33) and there was severe oxygen depletion from August to December 1993 and October to December 1994 (values falling to 0.1 ml l^{-1}). The standing crop is low throughout the year (7–22 individuals 10 cm^{-3}). Again, with the exception of one month, the assemblages are dominated by *Stainforthia fusiformis* with *Elphidium magellanicum* abundant or co-dominant. Subsidiary species include *Leptohalysis catella*, *Elphidium albiumbilicatum*, *Elphidium incertum* and *Epistominella vitrea*. *Stainforthia fusiformis* shows a positive correlation with chlorophyll *a* (used as a measure of primary production by phytoplankton in the surface water, 0–15 m). There was a rapid increase in numbers following a phytoplankton bloom but standing crop is not strongly correlated with chlorophyll *a* and this suggests that degraded organic matter and bacteria also serve as food during times when primary production is low (Gustafsson and Nordberg, 2000) but may alternatively indicate a lag in response. The more prolonged and more severe periods of dysoxia at 40 m is responsible for the low standing crop and low diversity compared with the 30 m site. Recovery from prolonged anoxia can be a slow process as documented in Drammensfjord, Norway, by Alve (1995a).

In Gullmar Fjord over a 17-month period a site at 117 m had dominant *Stainforthia fusiformis* in the 0–1 cm (46–88%) and 0–3 cm (49–97%) sediment layer (Gustafsson and Nordberg, 2001). In the 0–1 cm layer, the subsidiary species include *Saccammina* sp. (1 month) and *Bolivinellina pseudopunctata* (7 months). The standing crop from 0–1 cm ranges from 336–1590 and for the 0–3 cm interval 261–1292 individuals 10 cm^{-3}. However, the number of living tests from 0–3 cm in the sediment over an area of 10 cm^2 is 782–3875. *Stainforthia fusiformis* and some minor species (*Nonionella turgida*) show a positive correlation with chlorophyll *a* from the surface waters (2–15 m). *Stainforthia fusiformis* may have a life cycle of less than one month as it increased its numbers so rapidly (factor of 7) in response to food input.

Summary. Temperate fjords have a surface brackish layer above somewhat more marine deeper waters. The foraminiferal faunas of the surface layer are like those of estuaries while the deeper-water faunas are mainly modified shelf assemblages. There is a clear biogeographic pattern. Arctic Novaya Zemlaya and Svalbard stand apart from the temperate examples in the absence of southern taxa such as *Ammoscalaria runiana*, *Eggerelloides medius*, *Eggerelloides scaber*, *Miliammina fusca*, *Ammonia* group, *Elphidium macellum*, *Elphidium magellanicum*, *Elphidium williamsoni* and *Haynesina germanica*. The occurrence of so many of the southern taxa in the surface water layer of western Norway fjords (Figure 4.14) shows that summer temperatures must be sufficient for reproduction.

4.5.2   Canadian fjords

In Saanich Inlet, a silled fjord with restricted water circulation below 70 m depth and anoxic conditions from 70–150 m, the abundance of live individuals decreases with increasing water depth ($r = -0.533$) and few are encountered below 70 m (Blais-Stevens and Patterson, 1998). In Effingham Inlet the sill lies at 65 m and separates an anoxic inner basin from the oxic outer basin. Only small numbers of stained forms were encountered and Patterson *et al.* (2000) suggest that caution should be exercised in treating these as living occurrences; under anoxic conditions decay of dead protoplasm might be slow. Nevertheless, there are few stained individuals in anoxic waters and no evidence for transport in this quiescent environment. *Stainforthia feylingi* and *Bolivinellina pacifica* occur in dysoxic water. These may be analogues of *Stainforthia fusiformis* and *Bolivinellina pseudopunctata* in northwestern Europe. There are no stained *Leptohalysis catella* although this species is common dead.

4.6   Isolation basins

In Scotland, isostatic uplift has raised rock basins above the reach of modern sea level. Most examples are normal marine (salinity 33–38) with a few being brackish (salinity 17–30; Lloyd and Evans, 2002). The latter are dominated by *Miliammina fusca* while the normal marine examples have dominant *Elphidium williamsoni* alone or co-dominant with either *Miliammina fusca* or *Haynesina germanica*. In one example, *Eggerelloides scaber* is dominant.

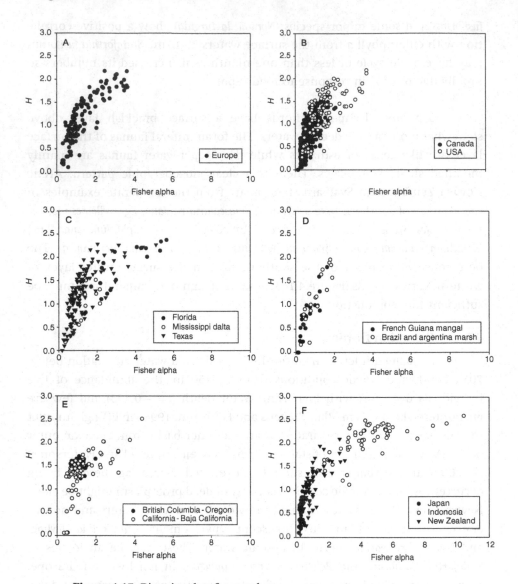

Figure 4.15. Diversity plots for marshes.

4.7   General attributes of faunas

4.7.1   Species diversity

With the exception of Indonesian mangal (low to moderate diversity), regardless of salinity, all marshes and mangals have low diversity (Figure 4.15). Estuaries and lagoons also have low diversity but some normal marine and hypersaline assemblages have moderate values (Figure 4.16). Fluvial-dominated Mississippi delta has low diversity whereas normal marine Ebro delta has low

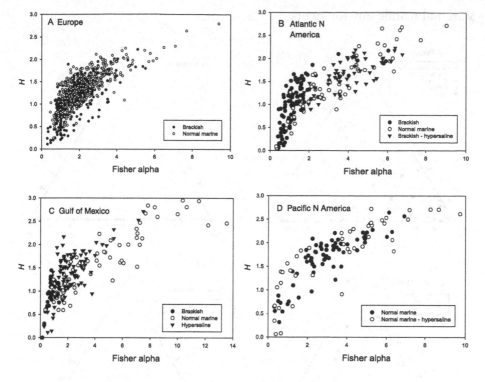

Figure 4.16. Diversity plots for estuaries and lagoons.

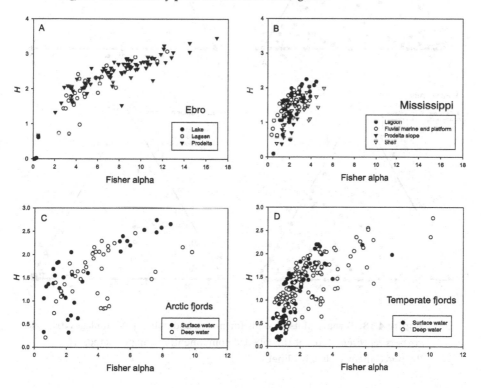

Figure 4.17. Diversity plots for deltas and fjords.

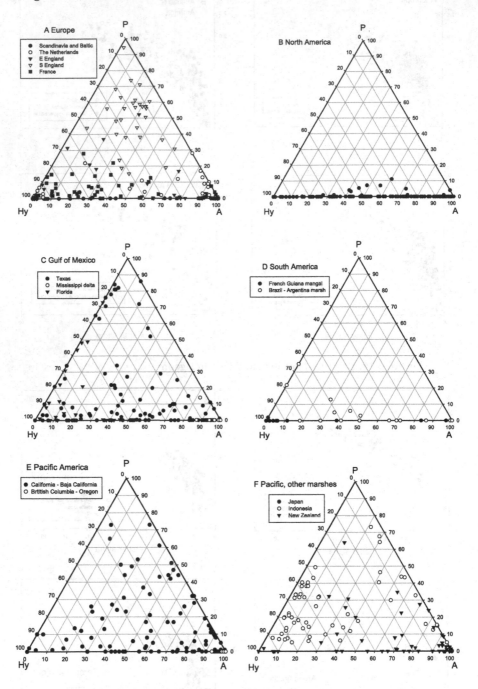

Figure 4.18. Ternary plots of walls for marsh and mangal assemblages (triangle corners represent 100% of the labelled component: A = agglutinated, P = porcelaneous, Hy = hyaline).

to moderate diversity with very low values in the almost freshwater lake (Figure 4.17A, B). The surface waters of fjords (variable salinities and temperatures) have low diversity but the more stable deep waters range from low to moderate (Figure 4.17C, D) depending on the availability of oxygen.

4.7.2   Wall structure

Marsh and mangal assemblages plot throughout the ternary diagram (Figure 4.18). Many have 100% agglutinated walls, especially those from high marshes and from marshes furthest from the open sea. Hyaline forms appear on low marshes and marshes closer to the open sea. Brackish marshes plot mainly along the agglutinated–hyaline side and hypersaline marshes along the hyaline–porcelaneous side. Marshes that vary seasonally from brackish to hypersaline (e.g., Texas) plot along both these sides. Brackish estuaries and lagoons plot mainly along the agglutinated hyaline side with only a small porcelaneous component (Figure 4.19). Normal marine and hypersaline examples plot along the hyaline–porcelaneous side with varying amounts of agglutinated walls. The same applies for deltas (Figure 4.20). In fjords, the normal marine deep-water assemblages plot along the agglutinated–hyaline side while the variable surface waters often have some porcelaneous forms.

4.7.3   Standing crop in 0–1 cm sediment layer

Samples with a count of fewer than 50 have been excluded from consideration so it is not possible to calculate average values. There are some zero values in all environments. Two features characterise the data: extreme variability within a small area; maximum values often several thousand individuals $10 \, cm^{-3}$. For marshes (Table 4.1), with the exception of the Mediterranean (maximum 330 individuals $10 \, cm^{-3}$), maximum values exceed 1000 individuals $10 \, cm^{-3}$ throughout the world. The typical values are several hundred $10 \, cm^{-3}$. For estuaries and lagoons, maximum values in Europe are ~2000–3000 individuals $10 \, cm^{-3}$. However, values are typically low in the Arctic, Exe estuary and Venice (<50) but >200 in the Hamble and northern Spain. In Atlantic North America, maximum values are 260–756 with typical values of 50–200 individuals $10 \, cm^{-3}$. Pacific lagoons have maximum 956 and typically 100–200 individuals $10 \, cm^{-3}$. There is a marked contrast between the Ebro delta (maximum 240, typically 10–100) and the Mississippi delta (maximum 8240, typically 80–500 individuals $10 \, cm^{-3}$) reflecting the nutrient differences between the two areas. Fjords in Europe are very variable according to degree of oxygenation. Maximum values sometimes exceed 1000 but may be as low as 100 and typical values may be <50 or range up to several hundred individuals $10 \, cm^{-3}$.

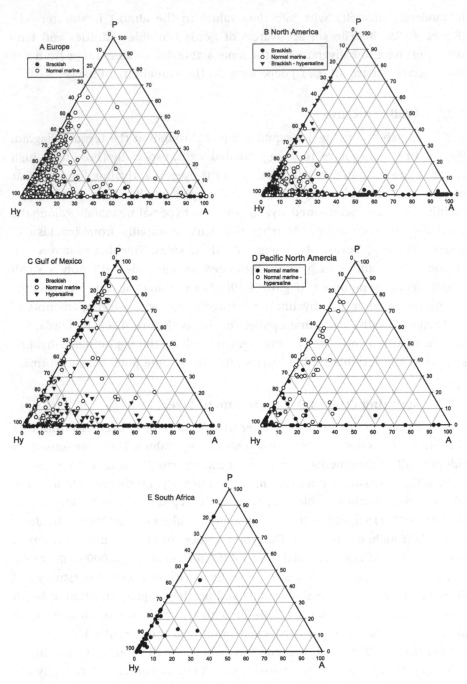

Figure 4.19. Ternary plots of walls for estuaries and lagoons (triangle corners represent 100% of the labelled component: A = agglutinated, P = porcelaneous, H = hyaline).

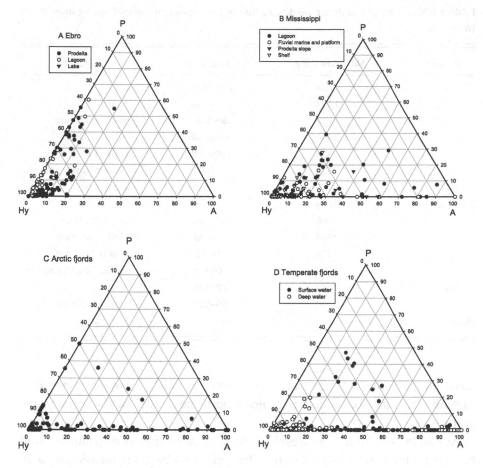

Figure 4.20. Ternary plots of walls for deltas and fjords (triangle corners represent 100% of the labelled component: A = agglutinated, P = porcelaneous, H = hyaline).

4.8   Summary

Although ecologists distinguish between estuaries, lagoons and deltas on geomorphological and sedimentological criteria, it is apparent that these distinctions are not important for the fauna. In general species diversity is low in extreme environments (brackish, hypersaline marshes, lagoons and estuaries) but ranges a little higher in normal marine lagoons. In terms of wall structure, whereas in marshes there are some assemblages that comprise mixtures of agglutinated and porcelaneous tests (with few hyaline forms), that is not the case for lagoons, estuaries, deltas or fjords. The general pattern is that brackish subtidal environments have assemblages with a mixture of agglutinated and hyaline walls. Hypersaline environments (found only in warm climates) have mixtures of porcelaneous and hyaline walls.

Table 4.1. *Summary of standing crop data for the 0–1 cm sediment interval (values individuals*
10 cm^{-3}).

Environment	Maximum value	Typical value	Location
Marsh	1798	200–1000	UK
	1252	100–200	Spain
	330	50–100	Mediterranean
	2464	50–1000	Atlantic USA
	2450	100–200	Gulf of Mexico
	1652	>100	Pacific North America
	10908	300–5000	Japan
Estuaries	3404	>200	Europe
	756	50–200	Atlantic USA
	2784	50–400	Gulf of Mexico
	376	10–75	South America
	1652	50–300	Pacific North America
Deltas	240	10–100	Spain
	8240	80–500	Gulf of Mexico
Fjords	2109	207–500	Europe

Environments that change seasonally from brackish to hypersaline have
assemblages that accordingly plot along both these sides of the ternary diagram.

As can be seen from the foregoing account, major species distributions and
biogeographic boundaries are controlled primarily by a combination of temperature and salinity. Species may be confined to a certain range of salinities
but there are no species found only in a single geomorphological setting such
as a lagoon, estuary, delta or fjord. However, there are certain species found
only in association with vegetated marshes and only rarely on adjacent mudflats (*Balticammina pseudomacrescens, Jadammina macrescens, Tiphotrocha comprimata, Trochammina inflata*). Local species distributions are determined by a
plexus of abiotic and biotic controls. In those studies where authors have used
multivariate and statistical measures to determine the relationships between
abiotic variables and species there is always a great deal of co-variance and
interdependence between variables. Consequently, it is difficult to know
exactly which controls are most important. It seems likely that the importance
of each variable changes with time as it comes close to the threshold for any
given species. Therefore, different variables are significant at different times.
Also some interdependent variables may reinforce one another in their influence on the fauna. Thus, salinity and temperature may interact so the response

of an organism to a change in salinity may be different at different temperatures, as these affect osmosis, concentration of dissolved oxygen, density and viscocity. There are differences between non-tidal and tidal settings. The former have a much more stable temperature and salinity structure on timescales ranging from days to weeks. Tidal areas are affected by currents and mixing so that changes in environmental parameters are continuous even if diurnally cyclic. Therefore some fjords and the Baltic have certain species adapted to constant conditions (*Ammotium cassis* living at the halocline). The standing crop is controlled by the availability of suitable food, dissolved oxygen (if dysoxic) and the degree of predation. Although maximum values sometimes reach several thousand individuals $10\,\mathrm{cm}^{-3}$ typical values are generally 50 to a few hundred. Several studies point to selectivity for particular types of food. In general, there is non-competitive feeding with each species taking a different part of the potential food supply.

In conclusion, it can be seen that although the general controls limiting marginal marine foraminiferal assemblages have been established there is still much to be learned about the operation of factors at a local level. Future progress in unravelling this will come from a combination of carefully controlled laboratory experiments to understand the physiological response to a forcing factor and field studies targeting specific themes.

5

Shelf seas

Wherever possible, citations of references and data tables are given in the captions to figures to save repetition and to avoid interrupting the flow of the text. There are more comprehensive versions of Figures 5.1 and 5.6 on the linked website (indicated by '+ web').

5.1   Introduction

The continental shelf includes the gently sloping sublittoral area from the open-sea shoreline to the shelf break where the inclination of sea floor increases with passage to the continental slope. The majority of shelves are shallower than 130 m but some reach depths of several hundred metres, as off Antarctica and in shelf basins. They are generally narrow along convergent plate margins where they border deep trenches (as around the Pacific Ocean), and wide along intraplate passive margins where they border a wide continental slope/rise (as around much of the Atlantic Ocean). Along the margin of California, USA, there is a transition zone from continental shelf to continental slope with deep elongated basins separated by ridges forming a continental borderland.

During the last glacial maximum (LGM) 18 000 years ago large areas of modern continental shelves were exposed as land and those at high latitudes were ice-covered. In some areas, for example, along the margins of the Norwegian–Greenland Sea, bordering mountains are cut by deep ice-gouged valleys that now form fjords. There are also large quantities of moraine deposits (e.g., off Newfoundland). In mid to low latitudes, during the LGM the shelves were crossed by rivers that disgorged their sediment load onto what is now the outer shelf and the adjacent continental slope. Sea level has risen by 120–130 m since the LGM. Except for deltas, sediment is now deposited mainly in marginal marine environments and the shelves are largely starved and non-depositional. Therefore shelf sediments are out of equilibrium with present sea level and are partially or entirely relict, relating to deposition in the early stages of the marine transgression. Such deposits cover around 70% of modern shelves (Kennett, 1982). In some cases, former alluvial deposits associated with a previous lower sea level remain exposed on the sea floor.

Phytoplankton and seaweeds are the main producers of carbon; bacteria are the main consumers in shelf regions. The rates of decomposition of organic matter under anoxic conditions are equal to or greater than aerobic decay. The main factors controlling the preservation of organic carbon in the sediment are primary production, source of organic matter (marine or terrigenous), sediment accumulation rate and oxygen content of near-bottom waters (Alongi, 1998). The physical energy on inner continental shelves comes from tides, currents, waves and storms. Even if storm conditions occur infrequently, they have great influence on the mobility of superficial sediment.

5.2   Europe and Africa

In general, the shelf slopes gently from the shore to the shelf edge but there may be local sea-floor topography that disturbs this pattern. The Kara Sea, the southeastern Barents Sea and the North Sea have wide shelves. The Baltic Sea, the Kattegat and the North Sea are epicontinental. The shelf is relatively narrow along much of the Norwegian and Iberian seaboards. The Kara and Barents seas border the Arctic Ocean and are partly ice-covered during the winter. From northern Norway to southern Iberia the margin is bathed in the North Atlantic Drift. This current warms the area and makes temperature-based faunal boundaries very diffuse. The Kattegat and Baltic are under continental influence and are partially ice-covered during the winter. The Kara Sea, the southern part of Scandinavia and the Baltic are microtidal, most of the open shelves are mesotidal and parts of the shelf around the British Isles are macrotidal. Storm waves affect the shelf from northern Norway to northern Spain. The western Iberian margin is affected by swell waves. The Barents and Kara seas are relatively sheltered from wave attack. The inner shelf in particular is a dynamic environment being affected by seasonal ice cover, tides or waves.

Primary productivity by phytoplankton in surface waters is limited by light in high latitudes (Kara Sea, Barents Sea) and the Baltic Sea, North Sea and Bay of Biscay are considered to be eutrophic (Alongi, 1998). Estimates of production range from 25–250 g C m^{-2} y^{-1} (70 Kara Sea; 25–96 Barents Sea; 75–150 Baltic Sea; 100–250 North Sea; 120 g C m^{-2} y^{-1} Bay of Biscay). Fontanier *et al.* (2002, p. 752) describe the Bay of Biscay as a typical 'temperate meso-oligotrophic environment' with an annual primary production of 145–170 g C m^{-2} y^{-1} but according to the Nixon classification it is mesotrophic (see Glossary). There can be change from oligotrophic to eutrophic according to season and development of thermohaline stratification or upwelling. Because shelf seas are invariably affected to some extent by currents it cannot be assumed that phytodetritus sinks vertically to the sea floor so quantification of food supply to the benthos is difficult.

5.2.1   Kara and Barents seas

The most important species progressively appear from north to south (Figure 5.1 + web). In the Skagerrak there are additional species present because this shelf basin is deep and therefore extends beyond normal shelf depths. The low-energy and seasonally ice-covered northern seas (Kara, Barents) and epicontinental Baltic Sea have some dominant species not found in other European shelf seas: *Adercotryma glomeratum, Ammotium cassis, Astrononion gallowayi, Cassidulina reniforme, Elphidium clavatum* and *Islandiella norcrossihelenae*. Hald *et al.* (1994) infer that there are three controls on the occurrence of *Elphidium clavatum*: fluctuating salinity and low-salinity water, turbid water, cold (<1 °C) and seasonally ice-covered water. Other assemblages also occur elsewhere: *Eggerella advena, Reophax scorpiurus* and *Saccammina atlantica*. The St. Anna Trough is a deep depression on the Barents–Kara shelf extending from 400 m in its southern part to 600 m in the north. It is ice-covered throughout most of the year and is bathed in Arctic bottom water (temperature *c.* −0.6 °C; salinity 34.85–35.00). Foraminifera >1 mm in size make up >90% of the foraminiferal fauna in the Barents Sea (Korsun *et al.*, 1998). The dominant species, *Reophax pilulifera*, responds to the summer peak of organic matter by increasing the volume of protoplasm. Biomass ranges from 0.06 to 1.66 mg cm^{-2} with the higher values in the deeper part of the trough and the lowest values on the sides. These values are low compared with other shelf seas and Korsun *et al.* consider this to reflect its oligotrophic setting; primary productivity 25–96 g C m^{-2} y^{-1} (Alongi, 1998). Elsewhere on the Barents shelf biomass is controlled mainly by large astrorhizids (*Rhabdammina abyssorum, Pelosina variabilis, Hyperammina nodosa*) and where these are present in muddy substrates biomass generally exceeds 1 mg 10 cm^{-2} compared with 0.1–1.0 mg 10 cm^{-2} on sandy sediments lacking these forms (Korsun, 2002). Miliolids are rare in cold northern seas.

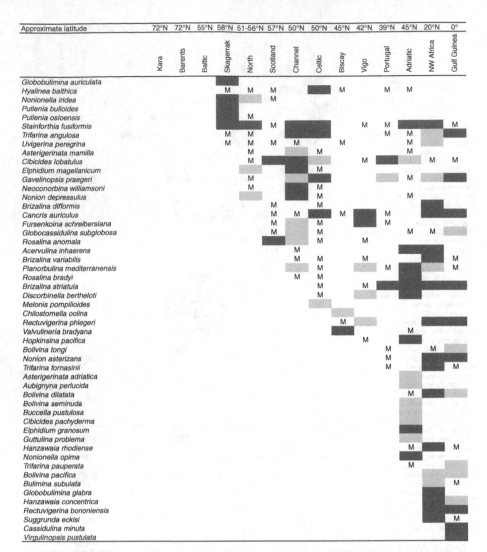

Figure 5.1. + web. East Atlantic shelf distributions. Black = dominant, grey = subsidiary, M = minor. Sources of data (not all references have data files): Kara Sea: Korsun, 1999 (WA-36); Barents Sea: Korsun *et al.*, 1994 (WA-111); Baltic Sea: Wefer, 1976a (WA-112); Skagerrak: Alve and Murray, 1995a, 1997 (WA-113); North Sea: Murray, 1985, 1992 (WA-114), Moodley, 1990b (WA-115); Scotland: Murray, 1985 (WA-114), 2003b (WA-116), 2003c; English Channel–Celtic Sea: Murray, 1965a (WA-40), 1970a, 1979, 1986 (WA-117), Sturrock and Murray, 1981 (WA-118), Scott *et al.*, 2003 (WA-119), Moulinier, 1967 (WA-120); Biscay: Fontanier *et al.*, 2002 (WA-121); Vigo: Diz *et al.*, 2004 (WA-122); Portugal: Seiler, 1975 (WA-123); Adriatic: Daniels, 1970 (WA-124), Haake, 1977 (WA-125), Barmawidjaja *et al.*, 1992 (WA-126), de Stigter *et al.*, 1998 (WA-127), Moodley *et al.*, 1998, Donnici and Serandrei-Barbero, 2002 (WA-128), Serandrei-Barbero *et al.*, 2003 (WA-129), Pranovi and Serandrei-Barbero, 1994 (WA-130); northwest Africa and Gulf of Guinea: Haake, 1980b (WA-131), Lutze, 1980 (WA-132), Lutze and Coulbourn, 1984, Coulbourn and Lutze, 1988, Schiebel, 1992 (WA-133), Timm, 1992 (WA-133), Altenbach *et al.*, 2003 (WA-133).

5.2.2 Baltic Sea

The Baltic Sea is linked to the North Sea via the Kattegat and Ska-gerrak. Because of the large freshwater input in the Baltic, there is a surface brackish layer overlying deeper more saline water so it is like a large stratified estuary. Most data come from a shallow area of Kiel Bay ('Hausgarten') from 5 m in the turbulent zone down to 28 m (Lutze, 1974a; Lutze *et al.*, 1983; Wefer, 1976a,b; Wefer and Lutze, 1976). In the summer and autumn there is a 10–15 m thick surface brackish layer with salinities of 12–19 above a halocline; below that salinities are 17–22. The surface layer covers sandy sediment with boulders that provide support for attached algae. The water below the halo-cline rests on a muddy substrate on the margin of a depositional basin. During the winter the two water layers become mixed. Time-series sampling was carried out at 6, 8, 11, 13, 23.5 and 27 m. At 6, 8, and 11 m most sediment assemblages are dominated by either *Miliammina fusca* or *Elphidium excavatum*. *Eoeponidella pulchella* is dominant in single assemblages at 8 and 13 m. *Ammotium cassis* becomes a subsidiary species at 11 m and is sometimes dominant at 13 and 23.5 m (spanning the halocline) but is not important at 27 m. At 23.5 and 27 m assemblages are dominated by *Elphidium incertum* and *Elphidium clavatum*. *Ammotium cassis* reproduces during the autumn and mature individuals are present by August. Reproduction takes place under high oxygen, temperatures < 8 °C and high food supply (>300 µg C l^{-1}). This species may be a relatively recent introduction as prior to Lutze's 1965 study Rottgardt (1952) had found only a few individuals and the species had not been recorded previ-ously. *Elphidium clavatum* lives in the upper few millimetres of sediment and reproduces at the greatest extent of the euphotic zone. *Elphidium excavatum* is euryhaline and lives in both sediment and on plants. *Elphidium incertum* favours a shallow infaunal habitat in sandy substrates. Reproduction seems linked to high oxygen values rather than any other parameter. Haake (1967) and Wefer (1976a) suggest that reproduction occurs two to four times a year. Algae are present from 6–13 m and their phytal assemblages are commonly dominated by *Ophthalmina kilianensis* but sometimes by *Elphidium excavatum* or *Elphidium gerthi* and occasionally by *Miliammina fusca*. Wefer and Lutze (1976) estimate biomass of 10–90 mg wet weight m^{-2}y^{-1} for the platform beneath surface water and 5411 mg wet weight m^{-2}y^{-1} in the basins. In addition, the phytal fauna has a biomass of 13 mg wet weight m^{-2}y^{-1}. Only three species are major contributors to biomass: *Ammotium cassis*, *Elphidium clavatum* and *Elphidium incertum*. There are similar differences in carbonate production: 10–35 mg m^{-2}y^{-1} on the platform and up to 3100 mg m^{-2}y^{-1} in the basin (Wefer and Lutze, 1978).

5.2.3   Skagerrak and North Sea

Temperate shelf seas extend from the Skagerrak to Portugal and include the Adriatic (Figure 5.1). The inner and mid shelves are affected by tidal currents and periodically by storm waves. The distribution of key species reflects these different energy conditions. High-energy areas have coarse sediment which may include gravel-grade material that provides a substrate for attachment of foraminifera and also for hydroids, bryozoa, etc., to which foraminifera may attach. Typical assemblages of these environments are dominated by *Textularia sagittula, Textularia truncata, Textularia bocki, Acervulina inhaerens, Cibicides lobatulus, Gavelinopsis praegeri* and *Rosalina anomala*. Sandy substrates commonly have *Eggerelloides medius, Eggerelloides scaber, Reophax fusiformis, Reophax scorpiurus, Quinqueloculina seminulum, Ammonia* group, *Cancris auriculus, Elphidium excavatum* and *Trifarina angulosa*. Lower-energy muddy substrates have *Bolivinellina pseudopunctata, Brizalina spathulata, Brizalina striatula, Bulimina gibba/elongata, Bulimina marginata, Cassidulina obtusa, Epistominella vitrea* and *Stainforthia fusiformis* (Figures 5.2–5.4). Some examples are discussed in more detail below.

The inner shelf and the western parts of the North Sea are vertically mixed throughout the year but the remainder is thermally stratified during the summer. Mixed and stratified areas are delimited by sharp tidal fronts. Nutrient concentrations are lower in the southern than in the northern part due to the influence of the Atlantic Ocean. A *Stainforthia fusiformis* assemblage is present in muddy substrates on the mid shelf and lives down to at least 11 cm (Collison, 1980). Although it makes up 57% of the whole-core living fauna it forms only 21% of the top 1 cm assemblage. Repeated sampling from September to June shows that silty fine sand from 25 m has an *Elphidium excavatum* assemblage (with subsidiary *Nonion depressulus*) throughout the year. Sand from a depth of 30 m has *Elphidium excavatum* in June but otherwise few living foraminifera and that from 31m has few living foraminifera throughout the year. Muddy fine sand from 52 m has an *Eggerelloides scaber* assemblage (with subsidiary *Stainforthia fusiformis*) in September and June and a *Stainforthia fusiformis* assemblage (with subsidiary *Elphidium excavatum* and *Bulimina gibba/elongata*) in December. Muddy fine sand from 63 and 81 m has a *Stainforthia fusiformis* assemblage (with subsidiary *Epistominella vitrea* in June and October at 63 m and *Reophax fusiformis* in December at 81 m). Thus, the assemblages are dynamic.

Long cores (10 cm, Murray, 1992; 25 cm, Moodley, 1990b) from 25–45 m have 0–1 cm assemblages, with one exception (*Ammonia* group), of *Elphidium excavatum* with subsidiary *Eggerelloides scaber, Buliminella elegantissima* and *Stainforthia*

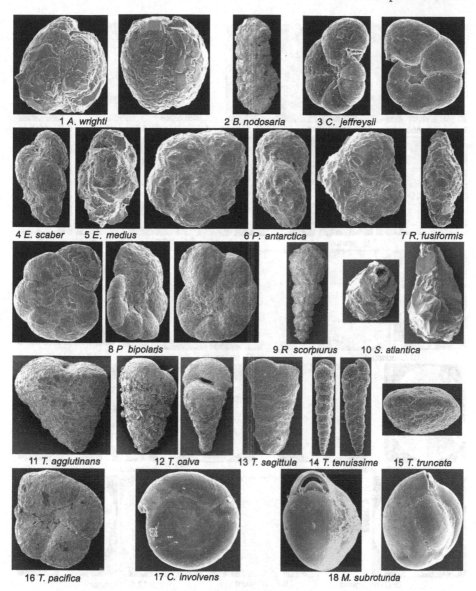

Figure 5.2. Scanning electron micrographs of shelf foraminifera (longest dimension, μm). 1. *Adercotryma wrighti* (160, 175). 2. *Bigenerina nodosaria* (1000). 3. *Cribrostomoides jeffreysii* (300, 300). 4. *Eggerelloides scaber* (480). 5. *Eggerelloides medius* (250). 6. *Portatrochammina antarctica* (315, 275, 265). 7. *Reophax fusiformis* (600). 8. *Portatrochammina bipolaris* (275, 240, 260). 9. *Reophax scorpiurus* (1800). 10. *Saccammina atlantica* (330, 220). 11. *Textularia agglutinans* (380). 12. *Textularia calva* (600, 600). 13. *Textularia sagittula* (830). 14. *Textularia tenuissima* (370, 370). 15. *Textularia truncata* (200). 16. *Trochammina pacifica* (500). 17. *Cornuspira involvens* (165). 18. *Miliolinella subrotunda* (200, 270).

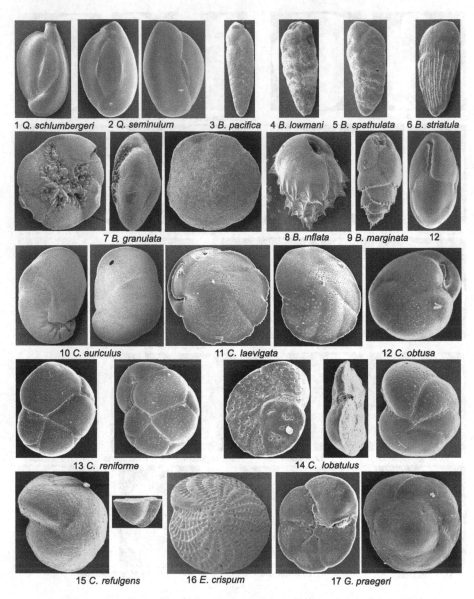

1 *Q. schlumbergeri*   2 *Q. seminulum*   3 *B. pacifica*   4 *B. lowmani*   5 *B. spathulata*   6 *B. striatula*

7 *B. granulata*   8 *B. inflata*   9 *B. marginata*   12

10 *C. auriculus*   11 *C. laevigata*   12 *C. obtusa*

13 *C. reniforme*   14 *C. lobatulus*

15 *C. refulgens*   16 *E. crispum*   17 *G. praegeri*

Figure 5.3. Scanning electron micrographs of shelf foraminifera (longest dimension, μm). 1. *Quinqueloculina schlumbergeri* (1000, 750). 2. *Quinqueloculina seminulum* (550, 360). 3. *Bolivina pacifica* (400). 4. *Brizalina lowmani* (200). 5. *Brizalina spathulata* (330). 6. *Brizalina striatula* (300). 7. *Buccella granulata* (300, 300, 360). 8. *Bulimina inflata* (500). 9. *Bulimina marginata* (330). 10. *Cancris auriculus* (1200, 1200). 11. *Cassidulina laevigata* (330, 300). 12. *Cassidulina obtusa* (230, 175). 13. *Cassidulina reniforme* (240, 220, 210). 14. *Cibicides lobatulus* (310, 400, 310). 15. *Cibicides refulgens* (900, 850). 16. *Elphidium crispum* (940). 17. *Gavelinopsis praegeri* (200, 310).

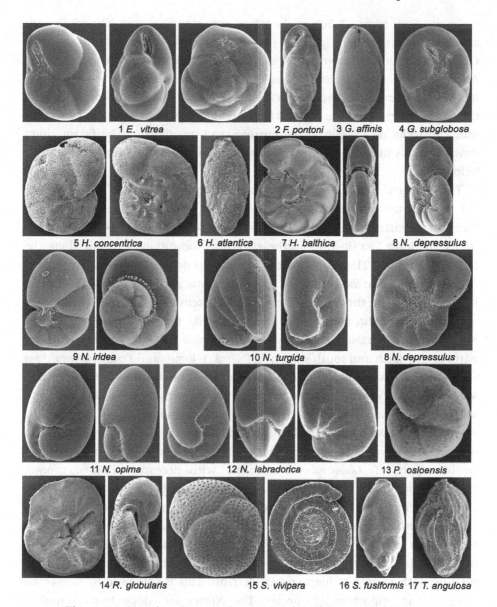

Figure 5.4. Scanning electron micrographs of shelf foraminifera (longest dimension, μm). 1. *Epistominella vitrea* (all 150). 2. *Fursenkoina pontoni* (230). 3. *Globobulimina affinis* (900). 4. *Globocassidulina subglobosa* (150). 5. *Hanzawaia concentrica* (300, 480). 6. *Hopkinsina pacifica atlantica* (200). 7. *Hyalinea balthica* (440, 370). 8. *Nonion depressulus* (280, 220). 9. *Nonionella iridea* (130, 130). 10. *Nonionella turgida* (300, 300). 11. *Nonionella opima* (280, 300, 210). 12. *Nonionellina labradorica* (630, 630). 13. *Pullenia osloensis* 200. 14. *Rosalina globularis* (330, 500, 420). 15. *Spirillina vivipara* (650). 16. *Stainforthia fusiformis* (280). 17. *Trifarina angulosa* (400).

fusiformis. They form 10–50% of the whole-core standing crop. The whole-core assemblages are either *Elphidium excavatum* or *Eggerelloides scaber*. Other species found infaunally down to 25 cm include *Reophax fusiformis*, *Ammonia* group, *Asterigerinata mamilla*, *Bolivinellina pseudopunctata*, *Bulimina gibba/elongata*, *Buliminella elegantissima* and *Stainforthia fusiformis* but there is no depth stratification. Moodley considers the distributions to reflect bioturbation by macrofauna introducing oxygen in the burrows and transporting the foraminifera downward, as the redox boundary lies at 4.5 mm and most foraminifera live below this.

The Skagerrak basin (700 m) is separated from the rest of the North Sea by a sill at ~270 m water depth and has an open connection to the Kattegat. The surface circulation is anticlockwise with coastal surface water from the southern North Sea flowing northeastward as the Jutland Current along the coast of Denmark. The deeper part of the basin is occupied by well-oxygenated Atlantic Water. The Skagerrak is a major depository for fine-grained sediment. In 1993 there were three main-factor defined assemblages: (1) *Haplophragmoides membranaceum*, *Epistominella vitrea* and *Nonionella iridea* restricted to depths >473 m; (2) *Globobulimina auriculata* and *Stainforthia fusiformis* on the shallower parts of the north and south slopes; (3) *Pullenia osloensis* and *Textularia tenuissima* with *Cassidulina laevigata* restricted to the north slope down to 473 m (Alve and Murray, 1995a). When this dataset was combined with that for 1994 (Alve and Murray, 1997) a new set of factor-defined assemblages was defined: (1) *Nonionella iridea* with *Melonis barleeanum* and *Pullenia osloensis* in the deep eastern part; (4) *Haplophragmoides membranaceum* and *Epistominella vitrea* in the deeper western part; (3) *Globobulimina auriculata* with accessory *Pullenia osloensis* and *Textularia tenuissima* covering most of the Norwegian slope; (2) *Stainforthia fusiformis*, dominates the Danish slope, under the influence of the Jutland Current, with a single occurrence on the Norwegian slope. The standing crops recorded in 1994 are significantly higher than those of 1993. There is a zone of high standing crop between 200 and 500 m on the Danish slope broadly correlating with an area of high organic detritus. This is colonised by *Stainforthia fusiformis*, an opportunistic species. The Norwegian slope lacks abundant organic debris and has a different assemblage (*Globobulimina auriculata*).

5.2.4   Western Britain

The west-facing continental shelf of the British Isles from Scotland to the English Channel is exposed to storm waves and is meso- to macrotidal with tidal currents being the dominant source of physical energy. There is strong seasonal forcing by surface heating and cooling leading to stratified areas separated from mixed areas by sharp fronts that vary only slightly in position due to tidal advection and the spring-neaps cycle. Enhanced biomass concentrations are found around fronts. There are no major rivers draining into the western seaboard of the British Isles so there is no major supply of terrigenous sediment except from the River Severn which drains into the Celtic Sea. The shelf is veneered with sand patches resting on a basal conglomerate. The sands are often carbonate-rich (>40% $CaCO_3$), being formed primarily of bioclasts from molluscs, serpulids and barnacles. Tidal currents induce large bottom shear stresses that are responsible for bedload transport of larger particles and for inducing regular resuspension of fine sediments in the water column. However, large, oscillatory shear stresses do not bring about transport of sediment; it is the asymmetry of flow that leads to net transport.

On the continental shelf to the north of Scotland there is a shell gravel sheet with coarse shell sands. Large shells and pebbles provide anchorage for hydroids and other attached epifauna that in turn provide sanctuary for attached benthic foraminifera such as tiny trochamminids, *Cibicides lobatulus* and *Rosalina anomala* (Murray, 1985). Cut into the shelf terrace are 'deeps' that are below the influence of storm waves and these are sinks for fine-grained sediment. Muck Deep reaches depths of >200 m and is cut in shelf with a depth of ~90 m while Stanton Deep (>160 m) is surrounded by shelf at ~130 m (Murray, 2003b). In Muck Deep replicate cores from 170 and 218 m have common *Eggerelloides medius*, *Bulimina marginata*, *Cassidulina laevigata*, *Epistominella vitrea* and *Nonionella turgida* (any of which may be dominant) with subsidiary *Textularia tenuissima* (170 m) or *Liebusella goesi* and *Recurvoides trochamminiforme* (218 m) in the top cm. The rank order of the common species differs between replicates for the 0.0–0.5 and 0.5–1.0 cm intervals indicating local patchiness. One replicate from 218 m has a *Pectinaria* tube in life position. This worm moves sideways through the sediment and appears to be responsible for mixing live individuals to a depth of several centimetres. In the other replicates >92% of the live individuals are in the top 1 cm even though the redox boundary is 2 to 5 cm below the surface. *Cornuspira involvens*, *Bulimina marginata*, *Cassidulina laevigata* and *Cassidulina obtusa* are almost equally abundant in the 0.0–0.5 cm interval. Below 1 cm there are few live forms. Correlation coefficients between foraminiferal abundance in size fractions and environmental factors give a confusing picture. *Eggerelloides medius* shows positive correlations with % TOC and to a lesser extent with per cent < 63 μm sediment and a strong negative correlation with grain size > 1000 μm. *Bulimina marginata* shows a strong positive correlation with % TOC and a negative correlation with per cent coarse sand. *Cassidulina laevigata* shows a strong positive correlation with per cent coarse sand and negative correlation with fine sediment and % TOC.

During the winter the water is vertically mixed in the English Channel but summer thermal stratification develops in the west. Except in sheltered areas such as coastal embayments where mud may accumulate, most of the sediments are sand or gravel, often with a high proportion of biogenic carbonate. Sand-grade sediment is mobilised during peak flood and ebb tidal flow. In the Celtic Sea, tidal currents and waves affect the inner shelf but the outer shelf is below storm-wave base and accumulates fine-grained sediments. In winter the waters are vertically mixed but during late spring and summer a thermocline develops over the mid and outer shelf. The sediment substrate exerts a strong control on the living foraminiferal assemblages (Murray, 1970a; 1979, 1986; Moulinier, 1972; Rosset-Moulinier, 1976; 1986; Sturrock and Murray, 1981). Based on observation of live material kept in small aquaria, the foraminifera have been grouped into three categories: I, attached immobile; M, attached mobile; and F, free (Sturrock and Murray, 1981). Attached immobile (sessile) forms have a flattened shape which may be deformed to match the substrate and are permanently or semi-permanently fixed to a stable substrate by cementation (glycoglue = glycosaminoglycans, Langer, 1993) or by a membrane; attached mobile forms are flattened and cling to stable substrates (macrofauna, pebbles) using their pseudopodia but they may also move freely in the sediment. Free-living foraminifera have a great variety of form and may be epifaunal and/or infaunal.

The living assemblages from the Channel and Celtic Sea can be grouped according to substrate and energy levels. Sturrock and Murray recognised six sediment groups but here they are reduced to three (A–C). Group A includes shell pavements and lag deposits around headlands (i.e., gravel-grade deposits) formed under high-energy conditions due to tidal currents. Group B includes mobile sand and well-sorted medium sand from high to medium energy tidal-current areas. Group C is low-energy muddy sand and mud where tidal currents are weak. In addition, Group D assemblages occur on organic substrates (such as hydroids attached to mollusc shells). Each has a distinctive field when plotted on a ternary diagram of free (F), attached-mobile (M) and attached-immobile (I) modes of life although there is some overlap between groups A and C (Figure 5.5). Group D organic substrates have mixtures of I and M. Group A has high proportions of I and M and is confined to the English Channel. The dominant species is *Cibicides lobatulus* with subsidiary *Rosalina anomala*. Group B is widely developed in the mid shelf of the western Channel. There are several dominant species but the ones forming most assemblages are *Spirillina vivipara* in the western Channel and *Textularia sagittula* in the Celtic Sea. Others include *Cassidulina obtusa*, *Portatrochammina murrayi* and *Trifarina angulosa*.

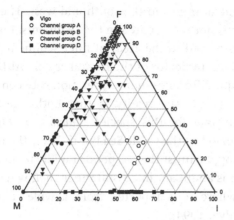

Figure 5.5. Ternary plot of mode of life (F, free; M, attached mobile; I, attached immobile) for medium to high energy environments of the English Channel–Celtic Sea, UK, and Vigo, Spain.

Group C has two geographic areas. On the inner shelf in the Channel there are assemblages dominated by *Ammonia* group, *Stainforthia fusiformis* or *Eggerelloides scaber*. In mid shelf in the Celtic Sea the principal assemblages are *Stainforthia fusiformis* and *Nonionella turgida*. There are also some dominated by *Eggerelloides scaber* or *Adercotryma wrighti*. In deeper water *Hyalinea balthica* dominates while in the entrance to the Bristol Channel *Bulimina gibba/elongata* is the principal form. The assemblages from the western Channel are broadly similar to those described by Rosset-Moulinier (1972; 1976; 1986) from coastal France. Three are dominated by I and M groups: *Cibicides lobatulus–Textularia truncata* (co-dominant) on coarse sediment including gravel, *Lepidodeuterammina ochracea* and *Remaneica plicata* assemblages on mobile sands. The fourth is the infaunal *Eggerelloides scaber* assemblage in muddy fine sands. In the eastern Channel, shelly sands from 10–24 m off the Seine, France, are dominated by *Elphidium* (*gerthi*, *magellanicum* or *selseyensis*) (Moulinier, 1967).

Multivariate analysis of their Celtic Sea data led Scott *et al.* (2003) to conclude that they had not measured the most important controlling environmental factors as only 30% of the species data are explained by the measured environmental variables. Partial canonical correspondence analysis shows that the first two axes are equally important: axis 1, mean grain size, skewness, latitude and per cent gravel; axis 2, temperature, depth and longitude; axis 3, per cent sand and sorting. These results confirm the IMF analytic approach presented above with I and M taxa on sites with a large proportion of gravel and F on fine-grained sediments. A frontal region of high productivity separates vertically mixed waters to the north and east from thermally stratified waters (during the late spring and summer) to the south and west. The high productivity of the frontal region is not represented in the sediments. Instead, organic debris appears to be transported to the areas of finer sediment. Scott *et al.* consider *Stainforthia fusiformis* an indicator of the frontal region. Although it might appear that there is a major difference in interpretation between Scott *et al.* and Sturrock and Murray (1981) the reality is that both stress the importance of a fine-grained substrate for this species. *Stainforthia fusiformis* may be an indicator of a frontal region in the Celtic Sea but if that is the case it is only because this species is an opportunist that responds rapidly to an input of food. In the English Channel it is abundant in coastal inner shelf where the water is vertically mixed. Elsewhere it is reported from a wide range of environments including fjords (Alve, 1994).

5.2.5 North Spain

The Ria de Vigo in northwestern Spain is an elongated embayment of the inner shelf subject to a macrotidal regime, which, combined with wind stress, can induce currents of up to 60 cm s^{-1} although normally they are much less than this. The sediments are quartz and bioclastic gravel and sands with a mean grain size of 0.35–2.2 mm. The sample from 28 m has some mud. Box cores from 6–42 m in January and September were sampled in 1 cm slices to depths ranging from 2–9 cm. The majority of the foraminifera are 63–125 μm (Diz *et al.*, 2004). The dominant and subsidiary taxa fall into two groups: (1) free-living: *Reophax scorpiurus, Bolivina pseudoplicata, Brizalina dilatata, Brizalina spathulata, Buliminella elegantissima, Cancris auriculus, Cassidulina obtusa, Nonion fabum, Nonionella stella*; (2) epifaunal, clinging or attached: *Lepidodeuterammina ochracea, Paratrochammina madeirae, Textularia conica, Cibicides refulgens, Eoeponidella pulchella, Gavelinopsis praegeri, Mullinoides* cf. *differens, Neoconorbina williamsoni, Patellina corrugata, Spirillina vivipara*. Free-living forms are dominant in assemblages from 28–41 m on sediments ranging from sand to gravel. Epifaunal taxa are dominant at depths of 6–42 m on coarse sand and gravels in exposed channels near the entrance or along the northern margin (Figure 5.5). Although there are minor differences in faunal composition between January and September samples from the same locality, the dominant forms principally have the same mode of life, either free living or epifaunal. There is a strong positive correlation between estimated current velocities and the abundance of epifaunal taxa ($r = 0.82$). However, there is no correlation between sediment grain size and the fauna. The authors calculated the average living depth (ALD) and found that infaunal taxa live at a wide range of depth. Epifaunal taxa also live down to several centimetres below the sediment surface and this may be because these high-porosity sediments are well flushed.

In spite of the high primary production, the sediments have a low organic content due to the high currents. The authors therefore suggest that bacterial films on the coarse grains may be a source of food.

5.2.6   Mediterranean

Compared with large oceans, enclosed seas such as the Mediterranean are relatively sheltered from wave attack. Salinities are slightly elevated due to the excess of evaporation over introduction of freshwater.

Phytal assemblages.   Along the south coast of France, seagrass (*Posidonia*) down to 20 m water depth provides habitats for phytal foraminifera particularly during the warm summer months (Blanc-Vernet, 1969, 1984; Vénec-Peyré and Le Calvez, 1981, 1988; Vénec-Peyré, 1984). The assemblages include fixed forms (*Hemisphaerammina bradyi, Hemisphaerammina crassa, Lituotuba lituiformis, Nubecularia lucifuga, Rhizonubecula adherens, Rosalina anglica, Rosalina posidonicola, Cibicides lobatulus* and *Planorbulina mediterranensis*). Among the free forms, miliolids are most abundant (*Quinqueloculina gualtieriana, Quinqueloculina vulgaris* and *Triloculina cuneata*). *Massilina secans* is abundant at 5 m, rare at 10 m and disappears at 15 m. Other species include *Elphidium crispum, Elphidium macellum, Rosalina globularis* and *Rosalina bradyi*. This area is just outside the northern limit of *Peneroplis* and *Sorites* (Murray, 1973). By contrast, the island of Volcano, to the north of Sicily, is within the limit of these forms. Detailed studies of its seagrass and seaweed assemblages show that phytal species can be divided into four morphotypes (A–D) according to their mode of life (Langer, 1993): (A) stationary, permanently attached (= I of Sturrock and Murray, 1981): these forms use secreted glycoglue to attach themselves firmly to their substrate: *Cyclocibicides vermiculatus, Daitrona* sp., *Miniacina miniacea, Cibicides lobatulus, Nubecularia lucifuga, Planorbulina mediterranensis, Sorites orbiculus*. Morphotype A occurs mainly on the broad leaves of *Posidonia oceanica* and *Sargassum hornschuchi*. It forms 5–62% of the assemblage at depths of 15–30 m. *Cyclocibicides vermiculatus* and *Planorbulina mediterranensis* are found on larger plants that have long life spans (>10 months). *Miniacina miniacea* is found only in shaded microhabitats. Permanently attached forms leave a cryptic record of the substrate on their attachment surface and this is of potential palaeoecological interest. (B) Temporarily attached (= M of Sturrock and Murray, 1981), including forms that are mobile when searching for food or during reproduction: *Asterigerinata mamilla, Cibicidella variabilis, Cibicides refulgens, Cymbaloporetta* spp., *Rosalina* spp. make up 25–85% of assemblages on seagrass and algae, and on microhabitats with a high sediment content (*Posidonia* rhizomes, algae) 10–45% of assemblages. (C) Suspension feeding, motile: keeled elphidiids and those with an acute periphery. These forms have a complex canal system and multiple apertural openings that give rise to a cobweb-like array of pseudopods. Their ideal habitat is within the three-dimensional network of algal blades (40–70%) and less so on seagrass leaves and smooth algae (15%). (D) Permanently motile, grazing epiphytes: small miliolids, peneroplids, *Spiroplectinella taiwanica*, *Textularia bocki*. Their highest abundance is in plant microhabitats that contain sediment (20–85%) and lower on arborescent algae (5–40%). *Peneroplis pertusus* occurs on *Dasycladus*, *Halopteris*, *Pseudolithophyllum* and *Posidonia* but not on rhizomes. *Peneroplis planatus* shows the opposite trend.

Comparison of epiphytic foraminifera from the eastern and western Mediterranean shows that *Amphistegina* does not occur west of Djerba (Tunisia), *Heterostegina* and *Calcarina* are restricted to the eastern part and *Amphistegina* is most frequent on *Halophila* in the east (Blanc-Vernet, 1984). The amount of light controls the presence or absence of symbiont-bearing foraminifera on plant substrates. Off Corsica, large granite boulders shade areas of sea floor. Plants living in the shade (sciaphile of Langer *et al.*, 1998) lack symbiont-bearing epiphytic foraminifera. Plants living in non-shaded areas have both symbiont-bearing foraminifera and higher species diversity. Thus, the controls on epiphytic foraminiferal assemblages are substrate, light, availability of plant substrates through time and food.

Sediment assemblages. The distribution of inner-shelf foraminifera on the Ebro Delta, Spain, is discussed in Chapter 4. In mud on the inner-mid shelf (0–90 m) off France, species occurring in >20% abundance include *Ammonia* group and *Nonion commune* and those >10% include *Nonionella turgida*, *Bulimina aculeata*, *Textularia bocki*, *Textularia sagittula*, *Sigmoilopsis schlumbergeri* and *Melonis barleeanum* (given as *Nonion parkerae*, Vénec-Peyré, 1984).

The Adriatic Sea is an elongated arm of the Mediterranean. A major feature of the oceanography is that the River Po discharges a large volume of freshwater. In summer there is 0–40 m slightly brackish surface water above an intermediate normal marine layer with a base at 150 m. In winter, mixing extends to ~250 m in the inner part. The surface layer has an anticlockwise circulation pattern with a northward flow along the eastern coast and a southward flow along the Italian coast. In winter and summer the freshwater is transported mainly along the Italian coast but in the summer there is also a freshwater plume that extends eastwards to Croatia. Mud deposition occurs along the Italian coast and this area is also enriched in nutrients and organic detritus. Since the 1970s phytoplankton blooms resulting from summer eutrophication have contributed phytodetritus to the sea floor along the Italian coast, resulting in oxygen depletion.

A single station on the northern slope at 16 m on fine sand sampled bimonthly from February 1991 to January 1995 gives a record of seasonal and inter-annual changes in the assemblages (Donnici and Serandrei Barbero, 2002). Overall, the dominant form throughout the sampling period in the > 125 μm assemblages is *Planorbulina mediterranensis*, an epifaunal species normally attached to a firm substrate. However, the 1991 and 1992 assemblages are dominated by *Vasiglobulina myristiformis*, *Rosalina bradyi* (both epifaunal attached), *Ammonia* group or *Quinqueloculina seminulum*. From November 1992 until January 1995, *Planorbulina mediterranensis* is dominant in 11 out of 14 sampling events. Thus, there is no obvious seasonality in the results. Of the numerous variables measured only suspended matter and biological oxygen demand appear to be significant. Serandrei Barbero and co-workers consider that phytoplankton abundance controls foraminiferal productivity in this area. Donnici and Serandrei Barbero (2002) used correspondence analysis to define three biotopes on the northern shelf. Biotope 1, 5–13.5 m, is nutrient-rich and characterised by *Ammonia* group. Biotope 2, 29.5–38.5 m, is a nutrient-rich zone of clay sediment occupied by *Nonionella opima*. Biotope 3, 21–46 m, is nutrient-poor sandy sediment and characterised by *Textularia conica*, *Miliolinella subrotunda* and *Elphidium granosum* assemblages.

Van der Zwaan (2000) suggests that in the clay belt associated with the Po discharge, stratification of the water column during the summer and high oxygen demand from decomposition of organic matter in the sediment leads to low oxygen conditions. The redox front is closer to the sediment surface during summer. Various foraminifera track the front with the tolerance to low oxygen increasing from *Ammonia* spp. > *Elphidium* spp. > *Textularia agglutinans* > *Nonionella turgida* to *Bulimina marginata* (*Nonionella turgida* of van der Zwaan is probably the same as *Nonionella opima* of Donnici and Serandrei Barbero, 2002, and Daniels, 1970). Box cores from 28–51 m on the northern slope have most living forms in the top 1 cm (Jorissen *et al.*, 1992). The lowest standing crops occur in the region with low oxygen values. Where the organic flux is high, the fauna is opportunistic and either epifaunal or mobile infaunal. These species benefit from high food supply and adequate oxygen levels after re-oxygenation during the autumn. At a single station from the mud belt, *Bolivina dilatata*, *Bolivina seminuda*, *Nonionella turgida*, *Stainforthia fusiformis* are essentially restricted to the top 2 cm of sediment and have a standing crop maximum in December. Predominantly infaunal species (i.e., evenly spread throughout 7 cm, *Ammoscalaria pseudospiralis*, *Eggerella advena*, *Eggerelloides scaber*, *Morulaeplecta bulbosa*, *Textularia agglutinans*) have a maximum in January and a minimum in May (Barmawidjaja *et al.*, 1992). Potential infaunal species (i.e., those restricted to the top 2 cm) do not vary greatly in abundance throughout the year (*Bolivina dilatata, Brizalina spathulata, Bulimina marginata, Epistominella vitrea, Hopkinsina pacifica*). They may track critical oxygen levels in the sediment. For the top 1 cm, the principal species are *Stainforthia fusiformis* and *Hopkinsina pacifica* with the former dominant in December and May and the latter in January. *Bolivina seminuda* is a subsidiary form.

Lim channel, Croatia, is an 11 km long inlet with a depth of 34 m over most of its length and salinities are 35–38 throughout the year. A thermocline is present from May until autumn and bottom temperatures range from 9–25 °C from winter to summer. The sediment is muddy silt burrowed by macrofauna to a depth of 30 cm. Repeat sampling over a 14-month period (Daniels, 1970; 1971) shows two main assemblages: *Nonionella opima* throughout most its length and *Ammonia* group in the innermost part. Because the assemblages are very diverse there are no subsidiary species consistently present. Minor assemblages are dominated by *Adercotryma* sp., *Eggerelloides scaber, Textularia agglutinans, Quinqueloculina planciana, Brizalina striatula, Discorbinella bertheloti, Epistominella vitrea* or *Hopkinsina* sp. Bottom-water oxygen is slightly lower in autumn or winter but never reaches a level that adversely affects the foraminifera and mainly there is 100% saturation. There is no information on the depth of the redox boundary in the sediment but foraminifera live to a depth of 20 cm probably in association with the macrofaunal burrows.

5.2.7   Africa

Northwest Africa and the Gulf of Guinea are mesotidal and experience swell wave disturbance. The data (Figure 5.1) are based on different sieve sizes and some data sets give only the common species. There is no orderly pattern of dominance on the shelf off northwest Africa. The only widespread living assemblages are: *Cancris auriculus* with subsidiary *Rectuvigerina phlegeri*; and *Rectuvigerina bononiensis* with subsidiary *Hanzawaia concentrica*. There are numerous different living assemblages occurring with one or a few occurrences including *Reophax calcareus, Textularia pseudogramen, Bolivina pseudoplicata, Brizalina difformis, Brizalina striatula, Brizalina variabilis, Globobulimina glabra, Hanzawaia concentrica, Nonion asterizans, Rectuvigerina phlegeri, Stainforthia fusiformis, Suggrunda eckisi, Trifarina fornasinii* and *Uvigerina peregrina*.

The Gulf of Guinea has a complex pattern of water masses and currents due to the interaction of the outflow of large rivers and the ocean. Shelf assemblages (down to 200 m) show a great deal of patchiness. The most widespread assemblages are dominated by one or other of: *Rectuvigerina phlegeri* (inner), *Cancris auriculus* (shelf to upper bathyal), *Siphotextularia caroliniana* (inner), *Gavelinopsis praegeri* (inner), *Brizalina striatula* (outer), *Cassidulina minuta* (mid). No subsidiary species is consistently present in these assemblages. Other

assemblages with one or two occurrences are *Ammoscalaria pseudospiralis,
Lagenammina difflugiformis, Nonion asterizans, Rectuvigerina bononiensis, Trifarina
angulosa* and *Virgulinopsis pustulata.* Altenbach *et al.* (2003) used the same datasets to differentiate >63 and >250 μm living assemblages and to analyse using
factor analysis. The inner-shelf assemblages (>63 μm) are dominated by *Rectuvigerina* cf. *phlegeri, Cancris auriculus* and *Gavelinopsis praegeri* or (>250 μm)
Cancris auriculus, Nouria polymorphinoides, Lagenammina difflugiformis or *Ammoscalaria pseudospiralis.* On the middle to outer shelf, the assemblages are dominated
by *Brizalina striatula, Nonion asterizans, Cancris auriculus* and *Ammoscalaria pseudospiralis.* Alterbach *et al.* relate faunal distributions to organic flux and the
shelf has the highest flux of organic material ($\sim$10 to >100 $gC_{org}m^{-2}y^{-1}$)
relative to the slope. However, they point out that flux rates are correlated with
water depth, sediment grain size, benthic oxygen respiration and chlorophyll
pigment equivalents. The authors conclude that correlation between foraminifera and any depth-related gradient, including fronts between water
masses, is just a matter of probability due to the interdependence of the
parameters.

5.2.8   Comparisons

Comparison of the various studies of the eastern margin of the
Atlantic suggests that some differences between areas may be due to differing
taxonomic usage. This may apply to species of *Globobulimina* and *Nonionella*.
There is a progressive southward increase in the number of species in
agglutinated, porcelaneous and hyaline groups (Figure 5.1). The arctic Kara
and Barents seas have few agglutinated species and none are indigenous as all
are also found further south. The most restricted form is euryhaline *Ammotium cassis,* which occurs as far south as the Baltic (and Scandinavian fjords;
see Chapter 4). Porcelaneous forms are rare in the arctic seas because of low
salinities. Among the hyaline taxa, *Elphidium subarcticum* and *Robertina arctica*
are confined to the Barents Sea; *Elphidium clavatum* extends as far south as the
Baltic (and southern Scandinavian fjords, see Chapter 4). *Ophthalmina kilianensis*
is restricted to the Baltic (but also occurs in shallow waters in the Kattegat
(see Chapter 4). The lack of data from the western shelf of Norway makes it
impossible to establish the northern limits of species found around the
British Isles and North Sea. Many agglutinated species are found here and
those restricted to this area include *Adercotryma wrighti, Eggerelloides medius,
Textularia tenuissima, Textularia truncata, Textularia bocki* and *Portatrochammina
murrayi.* The shelf off northwest Africa and the Gulf of Guinea has further
new appearances: *Reophax calcareus, Textularia pseudogramen* and *Siphotextularia
caroliniana* as well as dominant *Ammoscalaria pseudospiralis,* which occur in low

abundance as far north as the North Sea. Among the porcelaneous forms, although there are introductions around the British Isles, they are more obvious in the Adriatic Sea. There is a progressive change in the hyaline fauna from north to south. Forms confined to northwestern Europe include *Bolivinellina pseudopunctata*, *Cassidulina obtusa*, *Nonionella iridea*, *Pullenia bulloides*, *Pullenia osloensis*, *Elphidium magellanicum*, *Neoconorbina williamsoni*, *Fursenkoina schreibersiana* and *Rosalina anomala*. The Adriatic has several species not yet recorded elsewhere and this is true also for the shelf off northwest Africa and the Gulf of Guinea.

Although many factors control local distributions of foraminifera, temperature must be a primary control on biogeography. The eastern margin of the Atlantic does not have sharp temperature boundaries; instead there is a gradient of increasing bottom temperatures from north to south. Nevertheless, there is a marked difference in the bottom-water temperatures between the Barents Sea (-1.5 to $2\,°C$), the North Sea (4 to $15\,°C$), the Celtic Sea (9 to $12\,°C$), Adriatic (13 to $22\,°C$) and the Gulf of Guinea (14 to $25\,°C$). The development of a thermocline over most mid to outer-shelf seas means that the sea floor experiences a narrower range of annual temperature variation. Bottom temperatures are fixed in the winter when storms mix the surface and bottom waters. In spring and summer, the surface layer is warmed but below the thermocline the temperature remains at the winter level. In the autumn, storms again start to mix warm water down to the bottom. Excluding the Baltic because of its brackish salinities, there is a major faunal difference between the arctic seas and the North Sea region and there are further differences between this and the somewhat warmer western English Channel and Celtic Sea. There are also significant differences in assemblage composition due to tidal, storm and wave energy with a high proportion of attached epifaunal species in areas of coarse sediment and high energy (Scotland, western English Channel and Vigo, Spain).

5.3   Atlantic North America

Arctic Canadian waters are generally ice-covered from October to March and ice-free from August to September but there is local variability. Ice cover restricts primary production and exchange of gases with the atmosphere. However, unlike the Kara and Barents seas, the bottom waters here do not appear to be corrosive with respect to $CaCO_3$ (Hunt and Corliss, 1993). Nova Scotia is mesotidal to macrotidal whereas from Cape Cod southward is microtidal. Storm waves affect the area from Nova Scotia to New Jersey and south of this the shelf is affected by currents of the Gulf

Stream. Thus, throughout the length of this seaboard the inner-mid shelf is physically disturbed to some extent and ridge and swale topography, with coarse and fine sediments, respectively, is widely developed. The macro-faunal biomass and community structure is a function of the response to this (Alongi, 1998). Primary production is greatest in the coastal zone and espe-cially from May to September. There is resuspension of phytodetritus and local transport causing heterogeneity. The distribution of the main species and sources of data are listed in Figure 5.6 + web. It is immediately obvious that there are faunal changes between the Arctic and Grand Banks–Scotia shelf and between New Jersey and North Carolina. Therefore, individual species are not dominant throughout the whole of this large latitudinal range. Species occurring only in the Arctic include dominant *Recurvoides turbinatus* and *Islandiella helenae* and subsidiary *Paratrochammina bipolaris*, *Trochammina nana* and *Melonis barleeanum*. Species dominant from the Arctic to Grand Banks are *Adercotryma glomeratum* and *Nonionellina labradorica*. *Saccammina atlantica* and *Eggerella advena* are dominant from the Grand Banks to North Carolina. In the Arctic Hunt and Corliss (1993) found forms living down to 17 cm but with no obvious depth preferences. Overall, the cores are dominated by *Adercotryma glomeratum* and *Islandiella helenae* and these are also the dominant forms in the top 1 cm.

From the Grand Banks to New Jersey, the widespread dominant forms are *Eggerella advena, Cibicides lobatulus* and *Elphidium clavatum*. The tail of the Grand Banks south of Newfoundland is under the influence of the Labrador Current that brings cold water from the north; however along the outer part this mixes with warm water from the Gulf Stream flowing over the adjacent continental slope. The commonest assemblage is *Islandiella islandica* often with subsidiary *Elphidium clavatum* (Sen Gupta, 1971). Other assemblages are dominated by *Elphidium clavatum, Cibicides lobatulus, Nonionellina labradorica, Stainforthia loeblichi, Eggerella advena* or *Cassidulina teretis*. Recurrent group analysis of the results based on presence/absence shows that there are four recurrent groups (Sen Gupta and Hayes, 1979). Group 1, widely distributed in summer: *Astrononion gallowayi, Cibicides lobatulus, Eggerella advena, Elphidium clavatum, Islandiella islandica* and *Trochammina squamata*; group 2, generally on fine substrates >70 m, *Elphidium subarcticum, Stainforthia loeblichi, Globobulimina auriculata* and *Nonionellina labradorica*; group 3, restricted to 60–80 m: *Pseudopolymorphina novangliae, Saccammina atlantica* and *Trifarina angulosa*; group 4, confined to 70–90 m, *Elphidium bartletti*. Cluster analysis confirms groups 1, 2 (without *Elphidium subarcticum*) and 3. Group 4 includes *Quinqueloculina stalkeri* and *Recurvoides turbinatus*. The shelf off Nova Scotia is also influenced by the Labrador Current. Bank assemblages are dominated primarily by *Cibicidoides*

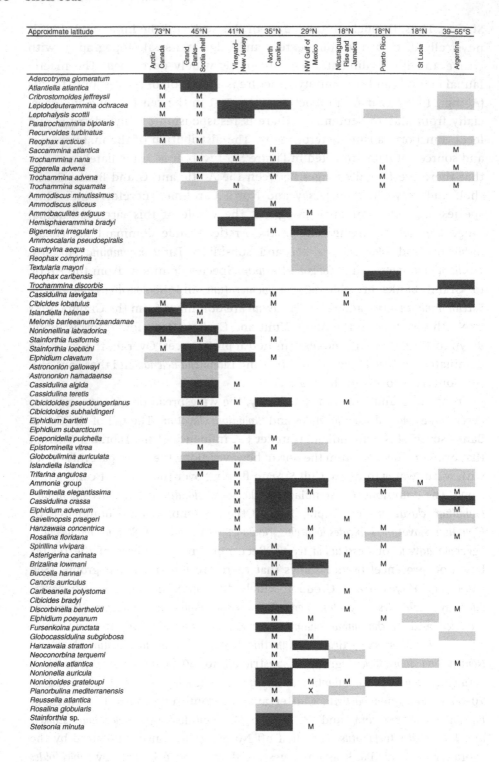

Approximate latitude	73°N	45°N	41°N	35°N	29°N	18°N	18°N	18°N	39–55°S
	Arctic Canada	Grand Banks–Scotia shelf	Vineyard–New Jersey	North Carolina	NW Gulf of Mexico	Nicaragua Rise and Jamaica	Puerto Rico	St Lucia	Argentina
Adercotryma glomeratum									
Atlantiella atlantica	M								
Cribrostomoides jeffreysii	M	M							
Lepidodeuterammina ochracea	M	M							
Leptohalysis scottii	M	M							
Paratrochammina bipolaris									
Recurvoides turbinatus		M							
Reophax arcticus	M	M							
Saccammina atlantica				M					M
Trochammina nana		M	M	M					
Eggerella advena						M			M
Trochammina advena		M		M		M			
Trochammina squamata			M			M			M
Ammodiscus minutissimus									
Ammodiscus siliceus				M					
Ammobaculites exiguus					M				
Hemisphaerammina bradyi				M					
Bigenerina irregularis									
Ammoscalaria pseudospiralis									
Gaudryina aequa									
Reophax comprima									
Textularia mayori									
Reophax caribensis									
Trochammina discorbis									
Cassidulina laevigata				M		M			
Cibicides lobatulus	M			M		M			
Islandiella helenae		M							
Melonis barleeanum/zaandamae		M							
Nonionellina labradorica									
Stainforthia fusiformis	M	M		M					
Stainforthia loeblichi	M								
Elphidium clavatum				M					
Astrononion gallowayi									
Astrononion hamadaense									
Cassidulina algida			M						
Cassidulina teretis									
Cibicidoides pseudoungerianus		M	M			M			
Cibicidoides subhaidingeri									
Elphidium bartletti									
Elphidium subarcticum									
Eoeponidella pulchella				M					
Epistominella vitrea		M	M	M					
Globobulimina auriculata									
Islandiella islandica		M							
Trifarina angulosa		M							
Ammonia group		M						M	
Buliminella elegantissima		M			M		M		M
Cassidulina crassa		M							M
Elphidium advenum		M			M				M
Gavelinopsis praegeri		M			M	M			
Hanzawaia concentrica		M			M		M		
Rosalina floridana					M	M			M
Spirillina vivipara				M			M		
Asterigerina carinata						M			
Brizalina lowmani				M			M		
Buccella hannai				M					
Cancris auriculus									
Caribeanella polystoma						M			
Cibicides bradyi									
Discorbinella bertheloti						M			M
Elphidium poeyanum				M	M		M		
Fursenkoina punctata									
Globocassidulina subglobosa				M	M				
Hanzawaia strattoni				M					
Neoconorbina terquemi				M					
Nonionella atlantica				M					M
Nonionella auricula									
Nonionoides grateloupi					M	M			
Planorbulina mediterranensis				M	X				
Reussella atlantica				M					
Rosalina globularis									
Stainforthia sp.									
Stetsonia minuta					M				

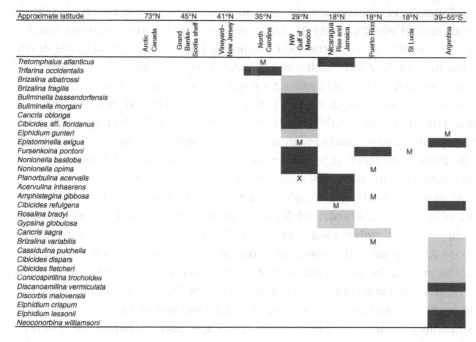

Figure 5.6. +web. Distributions in west Atlantic shelf seas. Black = dominant, grey = subsidiary, M = minor. Sources of data (not all references have data files): Arctic Canada: Schröder-Adams *et al.*, 1990b (WA-134), Hunt and Corliss, 1993 (WA-135); Grand Banks: Sen Gupta, 1971, Sen Gupta and Hayes, 1979 (WA-136), Miller, 1996 (WA-137); Vineyard Sound–Long Island: Murray, 1969 (WA-138); New Jersey: Poag *et al.*, 1980 (WA-139); North Carolina: Murray, 1969 (WA-140), Lueck and Snyder, 1997 (WA-141), Murosky and Snyder, 1994 (WA-142); Louisiana: Locklin and Maddocks, 1981 (WA-143), Sen Gupta *et al.*, 1996, Platon and Sen Gupta, 2001 (WA-144); Texas: Phleger, 1951 (WA-145), 1956 (WA-146), Hueni *et al.*, 1978, Poag and Tresslar, 1979, 1981; Nicaragua Rise: Peebles *et al.*, 1997 (WA 147); Jamaica: Marshall, 1976; Puerto Rico: Seiglie, 1974 (WA-148); St. Lucia: Sen Gupta and Schafer, 1973 (WA-149); Argentina: Boltovskoy *et al.*, 1983, Boltovskoy and Totah, 1985 (WA-150), Thompson, 1978 (WA-151).

subhaidingeri with subsidiary *Cibicides lobatulus*, *Cassidulina algida* and *Eggerella advena* but there are three assemblages dominated by one or other of these subsidiary species (Miller, 1996). On both the Grand Banks and the Scotia shelf, cibicidids are most abundant on coarse sediments which provide substrates for attachment. By contrast, in Vineyard Sound off southern Cape Cod where there are strong tidal currents there is an *Eggerella advena* assemblage with subsidiary *Lepidodeuterammina ochracea*, the latter living attached to coarse sediment particles.

Cape Hatteras marks a major oceanographic boundary between cool Virginian coastal water (salinity 31–34) to the north and warm Carolinian coastal water (salinity 35–36) to the south. Although some species found in the north continue south of Cape Hatteras (*Lepidodeuterammina ochracea, Eggerella advena, Ammobaculites exiguus* and *Cibicidoides pseudoungerianus*) many new forms appear (Figure 5.6+ web). An inner-shelf traverse across this boundary shows some of the faunal differences (Murray, 1969). To the north *Brizalina* sp., *Buliminella elegantissima* and *Nonionella* sp. are dominant in single assemblages. To the south the dominant species is *Stetsonia minuta* but in addition the following species are dominant in single assemblages: *Eggerella advena*, *Ammonia* group, *Cancris auriculus, Nonionoides grateloupi* and *Stainforthia* sp. The inner shelf off North Carolina is a sediment-starved, high-energy environment with a discontinuous thin veneer of mobile sand over rock ledges (Lueck and Snyder, 1997). Cores from 25–30 m off Wilmington have assemblages in the 0–1 cm interval dominated mainly by *Asterigerina carinata* but sometimes by *Lepidodeuterammina ochracea, Gavelinopsis praegeri, Nonionella auricula* or *Trifarina occidentalis*. With the exception of the latter two species, the others live attached or clinging to hard substrates. Similar conditions exist off Cape Fear at depths of 24–31 m (Murosky and Snyder, 1994). The dominant species are *Asterigerina carinata, Discorbinella bertheloti* and *Lepidodeuterammina ochracea* but there are single assemblages dominated by *Ammobaculites exiguus, Fursenkoina punctata* and *Stetsonia minuta*. In both areas living individuals occur down to 8 cm (limit of sampling) in the sediment but there are no distinct depth preferences for individual species. This may be because the sediment is mobilised during high-energy conditions.

5.4   Gulf of Mexico and Caribbean

This area is microtidal and of low to moderate energy due to storm waves. In the Gulf of Mexico the major source of freshwater is the Mississippi. This discharges a large amount of fine-grained sediment westwards onto the inner shelf and it is mainly transported westwards by currents and the prevailing wind. Close to the delta the most turbid plume restricts primary productivity through light limitation. Beyond this primary production is high (up to 5 $\mathrm{g\,C\,m^{-2}\,d^{-1}}$, Lohrenz *et al.*, 1999). During the last half of the twentieth century there were significant changes in the nutrient concentration of the discharge, particularly a doubling of the annual nitrate concentration. The latter enhanced primary production and this in turn has caused oxygen depletion of the lower water column. The most extensive area of oxygen-depleted waters in the Gulf of Mexico lies off Louisiana–Texas (Rabalais and

Turner, 2001). Dysoxic waters (<90 μM, i.e., $< 2 \, \text{ml} \, l^{-1} \, O_2$) averaged 8000–9000 km^2 in 1985–92 and 16 000–20 000 km^2 in 1993–2000. Hypoxia extends from 4 to 60 m but is typically between 5 and 30 m and occurs from late spring to late summer. The northern shelf has relict sediment related to previous sea levels. On the outer shelf there is a mixture of old and modern sediments so only a study of living assemblages will give a reliable picture of their ecology and distribution (Phleger, 1960c).

The Gulf of Mexico faunas have some similarity with those off North Carolina but because the area is one of low to moderate energy with fine-grained sediments and periodic dysoxia there are additional forms such as *Brizalina lowmani*, *Buliminella bassendorfensis*, *Buliminella morgani*, *Fursenkoina pontoni*, *Nonionella basiloba* and *Nonionella opima* (Figure 5.6 + web). The Louisiana shelf west of the Mississippi delta was sampled prior to the major increase in nutrient input (Locklin and Maddocks, 1981). The two most abundant species at that time were *Nonionella basiloba* and *Buliminella bassendorfensis* together forming 76% of the live assemblages. Other species included *Brizalina lowmani* and *Fursenkoina complanata*. Many juveniles were present especially during the spring. Together with *Ammonia* group, these four species showed seasonal changes from dominance to subsidiary. Sampling after the increase in nutrients found *Ammonia parkinsoniana* from 9 and 21 m under dysoxic conditions ($O_2 < 22$ μM, i.e., $<0.5 \, \text{ml} \, l^{-1}$; Sen Gupta *et al.*, 1996). *Brizalina lowmani* is dominant at 11 m ($25 \, \text{μM} = 0.5 \, \text{ml} \, l^{-1} \, O_2$). At 35 m in the oxic water adjacent to the margin of dysoxia there is a *Buliminella morgani* assemblage with subsidiary *Brizalina lowmani* and *Nonionella basiloba* and at 80 m from oxic conditions *Buliminella morgani* with *Brizalina albatrossi* and *Nonionella basiloba* (Platon and Sen Gupta, 2001). Foraminifera live down to 14 cm: *Buliminella morgani* is abundant down to 3–4 cm while *Brizalina lowmani* is present throughout. *Brizalina lowmani* is also found in the water column and Hueni *et al.* (1978) suggest that it might have a planktonic stage in its life cycle (meroplanktonic). However, because it is the most abundant small species in the sediment it is the most likely to be thrown into suspension during storms.

Hard substrates such as coral banks and algal-sponge bioherms off Texas and in the Caribbean have distinctive assemblages of epifaunal taxa. They include immobile *Placopsilina confusa*, *Acervulina inhaerens*, *Carpenteria utricularis*, *Cibicides refulgens*, *Gypsina plana*, *Homotrema rubrum*, *Neoconorbina terquemi*, *Planorbulina acervalis* and *Rosalina bradyi* and temporarily mobile *Amphistegina gibbosa*, *Asterigerina carinata*, *Cassidulina laevigata*, *Cibicides lobatulus*, *Gavelinopsis praegeri*, *Planorbulina mediterranensis* and *Rosalina floridana*. Soft substrates in the Caribbean are poorly known. The inner shelf off Puerto Rico has assemblages from 6–7 m dominated by *Ammotium salsum* and from 6–11 m by *Ammonia* group with subsidiary *Fursenkoina pontoni* and *Nonionoides grateloupi*. These are from near the outlet of a river. The majority of assemblages are *Fursenkoina pontoni* with subsidiary *Nonionoides grateloupi* with the rank order occasionally being reversed. Other common subsidiary species include *Reophax caribensis* and *Cancris sagra*. Seiglie (1974) suggests that the abundance of *Fursenkoina pontoni* is related to a recent environmental change which he considers to be the increase in pollution. He also attributes the loss of living *Fursenkoina spinicostata* to the same cause. A single assemblage from muddy sediment at 4 m off St. Lucia, West Indies, is *Nonionella atlantica* with subsidiary *Buliminella elegantissima* and *Elphidium poeyanum* (Sen Gupta and Schafer, 1973).

5.5   South America

A traverse across the macrotidal and wave-influenced shelf off Argentina, spanning the boundary between the coastal and Malvin Current zones, has two assemblages: *Neoconorbina williamsoni* with subsidiary *Ammonia* group at 40 m in the coastal zone and *Epistominella exigua* from 68–120 m (Boltovskoy and Totah, 1985). In Tierra del Fuego, the inner-shelf assemblage is dominated by attached trochamminids (Boltovskoy *et al.*, 1983).

5.6   Pacific Americas

This is a region of oceanographic contrasts. From Alaska to northern California the inner shelf is subject to storm waves and from California to northern Chile there is west-coast swell. Macrotidal and mesotidal areas include the inner shelf off British Columbia and Central America, but from the Canada–USA border to southern Mexico and off Peru the shelf is microtidal. The Californian Current flows slowly southwards along California causing water temperatures to be temperate (~15 °C in summer). In addition, there is upwelling from ~200 m that introduces cold nutrient-rich water from March to July. There is foraminiferal data for the inner shelf from Oregon to California, for some of the continental borderland basins, Todos Santos Bay and Baja California, Mexico, and for a few localities to the south.

5.6.1   Northern inner shelf

Beaches from Oregon to southern California are dominated by *Trochammina kellettae* (mainly south) and *Elphidium microgranulosum* (mainly north) with subsidiary *Elphidiella hannai*. One assemblage has dominant *Glabratella ornatissima*. In tidepools, the dominant species are *Discorbis monicana* and *Elphidium microgranulosum*. Subsidiary species include *Trochammina kellettae*,

Cibicides fletcheri, *Elphidiella hannai* and *Planoglabratella opercularis*. Other subsidiary species with single occurrences include *Alveolophragmium columbiense*, *Haplophragmoides canariensis* subsp. *mexicana*, *Buccella tenerrima*, *Eponides columbiensis*, *Glabratella ornatissima* and *Trochulina lomaensis* (Cooper, 1961, WA-152). *Glabratella ornatissima* lives epifaunally both on sand and on a variety of firm substrates including rocks, plants and other animals (Erskian and Lipps, 1987). This area is subject to strong wave activity from December to April which mobilises the sediment but *Glabratella ornatissima* withstands bedload transport because of its robust test. Some become trapped in the lower parts of algae in areas where the sediment is swept away. There is a cycle with a more-or-less dormant, sparse population of agamonts and plastogamic pairs sheltering in plants in the winter during high-wave-energy conditions. Reproduction occurs during spring and summer with an increase in numbers associated with increased food due to upwelling. By the end of the summer most individuals live on the sediment. Then the wave regime increases, the sediment becomes mobile, and the foraminifera seek the sanctuary of plants once again.

The turbulent inner shelf to a depth of ∼40 m has been sampled from Washington, USA, to Baja California, Mexico (Lankford and Phleger, 1973, WA-153). Of more than 100 species only 27 are common (>10%) at one or more localities. None of the samples off Oregon yielded assemblages of >50 live individuals. *Buliminella elegantissima* is dominant from Washington to southern Baja California and from 4.5–40 m on fine and coarse sand. *Elphidiella hannai* and *Elphidium lene* are confined to northern California. *Brizalina vaughani* and *Rosalina columbiensis* are dominant or subsidiary off California and Baja California. *Nonionella basiloba* is subsidiary off California and dominant off Baja California. *Reophax nana* and *Trochammina kellettae* are common only off Baja California. *Nouria polymorphinoides*, *Textularia schencki*, *Trochammina pacifica*, *Miliolinella californica*, *Quinqueloculina ackneriana* var. *bellatula*, *Quinqueloculina elongata*, *Quinqueloculina lamarckiana*, *Triloculina inflata*, *Triloculina inornata*, *Buccella parkerae*, *Cassidulina delicata*, *Dyocibicides biserialis*, *Elphidium sandiegoensis*, *Elphidium translucens*, *Neoconorbina parkerae*, *Rotorbinella campanulata* and *Trichohyalus ornatissima* have single or a few common occurrences probably representing local blooms at the time of sampling. Lankford and Phleger recognise depth distributions and distributions related to substrate using total assemblages but these are not evident from the live data as this is a much smaller dataset.

5.6.2 Baja California

Todos Santos Bay, Baja California, Mexico, has a broad shelf extending to ∼55 m. There is a deep channel (>370 m) to the southwest. The sediments fall into three groups: (1) coarse pebble and cobbles with epifaunal bryozoans and foraminifera, shell debris and some fine detrital material (<40 m); (2) fine sand and silt (0–55 m); (3) muds with faecal pellets (>180 m). Summer-upwelled cold water extends a short distance into the bay. Walton (1955, WA-154) took samples from the whole area during February 1952 and repeated sampling at some stations from March to November. His paper gives the locations of samples only for February. For the remainder, only the depths are given. The main assemblages in the bay are *Cibicides fletcheri* (5–48 m, all months), *Nonionella stella* (18–176 m, all months), *Alveolophragmium* cf. *columbiense* (22–37 m, all months), *Nonionella basiloba* (26–40 m, February–June), *Trochammina pacifica* (33–37 m, February) and *Reophax scorpiurus* (40–55 m, February). There are single occurrences of assemblages dominated by *Quinqueloculina* sp. (15 m, August), *Ammotium planissimum* (29 m, February), *Saccammina atlantica* (29 m, February), *Elphidium translucens* (29 m, February and June), *Cassidulina tortuosa* (42 m, February), *Globocassidulina subglobosa* (48 m, February), *Reophax curtus* (51 m, February) and *Goesella flintii* (71 m, November). Thus, although Walton (1955) recognised inner-, middle-, outer- and marginal-bay faunas, the picture is more complicated with some species dominant throughout the year and others only at certain months. On the slope beyond the bay in muddy sediment beneath the zone of upwelling there are assemblages dominated by bolivinids: *Bolivina acuminata* (99 m, November), *Bolivina pacifica* (101–371 m, July-August), *Brizalina argentea* (503–595 m, February and June), *Bolivina spissa* (641 m, July), *Globobulimina* spp. (198–271 m, all months), *Bulimina denudata* (201 m, February) and *Chilostomella ovoidea* (686 m, March).

5.6.3   Continental borderland

The continental borderland off southern California, USA, is topographically transitional between continental shelf and bathyal continental slope. There are elongate deep basins separated by ridges. The surface waters are under the influence of the California Current which has a shallow southward flow and a northward flow close to the coast down to a depth of 200 m (Douglas, 1981). At depths > 300 m the water is relatively constant in salinity and temperature and in this respect it differs from the open ocean where temperature decreases with depth. Basin water is nutrient-rich and oxygen-deficient and is sourced from intermediate depths to the south. A distinctive feature of the deep basins is that slow exchange of bottom waters, due to sills separating basins, leads to widespread dysoxia in the deep waters with oxygen commonly <1 ml l^{-1}, i.e., <45 µM. By contrast, the shallower waters that cover the ridges are oxic.

The characteristics of each basin depend on structural controls (sill and basin depth), sedimentary rates and process, and water quality at depths greater than the sill (Douglas and Heitman, 1979). The basins are divided into inner (Santa Barbara, Santa Monica, San Pedro, San Diego) and outer (Santa Cruz, Santa Catalina, San Nicholas) by the first line of islands and ridges. The inner borderland basins are broad and relatively shallow (<900 m) due to higher sedimentation rates that have led to basin infilling. Because their sills are in the oxygen minimum zone, Santa Monica and San Pedro basins are dysoxic–anoxic throughout the year with 22 μM (0.5 ml l^{-1}) at 500 m and <5 μM (<0.1 ml l^{-1}) on the basin floor deeper than 850 m. The outer borderland basins are deeper (1300–1900 m) and have deeper sill depths (1000–1100 m). Sedimentation rates are lower and the sediments are hemipelagic and biogenic. Oxygen levels at sill depth are 22 μM (0.5 ml l^{-1}) and although the basins are dysoxic they are less so than the inner basins.

Santa Barbara Basin. This undergoes short periodic fluctuations in oxygen levels of the bottom water and is dysoxic except for brief periods of anoxia. In the centre of the basin there is a *Beggiatoa* (bacterial) mat at the sediment–water interface. When bottom-water oxygen is depleted and hydrogen sulphide is present in the pore water at the sediment–water interface, the bacteria contribute to the deposition of light bands poor in detritus and thus to the formation of lamination. Most of the basin species are thin-walled *Globobulimina hoeglundi*, *Suggrunda eckisi*, *Nonionella stella* and *Bolivina seminuda* (Phleger and Soutar, 1973) or *Chilostomella ovoidea* and *Nonionella stella*. Slope assemblages have common *Fursenkoina cornuta* and *Brizalina argentea* (given as *Bolivina spissa*; see Bernhard et al., 1997). *Textularia earlandi* is most abundant on the slope but also occurs on the basin floor. Even though *Nonionella stella* lives deeper than the photic zone, it husbands chloroplasts which may be used to assimilate inorganic nitrogen (Grzymski et al., 2002). Most living forms are in the surface 0.25 cm at all seasons, with highest standing crops when oxygen is present and lowest when it reaches a minimum (~1 μM, Reimers et al., 1990; Bernhard and Reimers, 1991). By using fluorescent labelling, Bernhard et al. (2003) were able to show that there is structuring of the microfaunas and bacteria on a very fine scale. A core taken in February when oxygen was 2.4 μM had living foraminifera down to 10 mm with most (73%) in the upper 3 mm and 47% in the top 1 mm. Another core collected in September with 0.1 μM oxygen had rare live forms down to 8 mm and 55% in the top 1 mm. Two slope sites with oxygen levels of 15.4 and 3.5 μM have a *Suggrunda eckisi* assemblage with *Brizalina argentea*, *Bolivina seminuda* and *Nonionella stella* at 431 m and *Textularia earlandi* with *Brizalina argentea* and *Bolivina seminuda* at 522 m respectively (Bernhard et al., 1997, WA-155). There is great complexity of distributions in dysoxic conditions as anaerobic ciliates occur at a level with aerobic foraminifera and an aerobic ciliate occurs in anoxic sediment. This suggests that there are microscale variations in pore water geochemistry that might be modulated by biological activity. Dysoxic Santa Monica Basin (<0.2 ml l^{-1} oxygen) has a >125 μm *Cancris inaequalis* assemblage with subsidiary *Globobulimina pacifica* at 0–1 cm but at 1–2 cm *Globobulimina pacifica* is dominant with subsidiary *Cancris inaequalis* and *Nonionella fragilis* (Mackensen and Douglas, 1989 WA-156).

   San Pedro Basin.   Pairs of box cores collected in April, July and October from 704–722 m reveal the seasonal changes in down-core distribution of foraminifera in a dysoxic setting (oxygen variation 2.5–17 μM over an 11-year period) in mud–silt sediment (Silva *et al.*, 1996, WA-157). All six cores were studied for the >150 μm fraction down to 20 cm and three were studied also for the 63–150 μm fraction down to 10 cm. For the >150 μm assemblages, the proportion of live forms in the top 1 cm varies from 12–36% and for the >63 μm assemblages from 15–24%. The depth at which the cumulative 90% is reached varies from 5 to 8 cm. There are few live forms deeper than 10 cm. Agglutinated forms are treated as a single group and they are common subsidiary forms. *Chilostomella ovoidea* and *Globobulimina pacifica* dominate the >150 μm assemblages with the former dominating from the surface to between 1.5 and 5 cm and the latter dominating deeper than this. *Chilostomella oolina* and *Fursenkoina bramletti* are sometimes subsidiary at depth. *Bolivina minuta, Bolivina pacifica, Buliminella tenuata* and *Epistominella smithi* are essentially confined to the 63–150 μm fraction. Here, the >63 μm assemblage has been calculated from the two datasets. Although the general pattern of dominance of *Chilostomella ovoidea* in the top few centimetres and *Globobulimina pacifica* below this is still evident, there are differences. In April, *Bolivina pacifica* is dominant in the top 0.5 cm and *Bolivina minuta* intermittently between 2.5 and 7 cm. Although *Globobulimina pacifica* is still common in July it is replaced by *Epistominella smithi* and *Bolivina minuta* as dominant forms. Kaminski *et al.* (1995, WA-158) considered just the agglutinated forms. There is one assemblage with high dominance of an organic-cemented *Textularia* sp. and another with *Rhizammina irregularis* and subsidiary *Reophax bilocularis*.

   San Catalina Island and Basin.   This is part of the ridge between the inner San Pedro and outer borderland Santa Catalina basin. Oxygen values are >270 μM, i.e., >6 ml l^{-1} at depths down to $\sim$30 m and decrease to 70 μM $= 1.58$ ml l^{-1} at 200 m. The sediments range from calcareous sands to sandy silt. The foraminifera are primarily hyaline with an abundance of bolivinids (McGlasson, 1959, WA-159). In the coastal zone (2–7 m) there is an *Elphidium rugulosum* assemblage with *Buliminella elegantissima, Bulimina denudata* and *Bolivina quadrata.* This overlaps with a *Bolivina quadrata* assemblage (2–38 m). There are single occurrences of a *Cassidulina minuta* assemblage with *Bolivina quadrata* and *Bulimina denudata* (24 m) and *Nonionella basiloba* with *Bolivina pacifica* and *Bolivina quadrata* (33 m). A *Cancris sagra* assemblage with various bolivinids (40–101 m) partly overlaps a widely developed *Bolivina pacifica* assemblage in which the key species commonly forms >50% (46–190 m). There are other bolivinids which form single or few assemblages: *Bolivina acuminata* (64–100 m), *Bolivina compacta* (40 m) and *Brizalina acutula* (58 m). In Santa Catalina Basin at 1200–1300 m, the sediment surface is covered in 2–4 cm high tree-like growths of agglutinated foraminifera (Smith and Hamilton, 1983). The bottom water has oxygen of 18 μM (0.4 ml l^{-1}) and the sediments are silty clays with up to 6% TOC. Five agglutinated species were collected by divers in a submersible but only one, *Pelosina* cf. *arborescens* was clearly visible. The others are *Pelosina* cf. *cylindrica*, a spherical mudball *Oryctoderma* sp., mud-walled astrorhizids and a smaller astrorhizid (Levin *et al.*, 1991). The authors believe that disturbance by mound builders is an important source of spatial heterogeneity in this basin. At 829 m the >125 μm surface assemblage is *Bolivina spissa* with subsidiary *Globobulimina pacifica* but below 3 cm *Bolivina spissa* is absent and the dominant form is *Globobulimina pacifica* (Mackensen and Douglas, 1989, WA-156). The agglutinated foraminifera show patchy distribution with no single main assemblage: *Reophax bilocularis* or *Reophax excentricus*, more widespread *Saccorhiza* sp. and *Verneuilinulla*/?*Matanzia* (Kaminski *et al.*, 1995 WA-158). They infer that *Textularia* sp. is infaunal and suggest that with increasing severity of dysoxia only infaunal taxa survive.

San Diego Trough.   This is the most southerly of the studied California borderland inner basins (Uchio, 1960, WA-160). Only a few species dominate widespread assemblages: *Nonionella stella* (mainly 37–168 m, one occurrence at 631 m), *Bolivina pacifica* (73–769 m), *Reophax gracilis* (71–287 m), *Goesella flinti* (77–428 m) and *Fursenkoina apertura* (293–1171 m). There are local assemblages dominated by *Buliminella elegantissima* (16–20 m), *Nonionella basiloba* (27–38 m), *Globocassidulina subglobosa* (55–331 m) and *Cibicides spiralis* (1025–1169 m). R-mode cluster analysis (without transformation of the data) shows that species do not have high co-occurrence as they join at Bray–Curtis similarities of <50%. R-mode principal component analysis plots all species superimposed on one spot except for *Reophax gracilis, Bolivina spissa, Fursenkoina apertura* and *Nonionella stella* already recognised as the dominant taxa. Q-mode cluster analysis (without transformation of the data) has main clusters linking subclusters at 25–40% similarity. These show some degree of bathymetric overlap: 16–26, 54–503, 73–453, 452–768 and 640–1162 m. Uchio also recognised a series of depth zones with boundaries at 24, 82, 183, 458, 640 and 825 m.

Summary. There is a marked contrast between the higher-energy, well-oxygenated inner-shelf assemblages and those of the deep, dysoxic basins. In each case essentially the same species occur over a large latitudinal range due to the absence of sharp temperature boundaries. In the basins, species of *Bolivina* and *Brizalina* show distributions related to oxygen. Abundant *Brizalina argentea* occurs where oxygen is <18 µM, i.e., <0.4 ml l^{-1} whereas *Bolivina spissa* occurs between 22 and <45 µM = 0.5 and <1.7 ml l^{-1}. *Bolivina pacifica* is most abundant in oxygenated water but can tolerate dysoxia (Douglas, 1981). There are also morphological differences. *Bolivina* from dysoxic environments have thin walls and less ornament (Harman, 1964; Lutze, 1964). Those from shelves rich in oxygen are small and prolate (e.g., *Bolivina vaughani*) whereas the forms from dysoxic deeper waters are lanceolate and large (e.g., *Brizalina argentea*, *Bolivina spissa*). *Loxostomum pseudobeyrichi* and *Brizalina argentea* show morphological variation with respect to bathymetry and oxygen levels (Lutze, 1964). *Loxostomum pseudobeyrichi* has a lower width:length ratio in the basin and higher above the sill. Some of the variants of *Brizalina argentea* have been given separate names (*Bolivina argentea* var. *monicana* Zalesney, *Bolivina subargentea* Uchio).

5.6.4   Gulf of California and El Salvador

The Pacific shelf off Baja California is dominated by *Bolivina seminuda* (Phleger and Soutar, 1973). In the Gulf of California there are many assemblages often having a single occurrence (Phleger, 1964, WA-161; re-interpreted by Streeter, 1972). The entrance is microtidal and the head macrotidal. Assemblages on the shelf are dominated by *Reophax nana* (11 m), miliolids (11–27 m) on sand, *Ammonia* group (11–18 m) confined to the northern Gulf, *Nonionella stella* (15–33 m) on silt and clay, *Textularia schencki* (18–33 m), *Trochammina kellettae* (18 m), *Nonionella basiloba* (18 m), *Cassidulina* sp. (22 m), *Elphidium incertum* (26 m), *Hanzawaia nitidula* (27–33 m) on sand, *Buliminella elegantissima* (31 m), *Cancris auriculus* (31 m), *Fursenkoina pontoni* (40 m), *Trifarina jamaicensis* (62 m), *Bulimina denudata* (64 m), *Bulimina marginata* (75 m), *Brizalina vaughani* (80 m) and *Cancris panamensis* (91 m). Three assemblages from the shelf off El Salvador at 50–82 m in muddy fine volcaniclastic sand are dominated by *Brizalina striatula*, *Cancris panamensis*, *Cancris sagra* and *Hanzawaia concentrica* (Smith, 1964, WA-162). This is an area with dysoxic bottom water (26–51 µM = 0.58–1.15 ml l^{-1}).

5.7   West Pacific

Shelf seas have received very little attention except around Japan. On the Japan Sea coast of Hokkaido, Ishikari Bay is influenced by the warm Tsushima Current and cold water either from upwelling or from the coast. The shelf break is ~150 m. An *Eggerella advena* assemblage with *Lagenammina difflugiformis* and *Trochammina charlottensis* occurs in various sediment types from 20–38 m (Ikeya, 1970, WA-163). *Goesella flintii* with *Reophax gracilis* and *Trochammina charlottensis* occurs in sand and mud at 40 and 149 m and *Quinqueloculina seminulum* with *Cibicides* cf. *refulgens*, *Hanzawaia nipponica* and *Pseudononion japonicum* on sand at 60 m. On the Japan Sea coast of Honshu around the Oga Peninsula (Matoba, 1976a, b, WA-164, WA-165) under the Tsushima Current there is dominant or subsidiary *Ammobaculites* sp., *Eggerella advena*, *Textularia earlandi*, *Quinqueloculina* spp., *Pararotalia nipponica* and *Rosalina* spp. at depths from 5–50 m in fine to coarse sand. At slightly greater depths on sands with some silt there are *Textularia parvula*, *Bulimina marginata* and *Nonionella stella* assemblages with subsidiary *Nonionoides japonicum* and *Rectobolivina raphana*.

On the Pacific side of Japan, the sea between Hokkaido and Honshu extends from the shelf to bathyal depths. Two currents influence the area: the warm Tsushima Current flowing out of Tsugaru Straits and a southward flowing extension of the subarctic Oyashiro Current. The shelf assemblages vary considerably in composition (Ikeya, 1971, WA-166). On the north side at 54 and 56 m there is a *Nonion* sp. assemblage with subsidiary *Discammina compressa*, *Reophax* sp. and *Textularia* cf. *earlandi* in mud and an *Eggerella advena* assemblage with *Elphidium clavatum* and *Nonionella stella* in coarse sand. At 85–135 m in mud or muddy sand there is a *Nonionella stella* assemblage with *Islandiella islandica* and *Nonionoides scaphum* and a *Nonionellina labradorica* assemblage with *Uvigerina peregrina dirupta*. All these assemblages are under the influence of the cold current. On the south side of the area, the assemblages are bathed in warm water all year round. At 70 m on sand there is a miliolid-dominated assemblage (*Pateoris hauerinoides*, *Pyrgo ezo*). At 100 m in muddy sand, a *Hanzawaia nipponica* assemblage with *Guttulina* cf. *yamzakii* and *Islandiella islandica*, and at 115 m on coarse sand a *Cibicides lobatulus* assemblage with *Cibicidoides pseudoungerianus*. Miyako and Yamada bays are rias on northeastern Honshu. In Miyako Bay, there is an *Elphidium incertum* assemblage with *Buliminella elegantissima* and *Pseudononion tredecum* at 10 m on sand, an *Eggerella advena* assemblage with *Ammoglobigerina globigeriniformis* at 8–49 m on mud and a *Cassidulina complanata* assemblage at 45 m on shell sand. In Yamada Bay the dominant species are *Hopkinsina pacifica* and *Nonionella pulchella* from 14–37 m on mud (Ujiié and Kusukawa, 1969, WA-167).

5.8   Australia

Over a distance of 1000 km on the southwest coast of Australia there are embayments sheltered from oceanic swell which provide habitats for meadows of *Posidonia australis*. Seagrass leaves show a heterogeneous distribution of epiphytic foraminifera with standing crops of 0 to 4 individuals cm^{-2}. Often the highest densities occur in epiphytic algal growth on the grass blades. Zonation on a microscale is also present with *Trochulina dimidiata* and *Crithionina* spp. in the basal 15 cm where detritus is abundant; *Marginopora vertebralis*, *Amphisorus hemprichii* and discorbids occupy the middle regions while the top 10 cm, often algal-encrusted part of the leaves, has miliolids, buliminids, glabratellids, spirillinids, cibicidids and encrusting rotaliids (Semeniuk, 2000, WA-168). Analysis of variance shows that for most species there is homogenous distribution within the seagrass meadows. There is heterogeneity on a regional scale in the three bays studied (from north to south, Dongara, Whitfords and Albany). Dongara has *Peneroplis planatus*, *Quinqueloculina* spp., *Vertebralina striata*, *Rosalina* sp. and *Trochulina dimidiata*. Whitfords has dominant *Angulodiscorbis quadrangularis* with *Miliolinella* sp., *Annulopatellina annularis* and *Rosalina* sp. and Albany has *Trochulina dimidiata* with *Rosalina* sp. There are seasonal differences in standing crop and species diversity with both being lower in the winter (Semeniuk, 2001). Winter temperatures are similar over the 1000 km extent of the study area and the north-to-south changes reflect the differences in the summer temperatures which are higher in the north. Larger foraminifera (*Amphisorus hemprichii* and *Marginopora vertebralis*) are restricted to the tropical north and there are fewer *Quinqueloculina* species in the cooler south.

5.9   New Zealand

Sandy gravel from between rocks in an intertidal pool in Northland has a *Nonionella parri* assemblage with subsidiary *Quinqueloculina seminulum* and *Rosalina bradyi* (Hayward, 1979, WA-169). Forms living attached to *Corallina* include *Rosalina bradyi* and *Ammonia* group. In the inner shelf by the Cavalli Islands off North Island, common species include *Elphidium charlottensis*, *Elphidium oceanicum*, *Elphidium novozealandicum*, *Cibicides marlboroughensis*, *Trochulina dimidiata*, *Planoglabratella opercularis*, *Cassidulina carinata*, *Globocassidulina canalisuturata*, *Bulimina submarginata* and *Pileolina zealandica* (Hayward, 1982).

5.10   Pacific Ocean

The lagoon on Scilly atoll has no open connection with the surrounding ocean but has water of normal salinity (35). Down to 30 m, the sediments have an assemblage of a tiny rotaliid attached to sand grains. This is monospecific down to 8 m. From 40–52 m the common forms are *Textularia candeiana* and *Quinqueloculina tropicalis* with subsidiary tiny attached rotaliids (Salvat and Vénec-Peyré, 1981, WA-170; Vénec-Peyré and Salvat, 1981). The green alga *Microdictyon* grows on rock surfaces and supports an assemblage of 25 species mainly of *Textularia* with small numbers of miliolids. There are few species in common between the sediment and *Microdictyon* assemblages.

Moorea, Society Islands, is surrounded by a barrier reef that encloses a lagoon but there is good exchange of water between the ocean and the lagoon. Living foraminifera can be recognised from their coloured protoplasm (Vénec-Peyré, 1991). There is a marked contrast between the miliolid-dominated faunas of the lagoon and the rotaliid-dominated faunas of the outer oceanic slope. In the lagoon and on the reef, hard substrates including various calcareous algae and dead coral have a dominance of miliolids, especially *Triloculina* spp., *Quinqueloculina* spp. and *Miliolinella subrotunda* together with *Vertebralina striata*. There are differences between the fringing reef assemblage of *Vertebralina striata*, *Triloculina* cf. *oblonga*, *Triloculina planciana* and *Quinqueloculina agglutinans* and the barrier-reef assemblage of *Triloculina planciana*, *Miliolinella subrotunda* and *Miliolinella baragwanathi*. Foraminifera attached to algal thalli are rare on the fringing reef but *Cibicides mayori*, *Planorbulinoides retinaculatus*, *Acervulina inhaerens* and *Planogypsina squamiformis* are common on the barrier reef. There are encrusting forms such as *Homotrema rubrum*, *Planorbulina acervalis* and *Planorbulina rubra* on coral fragments. On the outer oceanic slope of the reef, *Amphistegina lessonii* forms >50% of the assemblages. Mobile sediments and rubble exposed to wave attack have an assemblage of small forms (<150 μm) such as *Rotaliammina*, *Siphotrochammina* and *Cymbaloporetta*, attached to detrital grains. These taxa are embedded in small cavities that they have excavated out of the carbonate substrate (Vénec-Peyré, 1985b; 1988; 1991; 1993; 1996). They preferentially colonise particles >1 mm in size and they occur at water depths down to ~40 m. The excavated material may help in the construction of the agglutinated test wall. Existence in a cavity may offer protection in a high-energy environment.

5.11   Indian Ocean

There is very little information on living forms. Assemblages from the inner shelf on the east coast of India are dominated by the *Ammonia* group with subsidiary *Bolivina pseudoplicata*, *Elphidium translucens* and *Nonionoides boueanum* (given as *Florilus*; Rasheed and Ragothaman, 1977). Standing crops are highest prior to the monsoon. Rock pools with seaweeds support epiphytic foraminiferal assemblages that feed on the rich microflora on the algal surfaces. Most assemblages are strongly dominated by *Pararotalia nipponica* with subsidiary *Cymbaloporetta bradyi* and *Rosalina floridana* but in a few instances the latter species become dominant (Rao *et al.*, 1982, WA-171). *Pararotalia nipponica* attaches itself by the umbilical side to the alga and some are found in pits on the algal surface. Adult foraminifera are often surrounded by juveniles.

Four broad facies are recognised in the Arabian Sea on the Pakistan–Indian shelf: *Ammonia-Nonionoides* (given as *Florilus*) in sand and mica-rich mud, *Ammonia-Cancris* in mica-rich mud, *Cassidulina-Cibicides* in pteropod-rich mud, and Buliminacea in mud from 334 m (Zobel, 1973). In the Gulf of Khambhat, Nigam (1984) found that living forms are absent on an extensive sand bank due to the high tidal-energy conditions causing substrate mobility. On muddy substrates under lower-energy conditions there is a low diversity *Ammonia, Ammobaculites, Pararotalia* assemblage. Off Cochin, the inner shelf has an *Ammonia* group assemblage with subsidiary *Nonionoides boueanum* and *Ammobaculites persicus* (4–18 m, mud), *Nonionoides boueanum* with subsidiary *Ammonia* group (4–25 m, mud), or *Cancris auriculus* with *Brizalina striatula* (25–30 m, sand) (Seibold and Seibold, 1981, WA-88). There are also single occurrences of assemblages dominated by one of *Ammobaculites persicus, Asterorotalia dentata, Brizalina striatula, Cassidella panikkari* or *Murrayinella erinacea*.

5.12   Southern Ocean

The shallow-water areas from 60–78° S are very variable due to different degrees of exposure to scour by floating ice and to local differences of algal productivity. Sunlight controls biological activity both directly (primary production) and indirectly (temperature and sea ice). Whereas Anvers Island at 64° S on the Antarctic Peninsula does not experience total darkness for periods of more than hours, McMurdo Sound at 78° S is in darkness for nearly four months each year. The mean annual temperatures are −2.8 and −17.6 °C respectively (Lipps and DeLaca, 1980). In many areas the shore is ice-covered but in some areas, e.g., parts of the Antarctic Peninsula, the shore is ice-free although exposed areas are scoured down to ∼15 m by floating ice.

Submarine cliffs.   These have a macroflora of seaweeds and a macrofauna of encrusting invertebrates. They provide diverse habitats for attached foraminifera. On the Antarctic Peninsula forms found in association with tunicates include *Tolypammina vagans, Turritellella shoneana, Cibicides refulgens* and *Rosalina globularis*. Sponges also have these species together with *Cribrostomoides*

*jeffreysii, Lepidodeuterammina ochracea, Nodulina dentaliniformis, Trochammina mal-
ovensis, Pyrgo elongata, Astrononion stelligera, Cassidulina crassa, Cassidulinoides parkerianus* and *Pullenia subcarinata* (Temnikov, 1976). Epiphytic foraminifera from
cliffs at 12 m are dominated by *Hemisphaerammina bradyi, Lepidodeuterammina
ochracea* or *Saccammina sphaerica* with subsidiary *Trochammina malovensis, Cornuspira involvens, Cibicides refulgens, Patellina corrugata* or *Rosalina globularis* (Lipps
and DeLaca, 1980, WA-172). Cliffs at 30 m have *Saccammina sphaerica, Cibicides
refulgens* or *Patellina corrugata* with subsidiary *Lepidodeuterammina ochracea, Cassidulina crassa* or *Rosalina globularis.*

Antarctic Peninsula. Gravel with crustose algae, extending from the
intertidal zone to about 45 m has *Cibicides refulgens* and *Rosalina globularis.* Algae
that support large numbers of foraminifera are 15–20 cm tall, finely branched
and dense and either spherical or globular in shape. Forms such as *Plocamium
coccineum, Pantoneura plocamioides* and *Picconiella plumosa* have the highest concentrations of dissolved and particulate organic carbon and bacteria, all of
which are potential food for foraminifera. Lipps and DeLaca (1980) suggest
that in the plant mesh foraminifera are trapped by the high viscosity of
the water and have a plentiful supply of food. Sediments from 2–4 m have
*Cribrostomoides jeffreysii, Lepidodeuterammina ochracea, Turritellella shonea, Textularia
wiesneri, Tolypammina vagans, Quinqueloculina seminulum* and *Epistominella exigua*
(Temnikov, 1976). Muddy substrates have an agglutinated fauna comprising
*Cribrostomoides jeffreysii, Gordiospira fragilis, Hemisphaerammina depressa, Hippocre-
pinella hirudinea, Lepidodeuterammina ochracea, Psammosphaera fusca, Nodulina dentaliniformis* and *Trochammina malovensis* (Lipps and DeLaca, 1980).

Ross Sea. McMurdo Sound has dominant *Haplophragmoides canariensis*
and *Cassidulinoides porrectus* on sediment substrates (Lipps and DeLaca, 1980).
In Terranova Bay, bryozoans provide shelter for attached foraminifera such
as *Psammosphaera fusca* forma *adhaerescens, Haplophragmoides canariensis,
Portatrochammina antarctica, Trochammina arctica, Cibicides refulgens* and *Rosalina
globularis* (Zampi *et al.,* 1997). *Haplophragmoides canariensis* is attached by agglutinated material to the trabeculae of the bryozoa, and encrusted with a fine
agglutinated film like that also seen in association with *Portatrochammina
antarctica.*

Explorers Cove is oligotrophic. Large (>1 mm) agglutinated foraminifera
morphologically similar to those in the deep sea can be sampled in shallow
water (down to ∼28 m) by divers, e.g., *Astrammina rara* and *Notodendrodes antarctikos* (Bowser *et al.,* 1995), *Astrammina triangularis* (Bowser *et al.,* 2002). Many
of the species are new to science and remain unnamed (Gooday *et al.,* 1996).

Crithionina delacai is abundant at 28 m water depth living infaunally in the top 1 cm of sediment feeding on diatoms and possibly on bacteria (Gooday *et al.*, 1995a). Some of the large species live both infaunally and also epifaunally on the sediment surface and on objects rising above it. The latter mode of life is thought to be an adaptation to suspension feeding. Because algal food is present for only a few months each year, most forms have more than one feeding strategy as a matter of survival. Also, species may be able to lower their metabolic rate during unfavourable periods. *In situ* feeding experiments show that *Astrammina rara* and *Astrorhiza* sp. ingest 2.5–3.5 ng C mg wet weight d^{-1} which is much lower than other elements of the meiofauna (Rivkin and DeLaca, 1990). However, the foraminifera synthesise two to five times more protein and three to five times less lipid than the metazoan meiofauna. The causes may be due to differences in selecting prey, in pathways and rates of digestion of the ingested prey algae or in metabolic and growth rates.

Gooday *et al.* (1996) compared the Explorers Cove faunas with analogues in Norwegian and Scottish fjords and the North Sea. The similarities may be largely due to similar sampling procedures: in all these studies relatively large volumes of sediment were washed over coarse sieves (>1 mm aperture). However, most ecological studies are based on samples having a small area (e.g., cores or small grabs) so large individuals are rarely encountered. That is the case with the cores collected by diving (Bernhard, 1987). She distinguished seven biotopes: boulder, open deep water, sponge mat, sediment below sponge mat (all with an average depth of 26 m), seasonally anoxic basin (18 m), shallow water (15 m) and anchor ice (4 m). On presence/absence the fauna of boulders, open deep water, sponge mat and sediment below sponge mat are very similar. Fewer species are present in seasonally anoxic basins, shallow water and anchor ice biotopes. Calcareous taxa are predominant on boulders and sponge mats; elsewhere, agglutinated forms predominate. Most of the living individuals in the anchor ice biotope (<7 m) were juveniles at the time of sampling and they were probably colonists from deeper water showing less seasonal stress. The distribution of foraminifera in sediment cores shows that most live forms are in the top 5 mm although some live forms are found in the anoxic zone deeper than 3 cm (Bernhard, 1989). The standing crops measured by ATP analysis, rose Bengal and Sudan Black B staining differ. Sudan Black B gives very low values. Rose Bengal gives somewhat higher values than ATP. Bernhard (1987) does not consider these shallow-water biotopes to be similar to the deep sea in terms of either standing-crop or species diversity. Furthermore, the list of species includes forms mainly known from relatively shallow water elsewhere. The only true deep-water form is *Epistominella exigua* and that is not common.

Deep shelf. The depth of the shelf break around Antarctica is commonly around 500 m and sometimes as deep as 700 m. Mikhalevich (2004) considers that the foraminiferal faunas are closely tied to water mass. In both east and west Antarctica the maximum number of species and individuals occurs between 180–300 m. The shallower shelf is bathed by the Antarctic Coastal Current with temperatures of −1 to −1.9 °C and has a largely endemic fauna. Along the shelf break, *Trifarina angulosa* is abundant. In the mid 1990s, the ice that covered the Larsen A shelf broke up and dispersed so that it was possible to sample the previously inaccessible sediments (Murray and Pudsey, 2004, WA-173). The assemblages are mainly dominated by *Nonionella iridea* with subsidiary *Globocassidulina subglobosa*. There is a single *Reophax subdentaliniformis* assemblage with subsidiary *Globocassidulina subglobosa*, a *Textularia earlandi* assemblage with *Reophax subdentaliniformis* and *Reophax subfusiformis* and an *Adercotryma wrighti* assemblage with *Pseudobolivina antarctica* and *Nonionella iridea*. By the ice margin there is a *Portatrochammina antarctica* assemblage with *Globocassidulina subglobosa*. Some species have green protoplasm even though they come from much deeper than the photic zone. This indicates feeding on fresh phytodetritus. Estimated primary production is 208–416 $g C m^{-2} y^{-1}$ so this is a eutrophic area. The standing crops are high (188–2430 individuals $10 cm^{-3}$) and show a negative correlation with water depth.

The large miliolid *Cornuspiroides rotundus* (maximum diameter 21.6 mm) occurs on the continental shelf around Antarctica on sand and sandy silt substrates left as a lag from the strong bottom currents. Its distribution is unrelated either to water mass or surface water productivity. It occurs with other cornuspirids with some individuals of *Cornuspira involvens* attaining a diameter of 10 mm (Schmiedl and Mackensen, 1993). The assemblages are dominated by one or more of *Trifarina angulosa*, *Globocassidulina biora*, *Globocassidulina rossensis*, *Cibicides lobatulus*, *Haplophragmoides canariensis* or *Cassidulinoides parkerianus*. Schmiedl and Mackensen speculate that the mode of life might be epifaunal, lying flat on the sediment, feeding on phytodetritus. Mikhalevich (2004) also comments on the large size of some other Antarctic taxa (*Pseudonodosinella, Glandulina, Pyrgo*). She also considers that there is a high proportion of endemic species that show circum-Antarctic distributions. Furthermore, she considers that forms found in deeper water elsewhere occur in shallower water around Antarctica due to the cold conditions. This is a good illustration of the fact that benthic foraminifera are not distributed with respect to water depth per se.

5.13   Stable isotopes

In the arctic Kara and Pechora seas, the oxygen stableisotopic composition of stained *Elphidium clavatum* tests from 15–20 m water depth is primarily controlled by mixing of seawater and river water. At greater depths, in normal marine water, temperature is the main control (Polyak *et al.*, 2003). For carbon isotopes, $\delta^{13}C$ is controlled by remineralisation of organic matter and by water mixing. *Elphidium clavatum* is infaunal and therefore has depleted values (3‰ lower than epifaunal *Cibicides lobatulus* from the same area). Estuarine *Haynesina orbiculare* is 2‰ heavier than *Elphidium clavatum* suggesting that it is shallow infaunal to epifaunal. Likewise, in the Laptev Sea, *Elphidium clavatum* shows a large negative offset in $\delta^{18}O$ where there is a freshwater influence. Bauch *et al.* (2004) suggest that this is a vital effect and that this species is especially sensitive to freshwater influence. *Elphidium groenlandicum* and *Haynesina orbiculare* also show negative offsets of $\delta^{18}O$ but they show a constant offset in a stable marine setting and in a seasonally, highly variable river-influenced environment.

In the Celtic Sea, UK, the values for *Quinqueloculina seminulum* and *Ammonia batavus* from stratified areas are more positive for $\delta^{18}O$ and less positive for $\delta^{13}C$ than those from areas where the water is vertically mixed throughout the year (Scourse *et al.*, 2004). For oxygen, *Ammonia batavus* and *Bulimina marginata* give values close to equilibrium and the latter calcifies during September at stratified localities and during spring or early summer at localities where the water is vertically mixed. *Quinqueloculina seminulum*, *Cibicides lobatulus* and *Bulimina gibba* show negative disequilibrium. *Quinqueloculina seminulum* calcifies during September when bottom-water temperatures are highest due to the downward mixing of warm water as the thermocline is destroyed by autumn storms.

The carbon isotopic composition of aragonitic and calcitic species that secrete their tests in isotopic equilibrium with respect to ^{18}O have been compared with that of dissolved inorganic carbon (DIC) to determine biological (vital effect) carbon isotopic fractionation on the California seaboard (Grossman, 1984a,b). Shallow-water *Cassidulina braziliensis*, *Cassidulina limbata* and *Cassidulina tortuosa* have similar $\delta^{13}C$ values and show an average enrichment factor of -0.2 ± 0.1‰ at temperatures of 8–10 °C. *Uvigerina curticosta*, *Uvigerina peregrina* and megalospheric *Brizalina argentea* from the slope are enriched by 0.7 ± 0.1‰ relative to ambient carbonate in the temperature range 3–9 °C. There is no temperature dependence for enrichment relative to ambient carbonate in any of these species. The enrichment values for *Cassidulina* (± 0.3‰) are close to the equilibrium value for calcite whereas for *Uvigerina* and *Brizalina* the values are 0.2–0.8‰ lower suggesting that these forms incorporate $\delta^{13}C$ depleted pore-water DIC in the form of metabolic CO_2. This also applies to

Globobulimina pacifica. All these forms show a microhabitat effect related to their infaunal mode of life. Other genera including *Pyrgo, Quinqueloculina, Triloculina* and *Lenticulina* secrete their tests not in ^{18}O equilibrium. Aragonitic *Hoeglundina elegans* is enriched in both ^{13}C and ^{18}O relative to equilibrium calcite. It is epifaunal and is therefore not influenced by the microhabitat effect (Grossman, 1987).

5.14   Summary

Data on live assemblages are sparse except for the North Atlantic and East Pacific North American shelves. Important controls on distribution patterns are temperature, salinity, energy levels, submarine vegetation and oxygen depletion. However, where several variables have been measured, correlation with the fauna gives ambiguous results showing that single factors are rarely the sole control (Scott *et al.*, 2003). Details for individual species are given in the Appendix.

Temperature is a major control on inner-shelf species with distinctive cold, temperate and warm assemblages. In addition, seasonal ice cover affects light penetration and primary productivity and thus standing crop. Moving ice can also abrade the sea floor down to depths of several metres as around Antarctica. Summer thermohaline stratification prevents warming of the bottom water until storms destroy the stratification and mix warm water down to the sea floor. The boundary between stratified and vertically mixed waters is normally a sharp front that may have high production and lead to distinctive assemblages (*Stainforthia fusiformis* in the Celtic Sea). Salinity is only locally important where it is brackish (as in the Baltic Sea). High energy from waves and tidal currents affects the grain size of the substrate (generally coarse), its mobility (especially for medium sand) and food supply (favours suspension feeders). High-energy areas have a significant proportion of epifaunal taxa attached to sessile benthos (bryozoa, hydroids, worm tubes, calcareous algae) or hard substrates (western seaboard of Europe, Caribbean). In low-energy areas with clear water and submarine vegetation, phytal foraminifera are common (Mediterranean, western Australia, oceanic atolls, Southern Ocean). Dysoxic bottom water is only locally developed (Adriatic Sea, northern Gulf of Mexico, California) so is not a major control in most shelf seas.

Species distributions are patchy and standing crop is highly variable in all environments ranging from 0 to 9300 individuals $10\,cm^{-3}$. However, maximum values $>1000\ 10\,cm^{-3}$ are found only in the Baltic, Skagerrak, Scotland, Adriatic, Grand Banks, California basins and Larsen shelf, Antarctica. Elsewhere, the maximum is <1000. Typical values on most shelves range from

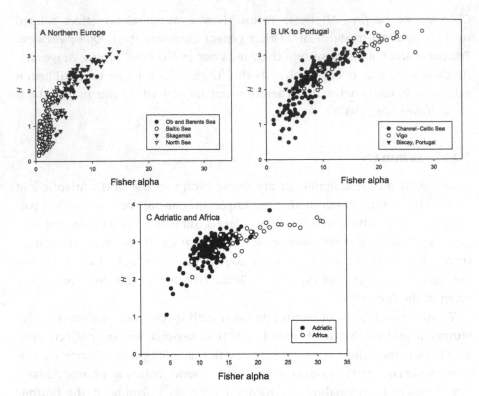

Figure 5.7. Species diversity of eastern Atlantic shelf seas.

50 to 500 10 cm^{-3}. However, some California borderland basins and Larsen shelf have higher typical values (up to 2000). The Santa Barbara Basin on the Californian continental borderland has lower standing crops above the sill where the sediments are bioturbated and well oxygenated (18–1890 individuals 10 cm^{-3}) and high values in laminated sediments deposited under oxygen-limited conditions where predators are excluded (1150 > 4000 individuals 10 cm^{-3}). Apart from seasonal variations in food supply (about which there is little reliable information) influencing factors are current and wave energy, sediment grain size, oxygenation and predation. Areas with high energy and sand substrates have lower standing crops than those with low-energy mud substrates (e.g., Scotland: sand, 1–87 cf. mud, 221–1750 individuals 10 cm^{-3}; Clyde Sea: sand, 200–500 cf. mud, 1000–4000; Iceland: sand, 15). For this reason values may be higher on the outer than inner shelf but the picture is not clear (e.g., Portugal: sand, 41 cf. mud, 98–420; Georgia Bight, USA: sand < 100 cf. mud > 300 individuals 10 cm^{-3}). Coarse sands have higher values than mobile sands because the latter is a very hostile environment.

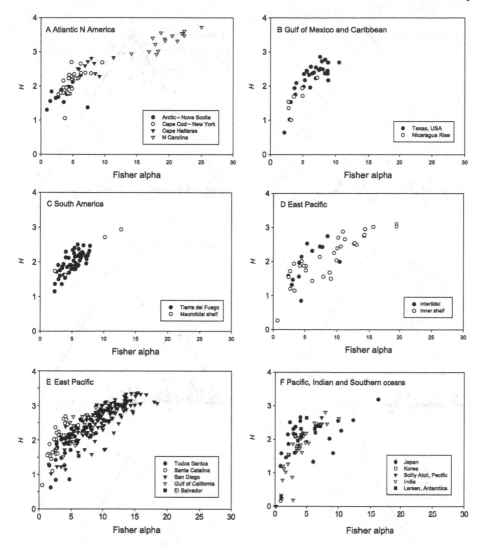

Figure 5.8. Species diversity of shelf seas.

Species diversity ranges from low in the North Sea and Baltic Sea (Figure 5.7A) to moderate in most European shelves (Figure 5.7A, B) and high in Vigo, Spain, and Africa (Figure 5.7B, C). In the Baltic, conditions are brackish and phytal assemblages have even lower diversity than the sediment assemblages. In the English Channel, the lowest values are off the Seine, France, and in the Celtic Sea, off the Bristol Channel. Both these areas are under the influence of rivers. With the exception of the North Sea, all the shelves having normal salinities have moderate to high diversity. On the USA seaboard, diversity is low to moderate from the Arctic to Cape Hatteras and moderate to high off

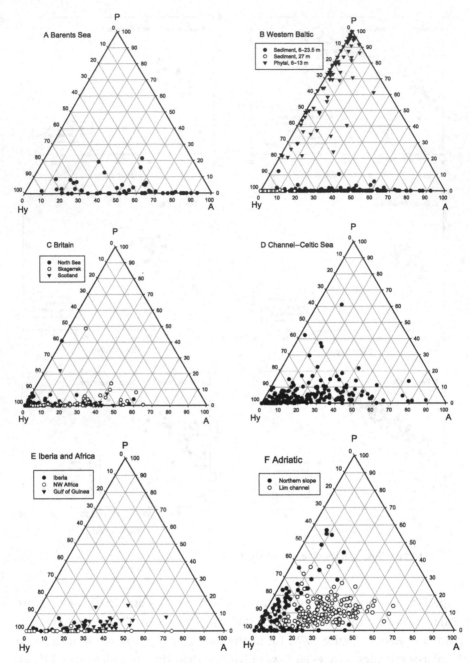

Figure 5.9. Ternary plots of walls for eastern Atlantic shelves (triangle corners represent 100% of the labelled component: A = agglutinated, P = porcelaneous, Hy = hyaline).

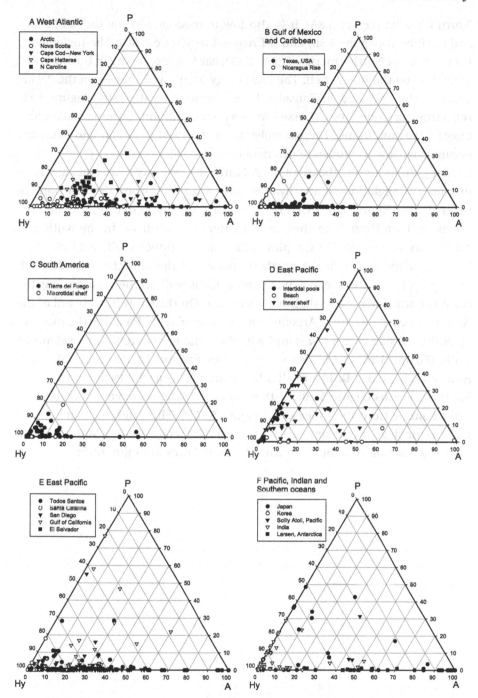

Figure 5.10. Ternary plots of walls for western Atlantic, Pacific, Indian and Southern oceans (triangle corners represent 100% of the labelled component: A = agglutinated, P = porcelaneous, Hy = hyaline).

North Carolina (Figure 5.8A). It is also low to moderate in the Gulf of Mexico and Caribbean (Figure 5.8B) and off Argentina (Figure 5.8C). The high-energy Pacific coast of North America (Figure 5.8D) has lower diversity in the intertidal zone than on the inner shelf. The low-energy basins and shelves on the Pacific coast of the USA and El Salvador have low to high diversity (Figure 5.8E) reflecting variation from stressed low-oxygen conditions to more favourable oxygenated conditions. The examples from the Pacific, Indian and Southern oceans have low to moderate diversity (Figure 5.8F).

Along the eastern seaboard of the Atlantic, the majority of assemblages have mixtures of agglutinated and hyaline walls with low proportions of porcelaneous walls (Figure 5.9). Assemblages with a greater proportion of porcelaneous wall are from deep shelf in the Barents seas, phytal in the Baltic and North seas, and scattered examples in the Channel (inner shelf) and Celtic Sea. In the Adriatic Sea both the northern slope and the Lim channel, although mainly hyaline, also have common porcelaneous walls as the salinity along the coast is normal marine to slightly hypersaline. On the western seaboard of the Atlantic from the Arctic to Argentina most assemblages plot along the hyaline-agglutinated side with a few samples having a modest content of porcelaneous walls (Figure 5.10, A–C). The same applies for the Pacific coast of North America (Figure 5.10D, E) and all other examples from the Pacific, Indian and Southern oceans (Figure 5.10F) with most of the samples having porcelaneous walls coming from the inner shelf (Figure 5.10D). Thus, although the overall pattern for shelf seas is similar to that originally recognised by Murray (1973), with the addition of a lot more data the boundaries are more diffuse.

6

Carbonate environments

6.1   Introduction

In the majority of sedimentary environments, the biogenic carbonate is diluted by clastic detritus transported in by rivers and forms only a small proportion of the sediment. The distinctive feature of carbonate environments is that they are areas not receiving clastic detritus from erosion of land. Therefore the biogenic material slowly accumulates to form carbonate sediment. The fundamental difference between cool–temperate carbonates and those in shallow tropical areas is that the latter have chemically deposited or cemented grains (ooids, aggregates, hardened pellets) and may have aragonite mud. All these are absent from cool and temperate examples where the characteristic mineral is calcite. Carbonate sediments composed of bioclasts of molluscs, echinoderms, bryozoans, barnacles, foraminifera and red calcareous algae are present in cool, temperate and tropical regions, but hermatypic corals and green calcareous algae are confined to shallow tropical areas. Two broad terms have been introduced to distinguish these differences: heterozoan and photozoan, the latter with corals and green calcareous algae (James, 1997). The term photozoan reflects the importance of light for the symbionts of the fauna and for the free-living algae.

6.2   Warm-water carbonate environments

6.2.1   Seagrass communities

Seagrasses are marine flowering plants that occur in shallow, subtidal environments in tropical areas, and extend into the intertidal zone in temperate areas. They are strongly influenced by light and cannot live in water depths where the benefits from photosynthesis are less than the loss from respiration. They influence the sediment with their roots and rhizomes, and by the variety of leaf form in the overlying water. The leaves are substrates for the attachment of a range of epiphytic small plants, including algae, and also for epiphytic animals, including foraminifera. In Florida, USA, *Thalassia* leaves are commonly covered with organic detritus and growths of epiphytic diatoms and these support foraminiferal faunas; but blades lacking diatoms have few living foraminifera (Grant *et al.*, 1973). Soritids gather detritus into a peripheral ring (Steinker and Steinker, 1976) while small *Discorbis candeianus* occur within patches of detritus. The abundance of *Thalassia* is seasonal from low in autumn/winter to a peak in spring/summer (Zieman, 1975) and this affects foraminiferal abundance. In Jamaica, biotic effects are considered to be more important than the physical environment in controlling the standing crops of foraminifera on *Thalassia* leaves although there is no evidence of partitioning of microhabitats or competition between species (Buzas *et al.*, 1977). During storms, *Thalassia* may be uprooted and transported over long distances and this provides a means of dispersal of attached foraminifera such as *Planorbulina acervalis*. However, clinging forms such as *Archaias angulatus* may be detached by waves and violent storms (Bock, 1970). Also anchored in the sediment there are rooted calcareous green codiacean algae such as *Halimeda* or *Penicillus* that also provide habitats for small organisms. There are three habitats on *Penicillus capitatus*. The rhizoids provide substrates for peneroplids, the stalk bears diatoms and supports *Archaias angulatus* and *Discorbis candeianus*, while the tuft traps fine-grained detritus and provides shelter for the latter two species together with *Quinqueloculina seminulum* and *Flintinoides labiosa* (Steinker and Steinker, 1976). Phytal miliolids have an ovate test with a thin wall, broad chambers, restricted aperture without a neck and lack ornament (Brasier, 1975a). *Peneroplis, Spirolina* and *Monalysidium* attach loosely and live in low-energy settings.

6.2.2   Larger foraminifera

The term 'larger foraminifera' is used for warm-water porcelaneous and hyaline taxa generally larger than 2 mm in diameter (Figures 3.1, 6.1). Porcelaneous larger foraminifera are commonly variable in size and form and this sometimes makes identification more difficult. Molecular genetics suggest that *Sorites marginalis* and *Sorites orbiculus* are morphotypes of one species (Holzmann *et al.*, 2001). Morphological variation in *Calcarina gaudichaudii* extends to the number of spines and stages of the life cycle (Röttger *et al*, 1990b). Larger foraminifera reproduce primarily by asexual reproduction with occasional sexual episodes (Röttger 1974; 1978; 1990; Röttger and

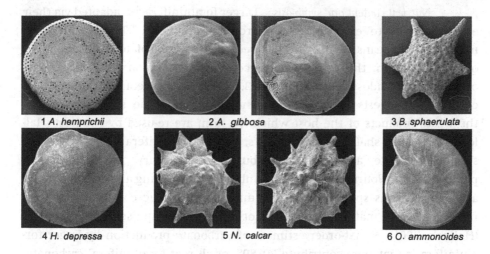

Figure 6.1. Scanning electron micrographs of tropical larger foraminifera (longest dimension, μm). 1. *Amphisorus hemprichii* (2200). 2. *Amphistegina gibbosa* (640, 1000). 3. *Baculogypsina sphaerulata* (2000). 4. *Heterostegina depressa* (3000). 5. *Neorotalia calcar* (700, 550). 6. *Operculina ammonoides* (2200).

Schmaljohann, 1976; Ross, 1972; Sakai and Nishihira, 1981; Lipps and Severin, 1985; Fujita *et al.*, 2000). The controls on global distribution patterns are thought to be water temperature, nutrient content, light intensity and hydrodynamic energy. Local distributions are also controlled by the intensities and interrelationships of these factors while abundance is controlled by interspecific competition. Overall, larger foraminifera are k strategists (slowly achieve populations that match the carrying capacity, k) but in shallow water they are opportunistic strategists (rapidly increase in numbers when conditions are favourable). In deep water with low light their numbers are low (Pecheux, 1994).

Temperature. Larger foraminifera are circum tropical in the Indian, Pacific and Atlantic oceans, mainly in oligotrophic areas. The latitudinal distribution is controlled primarily by temperature. The 15 °C isotherm for the coldest month limits their geographic distribution (Langer and Hottinger, 2000). Such low temperatures are tolerated by *Amphistegina* spp., *Amphisorus hemprichii* and *Sorites orbiculus*. For the majority of taxa, the lower temperature is ∼20 °C and for a small group (*Alveolinella quoyii, Heterostegina operculinoides, Operculinella venosus*) it is 22 °C. Laboratory experiments show that *Amphistegina* cease to move at temperatures of 12–16 °C (Zmiri *et al.*, 1974).

Nutrient content and symbiosis.   Larger foraminifera are adapted via their endosymbionts to exist under oligotrophic conditions. The central water masses of the oceans are oligotrophic because they are isolated from the input of nutrients from the land by the encompassing boundary currents of the oceanic gyres. Endosymbiosis (Section 3.2.11) is advantageous only if the host collects and digests particulate organic matter and the symbionts utilise the waste products of the host which, in turn, are re-used by the host (Hallock, 1981a). In shallow environments, larger foraminifera with their endosymbionts make a greater contribution to primary production than phytoplankton (Sournia, 1976). The combination of feeding and endosymbiosis allows organisms specialised to live under oligotrophic conditions to accumulate organic matter that would otherwise be dispersed into seawater (Hallock, 2001). A first-order estimate of carbonate production by larger foraminifera is that they contribute 70–80% of all reef foraminiferal carbonate, with an annual production of 44 million tonnes (43 million tons) representing 0.76% of modern oceanic carbonate production (Langer *et al.*, 1997). *Marginopora kudakajimensis* has a carbonate production of 5 $kg\,m^{-2}\,y^{-1}$ in a lagoon in the Ryukyu Islands (Fujita *et al.*, 2000).

Light intensity.   Irradiance levels decrease exponentially with increasing water depth and with transparency of the water (controlled by suspended particles including plankton and terrigenous detritus). The lower limit of the euphotic zone is 100–150 m in the Indo-Pacific and 50–100 m in the Caribbean because the latter is less oligotrophic so has more material in suspension in the water. The distribution of larger foraminifera is dependent on the light requirements of their endosymbionts. Shallow-dwelling forms have green or red pigment and include chlorophytes, rhodophytes and dinoflagellates. Alveolinids and all hyaline larger foraminifera have diatom symbionts and their depth range extends from shallow to the base of the euphotic zone.

Hydrodynamic energy.   Oceanic atolls and barrier reefs experience continual wave attack especially on the windward side and this is commonly reflected in faunal differences between the leeward and windward sides. Under extreme conditions, hurricanes and tsunamis may severely damage or destroy reef habitats. Currents are generated by tidal flow. To avoid being displaced by waves or currents, many larger foraminifera cling to the substrate, using pseudopodia or protoplasmic extensions. Even during reproduction, the gamonts of *Cycloclypeus carpenteri* are attached by elastic filaments (but not by pseudopodia; Krüger *et al.*, 1997). *Heterostegina depressa, Operculinella venosus* and *Operculina ammonoides* also have sheaths and filaments during reproduction. In

1972 a hurricane destroyed the reef biota of Funafuti Atoll. By 1995 much of the reef flat still had a smooth scoured surface but the area was being recolonised by foraminifera living epiphytically on algae. *Baculogypsina sphaerulata* was the dominant form and *Amphistegina lessonii, Amphistegina lobifera, Marginopora vertebralis* and smaller forms were present (Collen, 1996).

Substrate. All larger foraminifera are epifaunal on hard substrates such as coral rubble or other bioclasts and some are epiphytic on seagrass and calcareous algae.

Geographic distribution. The habitats of larger foraminifera are discontinuous in oceanic areas as they are confined to islands and submerged banks generally < 100 m deep. Dispersal mechanisms in oceanic surface circulation include transport on floating plant debris, fish, birds and propagules. The eastward decline in species diversity in the Pacific Ocean is attributed to the shoaling of the thermocline and the cooling of the Equatorial Undercurrent in the East Pacific. The propagules have to survive water temperatures below 18 °C and this may affect their survival (Belasky, 1996). *Marginopora vertebralis* living in the western Coral Sea and on the Great Barrier Reef, Australia, are genetically similar indicating that long-distance dispersal takes place (Benzie, 1991).

The Indo-Pacific is the area most favoured by larger foraminifera. Within this, there is an inner central Pacific province around the Philippines–Indonesia (characterised by *Baculogypsinoides spinosus*). Around this core is the central Indo-Pacific realm extending from Hawaii in the east to the Seychelles/Maldives in the Indian Ocean (*Alveolinella quoyii, Marginopora vertebralis, Amphistegina radiata, Calcarina* spp., *Cycloclypeus carpenteri, Operculinella venosus*). The western Indian Ocean, including the Red Sea and Arabian Gulf, lacks the species listed for the Indo-Pacific realm. The Caribbean realm has a different assemblage of larger foraminifera with few species present: *Archaias angulatus, Sorites orbiculus, Amphistegina gibbosa, Heterostegina depressa* (Langer and Hottinger, 2000).

Depth distributions of larger foraminifera. The only area for which there are detailed assemblage data is Okinawa, Japan (Hohenegger, 1994; Hohenegger *et al.*, 1999, WA-181). The reef flat has dominant *Calcarina gaudichaudii* in pools with *Peneroplis planatus* in moats and *Neorotalia calcar* on the beach. *Amphistegina lobifera* is a common subsidiary form in all these subenvironments. On the fore reef, hard substrates have dominant *Calcarina hispida* with *Amphistegina lobifera* (20–40 m) and *Operculina ammonoides* with *Heterostegina depressa* (50–70 m). On soft substrates, there is *Calcarina hispida* (20 m), *Operculinella venosus* (30–50 m), *Operculina ammonoides* (60–80 m) and *Heterostegina operculinoides* (90–100 m). A northern and southern transect was sampled in 1996. On hard substrates, the former has *Calcarina hispida* with varied subsidiary species (10–50 m) and the latter has *Amphistegina lessonii* and *Amphistegina radiata* (10–40 m) and *Operculina ammonoides* (60 m). On soft substrates the species distributions are less ordered. Details of species ecologies are summarised in the Appendix.

A coenocline is where several species have different tolerance ranges along an environmental gradient. In the case of water depth and the associated exponential decrease in light intensity, it is possible to model depth-related abundance (standing crop) of larger foraminifera (Hohenegger, 1995; 2000). However, local topographic effects create microhabitats that influence the abundance of the standing crop so that one simple model is not normally applicable. Also, differences in wave energy between environments affect the depth distributions of the same species. For instance, in the Maldives, the limits of depth-dependent species are shallower in lower-energy settings (Hottinger, 1980). Nevertheless, all depth distributions show left-side truncation at the water surface and there is a lower limit at the base of the euphotic zone. Distributions are unimodal and are best fitted by power-transformed (truncated) normal distributions (see Hohenegger, 2000).

Where patch reefs are developed on a continental shelf, conditions may be mesotrophic due to the input of nutrients from the adjacent land. Off Indonesia, periodically higher nutrient conditions may be unfavourable for larger foraminifera. Close to land, they live down to 6–9 m but in the outer-shelf zone they extend down to 40 m (Renema and Troelstra, 2001; Renema *et al.*, 2001). Species richness is reduced compared with Indo-Pacific oligotrophic ocean reefs. The largest species with long life cycles are absent (large size is an adaptation to oligotrophic conditions). On the leeward reef slopes at depths < 9 m there are *Neorotalia calcar*, *Amphistegina lobifera* and *Peneroplis planatus*. On exposed slopes at 6–9 m there is *Amphistegina lobifera*. The deep seaward slope (~12–27 m) has *Calcarina gaudichaudii*, *Amphistegina radiata* and *Heterostegina depressa* while the reef base (14 m near shore but 40 m offshore) has *Amphistegina papillosa*, *Operculinella venosus* and *Operculina ammonoides*.

Two conceptual models (Hallock, 1987; Figure 6.2) show possible responses by larger foraminifera to nutrient levels and light intensity. In model A, the species adapted to oligotrophic conditions do not occur in eutrophic regions. In model B, all species occur at the same irradiance level regardless of nutrient levels. From the observed depth distributions of larger foraminifera, model B does not apply. Model A predicts that loss of the most oligotrophic conditions

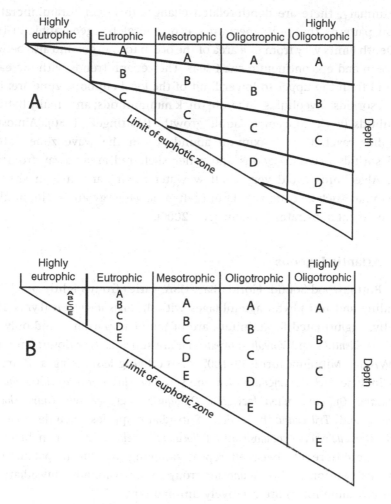

Figure 6.2. Models to explain depth distributions of larger foraminifera. Letters A–E represent species. A. Model A shows species related to nutrient levels. B. Model B shows species related to irradiance levels. Based on Hallock (1987) and Renema and Troelstra (2001).

will cause some species to disappear. This extinction can apply on a local or global scale. The evidence from Indonesia demonstrates the process on a local scale. Shallow nummulitids and amphisteginids occur at similar depths both close to shore (more nutrients, less transparent) and further offshore (fewer nutrients, greater transparency) and large deep-living taxa are absent (Renema and Troelstra, 2001). However, with the exception of *Baculogypsinoides spinosus*, calcarinids do not follow the model showing that they are more tolerant of higher nutrient levels.

In summary, there are depth-related changes in larger foraminiferal distribution patterns that are superimposed on the global geographic distributions. Depth limits vary from one area of the ocean to another and also between open ocean and epicontinental shelf seas. The general trends with increasing depth and from the upper to lower limit of the local euphotic zone are: large discoidal soritids; alveolinids; involute thick nummulitids; and finally flattened nummulitids (with concentric adult growth) (Hottinger, 1980). Almost all peneroplids reach their maximum abundance in the wave zone (<10 m). Discoid soritids occupy lagoons and other sheltered areas away from high energy. Alveolinids avoid very shallow water (<5 m) and live in sheltered microhabitats such as coral rubble in high- to low-energy areas. Nummulitids prefer sediment substrates (Hohenegger, 2000).

6.2.3   Atlantic lagoons

Buttonwood Sound, Florida Bay, USA, varies from slightly brackish to hypersaline and has phytal assemblages with *Archaias angulatus* (Lynts 1966). Generally, agglutinated foraminifera are of minor importance and only *Martinottiella occidentalis* and *Valvulina oviedoiana* are subsidiary or dominant (Lynts, 1971, WA-64). Miliolina form 50–100% of the living assemblages. Dominant species include *Archaias angulatus* (in one sample), *Hauerina bradyi, Quinqueloculina bosciana, Quinqueloculina laevigata, Quinqueloculina poeyana, Quinqueloculina subpoeyana* and *Triloculina bermudezi*. Subsidiary species include *Miliolinella circularis, Quinqueloculina sabulosa* and *Triloculina linneiana*. Lynts concludes that most of the differences between repeat sampling are due to patchiness in species distributions. The *Ammonia* group is occasionally subsidiary but otherwise hyaline forms are relatively unimportant.

In Laguna de Terminos, Campeche, Mexico, salinities are predominantly slightly brackish (25–30) but normal marine conditions are found near the entrances, and there is vegetation of seagrasses (Ayala-Castañares, 1963, WA-72). The sediments commonly have >50% CaCO₃. The single overall dominant form is the *Ammonia* group and because this often forms >70% of the assemblage there are few subsidiary species: miliolids, *Buliminella elegantissima, Elphidium gunteri* and *Elphidium* spp. Thus, the fauna is typical of a brackish lagoon.

Grand Cayman has a lagoon protected by a fringing reef. Living *Archaias angulatus* is most abundant on *Thalassia* banks in the inner lagoon and much reduced in abundance elsewhere. *Amphistegina gibbosa* and *Asterigerina carinata* are abundant on algal-veneered rubble in the fore-reef. *Trochulina rosea* is common on the reef top. These hyaline forms are rare in the lagoon (Li *et al.*, 1997).

Barbuda lies in the northern part of the Lesser Antilles. Because of its oceanic setting the waters are very clear, salinity is normal (34–36) and temperatures not very variable (26–29 °C). The lagoon is floored with carbonate mud but also has a flora of seagrass and calcareous algae. There is a phytal fauna that is found not only on the plants but also on substrates such as shells, coral fragments and pebbles (Brasier, 1975a,b,c, WA-76, WA-77). The phytal assemblages are commonly dominated by *Triloculina eburnea*, *Triloculina oblonga* or *Triloculina rotunda*. Other species that dominate one or a few assemblages are *Valvulina oviedoiana*, *Amphisorus hemprichii*, *Flintinoides labiosa*, *Hauerina bradyi*, *Quinqueloculina subpoeyana*, *Amphistegina gibbosa*, *Planulina acervalis*, *Rosalina floridana*, *Tretomphalus bulloides*, *Trochulina rosea* and *Valvulineria candeiana*. From cluster analysis, Brasier (1975b) found groups that relate to weed type. Group 1 includes *Triloculina eburnea*, *Triloculina rotunda*, *Triloculina oblonga*, *Flintinoides labiosa*, *Quinqueloculina subpoeyana*, *Quinqueloculina laevigata*, *Schlumbergerina alveoliniformis*, *Miliolinella circularis* and *Valvulina oviedoiana* and is primarily associated with fibrous algae. Group 2 comprises *Triloculina eburnea*, *Amphisorus hemprichii*, *Flintinoides labiosa*, *Valvulina oviedoiana* and *Discorbis* spp. and is from sheltered areas of mangrove. *Valvulina oviedoiana* prefers *Thalassia* rather than fibrous weeds. Group 3 includes a higher proportion of sediment associated with the plants and this is favoured by *Triloculina oblonga*. *Borelis pulchra* is a rare inhabitant of phytal assemblages. Lagoon sediment assemblages are dominated mainly by *Triloculina oblonga* but other species dominating one or a few assemblages are *Archaias angulatus*, *Cornuspira involvens*, *Miliolinella circularis*, *Peneroplis proteus*, *Quinqueloculina bidentata*, *Quinqueloculina laevigata*, *Quinqueloculina poeyana*, *Quinqueloculina subpoeyana*, *Triloculina eburnea*, *Brizalina spathulata*, *Cassidulina laevigata*, *Cibicides mollis*, *Discorbis mira*, *Discorbis orbicularis*, *Elphidium poeyanum*,*Trochulina rosea* or *Valvulineria candeiana*.

6.2.4   Atlantic shelf seas

In Bermuda, wave energy is an important ecological control. Living forms from shallow waters are rare on sediment, especially in wave-influenced areas, but more numerous on phytal substrates such as *Thalassia* and calcareous algae (Steinker and Clem, 1984). At the top of the intertidal zone in areas sheltered from wave attack, sediment loosely bound by algae has *Rosalina floridana* and *Triloculina rotunda*. *Planorbulina acervalis* and *Sorites orbiculus* are attached to *Thalassia* blades while among the accumulations of detritus on the leaves there are *Flintinoides labiosa*, *Rosalina floridana*, *Archaias angulatus* and *Peneroplis proteus*. In turbulent settings, *Halimeda* has *Planorbulina acervalis* but in quieter settings there are *Archaias angulatus*, *Peneroplis proteus*, *Pyrgo subsphaerica* and *Triloculina rotunda*, which cling to the alga with their pseudopodia. Mollasses Reef lies on the Atlantic side of the Florida Keys. The back-reef environment has a maximum depth of ~13 m and the bioclastic carbonate sediments are mainly vegetated with *Thalassia* and calcareous algae and thus resemble those of Florida Bay. There are relatively few foraminifera living on the sediment. Most live on the plants and the dominant species is *Rosalina candeiana* with subsidiary miliolids (Wright and Hay, 1971).

On the north coast of Jamaica, *Thalassia* habitats in Discovery Bay (3 m) and Pear Tree Bottom from a back-reef flat (10–15 cm at low water) have slightly different conditions (T 26–29 and 26–41 °C; salinity 33.5–34.5 and 28–34, respectively). Four replicate sediment samples from each area were taken monthly over a 12-month period (Buzas *et al.*, 1977, WA-74). Throughout most months, the assemblages at Pear Tree Bottom are dominated by *Brizalina striatula* with subsidiary *Cornuspira planorbis*, *Ammonia* group and *Bolivina subexcavata*. Occasionally one of these subsidiary species becomes dominant. In Discovery Bay asssemblages are dominated by: *Ammonia* group (7 replicates), *Amphistegina gibbosa* (3), *Bolivina subexcavata* (4), *Brizalina striatula* (21), *Discorbis mira* (2), *Rosalina globularis* (7), *Rosalina subaraucana* (1) or *Trifarina occidentalis* (9). Only one month has the same dominant form (*Brizalina striatula*) in all replicates and some months have a different dominant form for each replicate. Univariate and multivariate analysis of the data show that four species have similar monthly periodicities (*Brizalina striatula*, *Bolivina subexcavata*, *Trifarina occidentalis* and *Ammonia* group). Both stations have positive correlations between species and commonly the standing crops are very highly correlated. This shows that there is little competition between species. There is no statistical relationship between the 19 most abundant species and *Thalassia*, sediment grain size, organic carbon, temperature, salinity, turbidity, oxygen or pH. Therefore abiotic variables do not control the abundance of foraminifera in this environment. It is an example of a tropical environment with high species diversity, predictable stability and biotic regulation: '. . cropping by nondiscriminate predators reduces foraminiferal densities. When predation eases, the foraminifera quickly increase in density (logarithmic phase of growth), but before competition can ensue, their densities are again substantially reduced. In this way, species maintain similar proportions with time, show no evidence of competition, and maintain similar periodicities' (Buzas *et al.*, 1977, p. 61).

On the Barbuda shelf *Triloculina oblonga* is dominant in two assemblages. A range of species is dominant in single assemblages but only *Peneroplis proteus* dominates in several and these have no consistent subsidiary species. Unlike the protected lagoon, the shelf sediments are more affected by disturbance from waves and some are strongly winnowed. Brasier (1975c) notes that although *Homotrema rubrum* is abundant dead in the sediment it lives only on the underside of dead coral branches.

At the southern end of the Lesser Antilles is Tobago and, unlike Barbuda, it has significant relief so there are no real lagoons. At the eastern end of Tobago three small bays lie in the lee of small islands but the area is subject to heavy swell at all times. Single assemblages are dominated by one of the following: *Reophax comprima, Reophax scorpiurus, Textularia candeiana, Quinqueloculina poeyana, Triloculina oblonga, Amphistegina gibbosa, Reussella atlantica* or *Siphonina pulchra* (Radford, 1976a,b; 1995, WA-78). Two or more assemblages are dominated by *Cibicides antilleanus, Globocassidulina subglobosa* or *Trochulina rosea*. Forms associated with reefs include *Amphistegina gibbosa, Cibicides antilleanus* and *Trochulina rosea*. On the south coast a sheltered lagoon in Kilgwyn Bay has extensive banks of *Thalassia*. One sample has a *Sigmoilina distorta* assemblage with *Planoglabratella opercularis* while the other has a *Trochulina rosea* assemblage with *Triloculina oblonga* and *Miliammina fusca*. The Buccoo coral reef shelters a small bay from which two samples have yielded an *Elphidium poeyanum* assemblage and a *Cymbaloporetta squammosa* assemblage with *Amphistegina gibbosa*. The north coast of the island is subject to heavy swell throughout the year and is really inner shelf rather than lagoon. A single sample from off Courland River is dominated by *Triloculina oblonga*. Most assemblages are of *Nonionoides grateloupi* with subsidiary *Triloculina oblonga* or *Reophax comprima*. The latter is dominant in a single sample from deeper water less disturbed by wave activity. The common species occur throughout the depth range sampled (0–47 m) but some have restricted optimum depth ranges: <2 m, *Triloculina oblonga*; <10 m, *Quinqueloculina poeyana*; <15 m, *Trochulina conica, Trochulina rosea, Neoconorbina terquemi*.

6.2.5   Indian Ocean–Arabian Gulf

The Gulf is a land-locked epicontinental sea which connects via the Strait of Hormuz with the Indian Ocean. Because there is little freshwater input the main source of nutrients is from the Indian Ocean. Consequently, the area is oligotrophic and the water is very clear so light penetrates to considerable depths. The fauna and flora is linked to that of the Indian Ocean but is somewhat impoverished in diversity because the Gulf lies at the northern limit of tropical organisms such as corals and larger foraminifera (Basson *et al.*, 1977).

Lagoons.   The foraminiferal assemblages of the lagoon complex of the United Arab Emirates (formerly called the Trucial Coast of the Persian Gulf) have been described by Murray (1965b,c,d; 1966a,b,c; 1970b,c, WA-182). The

Abu Dhabi lagoon is connected with the open Gulf by a channel with an oolith tidal delta and connected with adjacent lagoons by channels. When it was sampled in the early 1960s the connection with the sea was restricted and salinities in the lagoon increased from 42 near the entrance to 50 in the outer lagoon and 70 in the inner lagoon. The area was highly stressed with poor development of seagrass and few living foraminifera mainly associated with vegetation (Murray, 1965c; 1970b). In the late 1960s the environment was changed through dredging the channels, leading to improved water exchange and less hypersaline conditions (maximum salinity 55). Also, the human population increased five-fold (and presumably there was a similar increase in sewage nutrients contributed to the sea). In 1969 seagrasses (*Halodule*) formed dense subtidal meadows in the outer-lagoon channels and living foraminifera were more numerous (Murray, 1970c). The high-energy oolith delta has an assemblage of scale-like *Rosalina adhaerens* or *Rotaliammina mayori* which live clinging to ooliths and larger bioclasts. The most widespread lagoonal assemblage is dominated by *Peneroplis planatus* with *Peneroplis pertusus*, *Miliolinella*, *Quinqueloculina* and *Triloculina* species living primarily in association with epiphytes on seaweeds and on seagrass. Dead coral has a *Triloculina* assemblage. *Ammonia* group and *Elphidium* cf. *advenum* are dominant on inner-lagoon carbonate muds. Sediment standing crop is low (1–47 individuals $10 \, \text{cm}^{-2}$) and biomass ranges from 0.01–4.06 mm^{-3} $10 \, \text{cm}^{-2}$ with the high values due to large *Peneroplis planatus*. It is not possible to quantify standing crop or biomass for epiphytic forms. Similar results were obtained by Ahmed (1991) from Tarut Bay, Saudi Arabia.

Shelf. The low intertidal zone of Bahrain (salinity 45) has an infaunal assemblage and in most months the *Ammonia* group is dominant with subsidiary *Elphidium advenum*, *Brizalina pacifica* and *Nonion* sp. During the two-year study period the standing crop significantly increased and *Brizalina pacifica* and *Nonion* sp. became much more important (Basson and Murray, 1995, WA-183). The range of standing crop is from 1–118 individuals $10 \, \text{cm}^{-3}$. The SHE analysis shows that the assemblages approximately follow a log series (Murray, 2003a).

On the inner shelf off the United Arab Emirates, living foraminifera are rare on the sediment (Murray, 1966c). The deeper basins of the Arabian Gulf have marl sediments with up to 1.5% organic carbon. Lutze (1974b, WA-184; Haake, 1970; 1975) sampled the Iranian side of the Gulf from 8–204 m but only three samples are deeper than 100 m. The principal assemblage is *Nonion asterizans* (8–143 m) followed by *Bulimina marginata biserialis* (31–63 m). Other assemblages are dominated by *Ammobaculites persicus* (8–17 m), *Ammonia* group (8–25 m), *Brizalina striatula* (16–41 m), *Nonionella opima* (21–45 m), *Cancris auriculus* (28–55 m)

or *Reophax calcareus* (66–98 m). None of the dominant species shows any correlation with sediment grain size. In shallow water and around islands at water
depths of 15–35 m, there are larger foraminifera including *Heterostegina depressa*,
Amphistegina madagascariensis and *Operculina ammonoides* (Lutze *et al.*, 1971).

6.2.6   Indian Ocean–Red Sea

The Gulf of Aqaba/Elat/Eilat has been intensively studied by Reiss and
Hottinger (1984) with many observations on the occurrence of living foraminifera but no numerical data. The northernmost occurrence of mangal is
20 km north of the entrance to the Gulf of Aqaba/Elat. No living foraminifera
are found on the aerial shoots of *Avicennia* but epiphytic assemblages of *Sorites
orbiculus* and *Sorites 'orbitolinoides'* are abundant on seagrasses (especially
Laurencia papilosa and *Halodule uninervis*) in June. Sediment assemblages
include *Neorotalia calcar*, *Amphistegina* and *Rosalina* in higher-energy channels.
Reiss and Hottinger point out that these mangal assemblages are similar
to the assemblages of shallow lagoons and reef platforms elsewhere in the
Red Sea.

The Gulf has steep sides that plunge to great depths. The euphotic zone
extends to around 130 m and there is submarine vegetation down to around
40 m. In a seagrass meadow (*Halophila*) extending from 10–40 m water depth
there is a sparse fauna of *Peneroplis planatus* on the rhizomes and horizontal
stems (Faber, 1991). Larger foraminifera, especially *Amphistegina lobifera* and
Amphisorus hemprichii, are much more abundant on both the vertical leaves
and horizontal rhizomes. Between 8–40 m, shore-parallel currents flow over
blocks of basement rock dislodged during winter storms. These blocks are
overgrown by the alga *Padina* during the calmer summer months and this
forms a suitable habitat for larger foraminifera such as *Amphisorus, Sorites,
Borelis, Amphistegina* and *Heterostegina depressa.* Coarse coral debris present
from 40–75 m is cemented by encrusting foraminifera including *Acervulina
inhaerens* and provides shelter for *Heterostegina depressa* which has its highest
abundance   at   70–80 m.   Other   foraminifera   with   symbionts   (*Operculina ammonoides, Heterocyclina tuberculata*) have their lower limit here but
Amphistegina extends to 90–100 m. In more sheltered areas < 40 m deep with
patches of seagrass (*Cymodocea*), *Miniacina* and *Acervulina* encrust the basal
parts of their stems. Amphisteginids show morphological changes with
respect to water depth and decreasing light. Thick-walled, subglobular
Amphistegina lessonii and *Amphistegina lobifera* occur in shallow waters but are
gradually replaced by thin-walled, lenticular *Amphistegina papillosa* and
Amphistegina bicirculata at 40–80 m.

Jiddah Bay, Red Sea, is a microtidal area protected by coast-parallel linear barrier reefs. Seagrass covers the muddy and sandy carbonate sediments on the flanks of the reefs and in the channels (depth 20–40 m) between them. Due to restricted circulation with the open Red Sea the area is hypersaline with salinities of 37–42 in winter and 39–48 in summer. Wave attack is most severe on the seaward side of reefs during the winter. However, the environment is not very variable and this is reflected in the foraminiferal assemblages (Bahafzallah, 1979, WA-185). These are dominated by porcelaneous forms such as *Peneroplis planatus*, *Quinqueloculina oblonga*, *Sorites marginalis* and *Spirolina arietina* with many other miliolids as minor species. A few assemblages are dominated by the *Ammonia* group, *Cymbaloporetta bradyi*, *Nonionoides grateloupi* or *Rosalina adhaerens*.

6.2.7   Pacific Ocean

The windward side of Apo Reef, Philippines, lacks seagrass and has sparse living assemblages (Glenn *et al.*, 1981). The reef wall has *Amphistegina lessonii*, *Marginopora vertebralis* and *Neoconorbina* spp. in patches of poorly sorted sediment. *Marginopora* and *Heterostegina* also occur on hard surfaces. The other environments generally have a high abundance of attached juvenile hyaline species together with the following adults: in the moat and outer-reef slope *Cymbaloporetta squammosa* and *Glabratella tabernacularis*; the outer-reef flat and algal crest have a high proportion of coral debris with *Calcarina spengleri*, *Cymbaloporetta bradyi* and *Cymbaloporetta squammosa*; turf-like algae growing on dead coral have *Calcarina hispida*, *Triloculina bermudezi*, *Cymbaloporetta bradyi* and *Miliolinella* spp.; inner-reef flat and back reef have *Calcarina hispida* and *Neorotalia calcar*; on the sandy sediments in the lagoon the common forms are *Cymbaloporetta squammosa*, *Glabratella pulvinata*, *Neorotalia calcar* and *Rosalina globularis*. *Calcarina spengleri* is rare or absent; in the deep lagoon at 12–15 m there are rare live *Elphidium* spp., *Sorites marginalis* and *Peneroplis* spp.; patch reefs in the lagoon have *Neorotalia calcar*, *Calcarina hispida*, *Bolivina* spp., *Planorbulina acervalis*, *Rosalina globularis*, *Angulodiscorbis quadrangularis* and *Cymbaloporetta squammosa*.

The volcanic island of Moorea is surrounded by a lagoon and barrier reef (Vénec-Peyré, 1985b). Three radial traverses give information on different aspects of the environment. The west side is protected from the dominant east–southeast winds. The lagoon has muddy sediments but the reef has coarse sediment due to the hydrodynamic regimes. There are few living foraminifera and those present are small *Rotaliammina*, cibicidids and planorbulinids attached to bioclasts. Coral fragments have miliolids. The northwest is also protected from the east–southeast winds. Forms epiphytic on algae include *Triloculina oblonga*,

Triloculina planciana and *Quinqueloculina agglutinans*. On the reef the main forms are *Triloculina planciana, Miliolinella subrotunda* and *Pyrgo* sp. The Afareaitu transect has a sparse microfauna attached to bioclasts. There is a diverse fauna in the channel at 25–32 m with dominant *Textularia* sp. The reef off Îlots has an assemblage of attached glabratellids with some miliolids. It is considered that *Rotaliammina* plays a role in bio-erosion (Vénec-Peyré, 1985a). This area also has juvenile *Tretomphalus*. Scilly atoll in the Society Islands also has few foraminifera living on the sediment in the outer margins of the lagoon (Vénec-Peyré and Salvat, 1981).

6.3   Cool–temperate carbonate environments

6.3.1   Atlantic Ocean

The inner to mid shelves of the western English Channel and parts of the Celtic Sea have sediments with >50% $CaCO_3$ due to the accumulation of mainly molluscan bioclasts. This area is one of high energy due to tidal currents and waves. There is little terrigenous sediment contributed by rivers and the carbonate-rich sediments form a thin veneer on the eroded bedrock of the shelf. Part of the foraminiferal fauna lives either permanently attached or clinging to stable substrates that are not moved by the tidal currents. The assemblages are described in Section 5.2.4. A characteristic feature is the abundance of attached species in this high-energy environment. There are no species that are confined to these carbonate sediments.

6.3.2   Balearic Islands, Mediterranean

Heterozoan carbonate sediments accumulate in sheltered bays on Mallorca and Menorca. Seagrass (*Posidonia*) provides a substrate for attached forms such as *Nubecularia lucifuga, Planorbulina mediterranensis, Planorbulina acervalis, Cyclocibicides vermiculatus, Discorbis, Cibicidella, Cibicides* and *Acervulina*. In the roots there are miliolids and *Miniacina*. Other forms clinging to the plants include *Amphisorus hemprichii, Peneroplis* and *Vertebralina* (Colom, 1974). These species are very close to the northern limit of their distribution (Murray, 1973).

6.4   Distinction between warm and cool–temperate carbonate shelf faunas

The key difference is the absence of larger foraminifera from cool-temperate areas. The small benthic foraminiferal assemblages are distinctly different with no species in common. *Quinqueloculina* species may be smooth, striate, costate, reticulate or undulose in tropical to warm–temperate

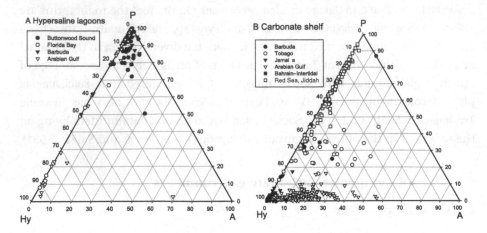

Figure 6.3. Ternary plot of walls. A. Hypersaline lagoons. B. Carbonate shelves (triangle corners represent 100% of the labelled component: A = agglutinated, P = porcelaneous, Hy = hyaline).

environments but only smooth and striate forms are known from cool–temperate environments (Murray, 1987a). Whereas *Cibicides* is often abundant in cool areas it is rare in warm waters.

6.5   Summary

Warm-water carbonate environments include reefs, lagoons and shelf seas whereas cool–temperate carbonates are confined to continental shelves.

Warm-water carbonate environments are circum tropical and mainly developed in oligotrophic settings. However, some tropical shelf seas are mesotrophic. Characteristic taxa in open normal settings are larger foraminifera, which vary in abundance and kind according to water temperature, nutrient content, light intensity, hydrodynamic activity and substrate. The majority of larger foraminifera live on hard substrates or on plants. Maximum diversity is found in the central Pacific and minimum diversity occurs at the edges of distribution in the Indo-Pacific. The Caribbean has an impoverished fauna.

In normal marine to slightly brackish lagoons hyaline taxa are dominant (Mexico) but the normal marine Barbuda lagoon has a hyaline–porcelaneous mix. Where salinities are hypersaline (Florida Bay, Arabian Gulf) porcelaneous forms dominate (Figure 6.3A). In the open ocean where environments are constantly under wave attack, there are many small forms attached to coral debris and other large bioclasts. Shelf seas of normal salinity have mainly hyaline assemblages (Jamaica) or mixtures of all three wall types (Tobago) as

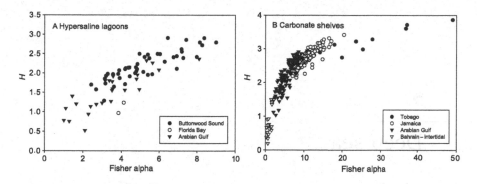

Figure 6.4. Diversity plots. A. Hypersaline lagoons. B. Carbonate shelves.

does the hypersaline shelf of the Arabian Gulf. However, the Red Sea, and to a lesser extent Barbuda, are hyaline–porcelaneous mixes (Figure 6.3B).

Species diversity is low to moderate in hypersaline lagoons (Figure 6.4A) and on a hypersaline shelf (Bahrain, Arabian Gulf, Figure 6.4B), but moderate to high on normal marine shelves (Jamaica, Tobago, Red Sea, Figure 6.4B). However, the extremely high values for Tobago may be due to oversplitting (Radford, S., 2004 personal communication).

Relatively little is known of cool–temperate carbonate assemblages. Those from low-energy shallow water in the Mediterranean have affinities with the Indian Ocean as several warm-water species reach their northernmost (and westward) limit of distribution. The English Channel high-energy shelf assemblages have a high proportion of attached forms such as *Cibicides, Rosalina* and textulariids. Most assemblages have a hyaline–agglutinated mix with <20% porcelaneous forms. Species diversity is low to moderate (Figure 5.7).

Cool–temperate and tropical carbonate shelves have similar diversity and similar patterns of wall structure. However, they are fundamentally different at species level.

7

Deep sea

7.1 Introduction

The deep sea is that part of the ocean beyond the shelf break and remains one of the least explored areas of the Earth. The topography varies depending on the geological setting. Passive margins such as those around most of the Atlantic have a continental slope which passes oceanwards into a continental rise beyond which there is abyssal plain. Active margins may have more complex topography with perched sedimentary basins and terraces on the slope and a trench at the base. Slopes may be cut by valleys, known as submarine canyons, which act as conduits for the transport of sediment by turbidity currents or mass flow into deeper water. In the centre of the Atlantic and Indian oceans there is a mid-ocean ridge system; in the Pacific it is in the east. Between the ocean ridge and the continental rises there are abyssal plains. There is no universal agreement on the terminology of topographic divisions of the oceans. In this book the term bathyal refers to the continental slope and rise from the shelf break to ~4 km, abyssal to areas > 4 km and hadal is used for trenches where depths reach as much as 11 km.

7.1.1 Oceanography

Good general overviews of oceanic circulation are given in Kennett (1982) and Gage and Tyler (1991). Each water mass has its own temperature–salinity characteristics, which define its density. Where water masses are layered, the densest is at the bottom (thermohaline stratification). The densest water forms in the Weddell Sea in the Southern Ocean and is known as Antarctic Bottom Water (AABW). The essential feature of water in the oceans is that it is constantly moving. The source of energy is from the sun (which heats the surface layer) and the rotation of the Earth. Surface-water masses flow in different directions from intermediate and bottom-water masses. Wind stress at the sea surface causes both horizontal and vertical movement of the water. At the ocean surface there are gyres in the northern and southern subtropical central areas. In the North Atlantic, the central gyre is bordered to the north by the Gulf Stream, a western boundary current. In the North Pacific the comparable current is the Kuroshio. The return flow is as eastern boundary currents such as the Canary Current in the North Atlantic. These currents make latitudinal transfers of heat. The subtropical gyres are occupied by central water masses that form under anticyclonic wind systems.

The basic pattern of deep-ocean circulation is that most bottom waters form at high latitudes due to cooling of water either in the Southern Ocean or the Greenland–Norwegian Sea. The water flows away from the source areas (thermohaline circulation) and is deflected by the Coriolis force – to the right in the northern hemisphere and to the left in the southern hemisphere. Most deep areas of the ocean have bottom waters derived from one or other of these source areas. Some deep bottom currents are powerful and exert a strong influence on the bottom fauna (e.g., the Mediterranean Outflow Water, MOW). In the Southern Ocean, there is the clockwise Antarctic Circumpolar Current (ACC) which extends down to the ocean floor. The circulation of the Antarctic waters causes aeration by exposing them to the atmosphere before they become deep waters and spread into the bottoms of the Atlantic, Pacific and Indian oceans. This system has a major role in equalisation of water characteristics of the major oceans. Once a water mass is removed from direct contact with the atmosphere it loses oxygen through the decomposition of sinking detrital organic matter. In some parts of the ocean there is an oxygen minimum layer where oxygen values are at their lowest. However, only rarely does oxygen become totally depleted to give anoxic conditions, but where this does happen it is ecologically very important (see Levin, 2003).

7.1.2   Primary production

The main source of food for organisms living in the deep sea is from primary production by phytoplankton in the surface waters. The important aspects are:

- Variations in primary production due to differences of fertility of the surface waters.
- Aggregation of organic material into phytodetritus, marine snow or faecal pellets, all of which settle at a much faster rate than is the case for individual particles.

- Exponential loss of labile organic material due to remineralisation as it sinks through the water column.
- Seasonality of input of phytodetritus in some parts of the ocean.

There is a lag of around four weeks between a surface phytoplankton bloom and the arrival of phytodetritus at 4 km on the sea floor (Lampitt, 1996) and the settling velocity is around 100 m d^{-1} (Gage and Tyler, 1991). Only 1–3% of surface primary production reaches the abyssal ocean floor and only a tiny proportion of this is incorporated into the sediment as the major part is consumed by the benthic fauna including foraminifera (Meyer-Reil and Köster, 1991). Comparison of sediment trap data and modelling of flux processes in the South Atlantic suggests that shallow traps underestimate downward flux while deep traps are affected by lateral advection and give overestimates (Schlitzer et al., 2003). This is because the calculation of flux is based on downward vertical movement only. Particulate food availability in the deep sea is generally a significant limiting factor despite large seasonal influxes (Gage and Tyler, 1991). Dissolved organic matter (DOM) resulting from decay represents the largest concentration of organic matter in the oceans. However, it is not known whether many benthic foraminifera can utilise this potential food source.

Where there is a quantitative estimate of surface-water primary production and where the organic flux is mainly vertical it is possible to estimate the amount that reaches the ocean floor using the following equation developed by Berger and Wefer (1990) and Herguera (1992) (see also Fontanier et al., 2002):

$$Jz = [2\sqrt{PP}(PP/z)] + [(5/\sqrt{PP})(PP/\sqrt{z})]$$

where Jz is the estimate of the annual flux of organic matter (J) that reaches the sea floor at depth z, and PP is primary production in g C m^{-2} y^{-1}. The first term refers to the labile component, which is more rapidly decayed in transit and more easily utilised as food by foraminifera, and the second term refers to refractory material, which is less abundant in plankton but less easily remineralised during descent; thus it decreases more slowly with depth but is less easily used as food by foraminifera. Most continental margins are affected to a greater or lesser extent by contour currents (Gulf Stream, MOW, Western Boundary Undercurrent). Gooday (2003) suggests that productivity estimates vary by a factor of two to three and do not take into account lateral variability. Although benthic–pelagic coupling is understood in a general sense, in detail there are major gaps in knowledge. The supply of food may be the ultimate controlling factor for benthic standing crop and biomass, but 'No one factor controls the benthic community. Nearly all factors are coupled and their relative importance at different sites determines benthic community structure' (Flach, 2003, pp. 351–2).

7.2   Foraminifera

7.2.1   General comments

In their review of deep-sea foraminifera, Douglas and Woodruff (1981) emphasised that our knowledge up to 1980 is almost entirely based on total assemblages. Unfortunately, that continues to be the case especially for areas away from continental margins. As there are relatively few deep-sea studies in relation to the area of the oceans, it might seem perverse to ignore those based on total assemblages. Yet because the sedimentation rate is so low, the surface 1 cm of oceanic sediment commonly represents hundreds of years of accumulation. On slopes where mass flow takes place, in the eroded source area the surface sediment may not even be modern. Thus, there is very good reason to be cautious.

Xenophyophores are giant deep-sea rhizopod protists now shown by molecular genetics to be foraminifera (Gooday and Tendal, 1988; Lee et al., 2000; Pawlowski et al., 2003). Komokiaceans are soft-walled, fragile agglutinated foraminifera. Some have no obvious aperture and pseudopodia emerge from the wall (Richardson, 2001). None of these forms is likely to survive fossilisation although there are similarities with some body and trace fossils (Maybury and Evans, 1994). Nevertheless, they are abundantly present in deep-sea settings and must make a significant contribution to deep-sea ecology. Some may have a facultative association with bryozoans (Gooday and Cook, 1984).

The deep ocean is the least-explored marine environment for foraminiferal ecology and most recent advances have been possible because of improved technology: invention of the multicorer allowing retrieval of the sediment–water interface (Barnett et al., 1984), and pressurised culture chambers for experimental work (Gooday, 1994). Examining deep-sea samples for living foraminifera presents some major challenges. Where dead planktonic foraminiferal tests are present in the sediment, they commonly outnumber the benthic tests by >100 : 1. Only a small proportion of the benthic tests are of live individuals. Some forms, such as *Cribrostomoides subglobosus*, show great fluctuations in protoplasmic volume (Linke and Lutze, 1993). Indeed, in many species the protoplasm occupies only a small part of the test, especially during periods of low flux of food. This makes it more difficult to detect individuals stained with rose Bengal. An additional complication is that some tubular tests are occupied by metazoans after the death of the original foraminiferal inhabitant (Gooday, 1984). Even where authors have separated living assemblages there are problems because of lack of consistent methodology. A wide range of different size fractions are used (>32, 63, 125, 150, 250, 300 μm). However, small key species, such as *Epistominella exigua*, are severely under represented unless >63 μm fractions are examined. Some workers distinguish live forms using ATP while others use rose Bengal stain. Some include both soft- and hard-shelled forms, others just 'fossilisable' forms. Data may relate to the top 1 cm with or without slices down to several centimetres subsurface, while others present summary data for whole cores (e.g., 0–10 cm). This makes meaningful comparisons between the different studies very difficult and sometimes impossible.

Early studies were aimed at establishing a relationship with water depth (e.g., Phleger, 1951; Bandy, 1956). However, it soon became clear that depth zones are ill-defined. From the 1970s various authors used multivariate analysis to relate deep-sea foraminiferal assemblages to bottom-water masses (e.g., Streeter, 1973). However, as pointed out for the Gulf of Mexico, there is only a general rather than a specific relationship between assemblages and water masses (Murray, 1991, p. 119). Over the past two decades the favoured influences are oxygen and flux of organic matter; role of foraminifera versus other meiofaunal (and sometimes macrofaunal) groups in carbon recycling; and patterns of biodiversity.

Sections 7.2.2–7.2.13 deal with the bathyal zone and section 7.2.14 with abyssal plains.

7.2.2   Foraminifera and organic flux

The ecological importance of the flux of organic matter to the deep sea is now widely recognised as a major control on deep-sea faunas and in many cases the fluxes are seasonal (Gage and Tyler, 1991). Seasonal fluxes may be consistent between years or there may be significant interannual variation in the timing and intensity of maxima and in the composition of the settling material. The monsoon controls such variations in the intensity of upwelling and primary productivity in the Arabian Sea while the interannual variation in ice cover off northwest Greenland influences primary production (see review by Gooday, 2002). In the northeast Atlantic, where seasonality of phytodetritus was first discovered, there are interannual differences in its composition (spring–early summer, predominantly diatoms; autumn, predominantly gelatinous with coccolithophorids). It must be universally true that phytodetritus shows temporal variation in floral composition. It also comprises labile and refractory components. Furthermore, apart from phytodetritus there are zooplankton faecal pellets that descend to the ocean floor and contribute to the pool of organic carbon.

Phytodetritus.   There is compelling evidence that benthic foraminifera show resource partitioning, that is, they feed on different types of food in a non-competitive way. Thus, although most authors seem to assume that C_{org} (as a measure of all organic inputs of carbon to the sea floor) is a single variable, it is much more likely that it must be a complex of variables as far as the benthic fauna is concerned. This is clearly shown by results from the northeast Atlantic where the importance of phytodetritus for benthic foraminifera has been documented by Gooday and co-workers in numerous papers. It is inferred that certain foraminifera feed on the phytodetritus because they have green protoplasm even though they are living in the aphotic zone (Thiel *et al.*, 1988; Gooday and Turley, 1990). However, the species that feed directly on phytodetritus are only a small proportion of the whole living assemblage. For the remainder, the seasonality of organic flux seems not to be important in causing short-term variations in abundance. Cores collected after a phytodetritus event show a layer of fluffy phytodetritus millimetres to centimetres thick overlying the sediment surface and bottom photographs show phytodetritus preferentially deposited in hollows. Foraminifera found living in the phytodetritus layer include tiny forms (>28 μm; Gooday *et al.*, 1995b) as well as >63 μm *Epistominella exigua, Alabaminella weddellensis* (given as *Eponides* by Gooday, 1988) and allogromiids (subsequently named *Tinogullmia riemanni* in Gooday, 1990); these forms are rare in the sediment. Other species, *Pyrgoella* sp. and *Fontbotia wuellerstorfi* (given as *Planulina*), also occur mainly in the phytodetritus whereas *Adercotryma glomeratum* occurs in the sediment as well. These species have been grouped into three categories (Gooday, 1993). 1. Abundant in phytodetritus and mostly <150 μm in size: *Nuttallides pusillus* (given as *Alabaminella weddellensis* but revised to *Eponides pusillus* by Gooday and Hughes, 2002), *Epistominella exigua, Tinogullmia riemanni.* 2. Uncommon but confined to phytodetritus (mainly small): *Fursenkoina* sp., *Pyrgoella* sp., *Morulaeplecta* sp., *Parafissurina fusuliformis.* 3. Uncommon species not confined to phytodetritus and ranging in size up to 300 μm: *Adercotryma glomeratum, Fontbotia wuellerstorfi* (given as *Cibicides*), *Gavelinopsis lobatulus, Globocassidulina subglobosa, Oridorsalis umbonatus, Trochammina* sp. (Figure 7.1). The diversity of phytodetritus assemblages is low (Fisher alpha 4–5, *H* 2.53–2.66) compared with that of the underlying 0–1 cm surface sediment layer (Fisher alpha 55–65, *H* 6.15–6.17; Gooday, 1996). The vertical distribution in the sediment is 52–71% live in the top 1 cm and 89–92% live in the top 2 cm (Gooday, 1986a,b). Two sites in Rockall Trough (1926 m with some phytodetritus, 3600 m trace phytodetritus) and one site in Hatton-Rockall Basin (1100 m no phytodetritus) are dominated by soft-shelled agglutinated species. The standing crops are 25, 19–103 and 50 individuals $10\,cm^{-2}$, respectively in the top 0.5 cm of sediment (Hughes

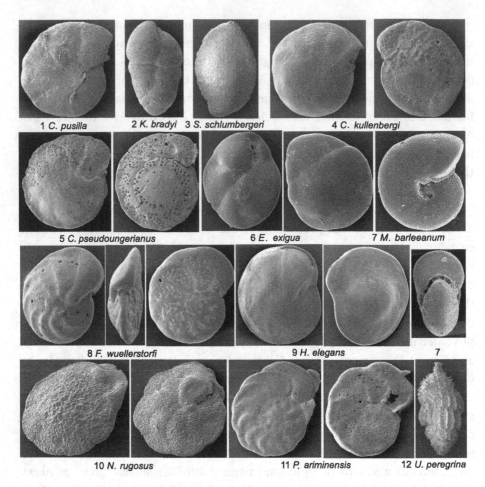

Figure 7.1. Scanning electron micrographs of deep-sea foraminifera (longest dimension, μm). 1. *Cyclammina pusilla* (800). 2. *Karreriella bradyi* (540). 3. *Sigmoilopsis schlumbergeri* (670). 4. *Cibicidoides kullenbergi* (650, 690). 5. *Cibicidoides pseudoungerianus* (700, 700). 6. *Epistominella exigua* (270, 210). 7. *Melonis barleeanum* (450, 450). 8. *Fontbotia wuellerstorfi* (360, 560, 450). 9. *Hoeglundina elegans* (700, 900). 10. *Nuttallides rugosus* (330, 300). 11. *Planulina ariminensis* (500, 400). 12. *Uvigerina peregrina* (860).

et al., 2000). Fisher alpha species diversity is 8.75, 5.33–10.65 and 5.13, respectively.

Comparison of samples from April (time of spring phytoplankton bloom at the surface) and July (after the arrival of phytodetritus at the bottom) show that the sediment samples from the two periods have an average of 385 ± 79 and 474 ± 216 individuals 3.46 cm^{-3}, respectively but in July the combined sediment and phytodetritus has 713 ± 281 individuals 3.46 cm^{-3} (Gooday and

Lambshead, 1989). Apart from changes of standing crop, there are also changes in faunal composition. In April, the sediment has common *Ovammina* sp. and *Nonionella iridea* whereas in July *Ovammina* sp. is absent, *Nonionella iridea* is common together with *Cassidulina teretis*, which may have been reproducing at that time (individuals small). Multivariate analysis confirms the role of phyto-detritus in structuring the assemblages (Lambshead and Gooday, 1990). A similar comparison at a site in Rockall Trough (Gooday and Hughes, 2002) reveals comparable trends. The May sample (pre-phytoplankton fall) has 830 individuals $10 \, cm^{-3}$ while the July sample (with phytodetritus) has 2379 indi-viduals $10 \, cm^{-3}$. Fisher diversities are 38.5 for May (sediment assemblage) and 33.0 for July (sediment and phytodetritus assemblages combined). The phyto-detritus species *Nuttallides pusillus* and *Nonionella iridea* are more abundant in July than May. Whereas the former is most abundant in the phytodetritus and less so in the 0.0–0.5 cm sediment layer, for the latter the case is reversed and it is never found embedded in phytodetritus aggregates but has an agglutinated cyst. It has been suggested that after the consumption of the new phyto-detritus, there is a faunal change to dominance of *Quinqueloculina* sp. and that this species returns to the sediment and possibly becomes dormant during the winter (Gooday and Rathburn, 1999).

Comparison of a Porcupine abyssal plain site (seasonal phytodetritus) with the Madeira and Cape Verde abyssal plains (no seasonal signal) reveals simi-larities and differences. The standing crop of complete individuals of the sediment is higher at Porcupine than elsewhere (266 compared with 189 and 220 individuals $10 \, cm^{-3}$) and even higher when the phytodetritus species are included (314). However, when fragile taxa such as komokiaceans and tubular agglutinated forms are included, the standing crop is higher on the Madeira abyssal plain (Gooday, 1996). On a smaller spatial scale, Gooday *et al.* (1995b) found that two adjacent cores have profoundly different assemblages. One core, with a large amount of phytodetritus, has an abundance of <63 μm soft-walled saccamminids and allogromiids whereas the other, without phytodetritus, has few of these forms.

Stable isotopes.   Carbon stable isotope records of live and dead *Cibici-doides* sp. and *Fontbotia wuellerstorfi* are closely comparable in the eastern South Atlantic and Weddell Sea (Mackensen *et al.*, 1993b). These are epifaunal taxa with no particular association with phytodetritus. Between the subtropical front at ~41° S and the southern boundary of the ACC at ~55° S, $\delta^{13}C$ values for *Fontbotia wuellerstorfi* and *Cibicidoides* spp. are generally lower than those of the bottom water $\delta^{13}C$ total dissolved inorganic carbon. Although these taxa are epifaunal and should not show any microhabitat effects, the authors suggest that in areas where there is seasonal input of phytodetritus, the $\delta^{13}C$ gradient in the sediment might extend up into the phytodetritus layer thereby creating a microhabitat, especially if growth and test calcification proceed mainly during such events. However, stable isotopes and Cd/Ca ratios of *Fontbotia wuellerstorfi* (given as *Cibicidoides* by Tachikawa and Elderfield, 2002) from the northeast Atlantic show no variability in the surface few centimetres of sediment and give a reliable record of bottom-water values even though two of the sites receive seasonal inputs of phytodetritus.

Flux rates. The occurrence of foraminifera >250 μm in the surface 1 cm of sediment from the Gulf of Guinea to the Arctic Ocean and the 63–250 μm fraction from just the Gulf of Guinea and the Arctic Ocean have been related to organic flux (Altenbach *et al.*, 1999). The results indicate a broad tolerance to flux rates (one to three orders of magnitude) and different mean values for different species, although there is considerable overlap (Figure 7.2). Because of this variability, individual species cannot be used to predict flux

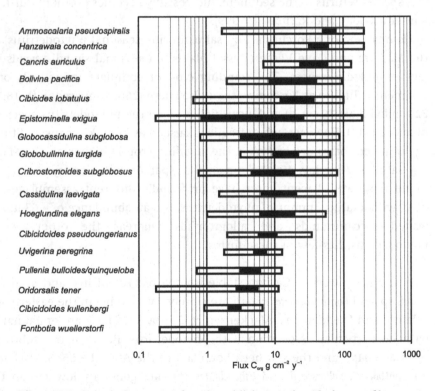

Figure 7.2. Organic flux ranges for selected living species. Black part of bar indicates standard deviation about the mean (based on Altenbach *et al.*, 1999).

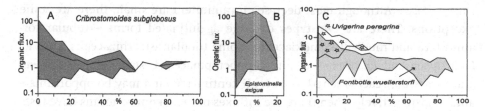

Figure 7.3. Range of organic flux (g C_{org} cm^{-2} y^{-1}) for different species. The median line in A and B gives mean values for the intervals from 0-2.5% and thereafter in steps of 5% (based on Altenbach *et al.*, 1999).

rates. As might be expected, a plot of percentages of both live and dead counts against organic flux shows the broadest range of flux at the lowest percentages (Figure 7.3). The only species that show increasing abundance with decreasing flux are those from oligotrophic environments. In the case of *Uvigerina peregrina* and *Fontbotia wuellerstorfi*, there is considerable overlap in flux rates at their lower abundances; it is only where each species becomes dominant that they show separation of flux requirements. Altenbach *et al.* conclude that at depths <1000 m, organic flux is not a primary control (partly because there is resuspension of material that makes flux calculations incorrect). Below 1000 m, flux rates >3 g C m^{-2} y^{-1} are restricted to areas beneath regions of high primary productivity. There is a transitional zone with fluxes of 2-3 g C m^{-2} y^{-1} and oligotrophic abyssal environments below 1-2 g C m^{-2} y^{-1}. They consider the 2-3 g C m^{-2} y^{-1} limit to be a region where several species reach their lower or upper thresholds. However, they also point out that there are exceptions where species have particular requirements, such as the preference for freshly deposited organic matter (as for *Cribrostomoides subglobosus*, *Epistominella exigua* and *Melonis barleeanum*). Finally, annual flux rate does not reflect seasonal pulses such as those of phytodetritus.

The boundary at 1000 m also affects foraminiferal biomass. Above this, the flux rates vary within one order of magnitude while the biomass is randomly distributed within three orders of magnitude. Below this, flux rate and biomass are significantly correlated ($r = 0.7$, error probability < 0.001%, for $n = 108$; Altenbach and Struck, 2001). Nevertheless, there is a very high standard deviation (0.65, a factor of 4) for this exponential relationship. The authors attribute this to biological variability and 'methodological perturbation'. These include the problems of adequately sampling large individuals, recording biomass of only the surface 1 cm when individuals live infaunally to greater depths, the rapid growth response of certain species to a pulse of food, and spatial and temporal variability of standing crop.

Size.   Although most deep-sea foraminifera are small, there are some exceptions. There are two types of large agglutinated forms – tubular foraminifera and mudball komokiaceans. Some tubular astrorhizaceans, such as *Bathysiphon* and *Rhabdammina*, and hippocrepinaceans, such as *Hyperammina* and *Saccorhiza*, reach a length of several centimetres and may be opportunists able to grow actively when there are influxes of food; komokiaceans have tests 1–5 mm in size and are probably not opportunistic with respect to pulsed food supplies. Tubular astrorhizaceans and hippocrepinaceans occur in abundance in eutrophic areas where there is a good supply of food from surface primary production, bottom currents or down-slope transport (e.g., Oman margin). Komokiaceans are more common on oligotrophic abyssal plains. They have small biomass in relation to the quantity of detrital material that they use to form their test. Gooday *et al.* (1997) speculate that the tubular forms feed on labile organic matter whereas the komokiaceans feed on refractory organic material, bacteria and possibly on dissolved organic material.

7.2.3   Atlantic Ocean: Europe–Africa

The Norwegian–Greenland Sea lies between the Arctic Ocean and the North Atlantic from which it is separated by the Greenland–Iceland–Scotland ridge over which there are powerful currents from the Norwegian Sea. Contour currents affect both the Norwegian and northwestern European upper slopes. The empirical relationship between primary production and flux of organic material to the sea floor does not apply where there are bottom currents that laterally advect food.

Assemblages under the influence of currents

Water masses.   On the northern flank of Iceland, >106 μm assemblages vary with water mass (Jennings *et al.*, 2004; WA-191). At 165 m, under the seasonally warmed surface-water layer, there is a *Globobulimina auriculata* assemblage with *Nonionellina labradorica*; at 385–450 m, beneath cooled Atlantic Intermediate Water (AIW) of the North Iceland Irminger Current, *Nonionellina labradorica* with *Globobulimina auriculata*; and at 656 m, in Arctic Water, *Melonis barleeanum* with *Globobulimina auriculata*, *Silicosigmoilina groenlandica* and *Adercotryma glomeratum*. Standing crop is low (12–63 individuals 10 cm^{-3}). Two transects across the Greenland–Iceland–Scotland ridge show contrasts between the southern Norwegian Sea and the North Atlantic (Mackensen, 1987, WA-192, >125 μm). The assemblages in the Norwegian Sea are more varied with almost every sample having a different dominant species: *Melonis zaandamae* 500–1101 m, *Cibicides refulgens* 605 m, *Cassidulina teretis* 997–1356 m, *Triloculina frigida* 1224 m, *Cribrostomoides subglobosus* 1601 m, *Fontbotia wuellerstorfi*

1798 m and *Hormosinella guttifer* 2305 m. Other subsidiary species include *Adercotryma glomeratum, Ammoglobigerina globigeriniformis, Lagenammina difflugiformis, Epistominella exigua, Globobulimina turgida, Globocassidulina subglobosa, Oridorsalis umbonatus, Pullenia bulloides* and *Rupertia stabilis.* Two samples on the Atlantic side are dominated by *Bolivinellina pseudopuncata* with *Cibicides lobatulus, Melonis zaandamae* and *Trifarina angulosa* at 406 m and by *Cibicides lobatulus* at 596 m. Both have sediment dominated by coarser material, some of which must be suitable as a substrate for *Cibicides lobatulus.*

Substrate. The upper part of the Norwegian continental slope is influenced by the north-flowing Norwegian Current. Along the southern shelf edge at 200–600 m there is a *Trifarina angulosa* assemblage on sandy sediment (Mackensen *et al.*, 1985, WA-193). From 600–800 m in the north there is an abundance of attached suspension-feeding *Rupertia stabilis* (Lutze and Altenbach, 1988; Linke and Lutze, 1993). Distributions are patchy with standing crops reaching a maximum of 100 individuals $10 \, \mathrm{cm}^{-2}$ on exposed rock substrates. Sponge spicules are used to support the pseudopodia in the current to aid feeding from material in suspension. Under experimental conditions, individuals rearrange the sponge spicules in response to varying current directions in order to maximise exposure. This form is well adapted to an environment where food particles are transported laterally by currents giving a relatively uniform food supply, stable protoplasmic volume and ATP content (153 ± 23 ng ATP per individual) with a reduced turnover rate of ATP ($0.008 \, \mathrm{s}^{-1}$) throughout the year (Heeger, 1990; Linke, 1992). Other species, such as *Cribrostomoides subglobosus, Rhabdammina abyssorum* and *Pyrgo rotalaria* experience large seasonal fluctuations of ATP but have a relatively constant ATP turnover rate. The latter allows them to respond opportunistically to sudden inputs of food. However, it does mean that there are major fluctuations in protoplasm volume between periods of starvation and plenty. Experimental studies on *Cribrostomoides subglobosus* show that following a pulse of food this species increases its body mass by 89% within three days but as food declines so too does body mass (Altenbach, 1992).

Contour currents. Bottom water discharged from the Mediterranean through the Strait of Gibraltar into the Gulf of Cadiz forms a contour current (MOW) which turns right due to the Coriolis Effect and flows along the upper slope past Portugal and France. Mediterranean Outflow Water enters the Gulf of Cadiz at 150–280 m, flows down to 700 m mixing with North Atlantic Central Water (NACW) before becoming a broad contour current along the slope at depths of 450–1500 m. There is a decrease in bottom currents from east to west

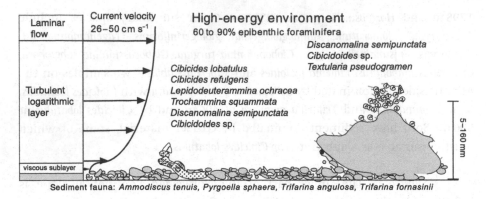

Figure 7.4. Epifaunal colonisation of elevated hard substrates in the high-energy
Gulf of Cadiz (modified from Schönfeld, 2002, with his permission).

in the Gulf of Cadiz with proximal rock outcrops and gravel, sand and sandy
silt beneath the main flow paths, and clay and clayey silt in the quieter zones
in between. The Faro Drift is a contourite resulting from this flow. The
>250 μm living assemblages (Schönfeld, 2002; WA-194) show no one species to
be widely dominant: 396 to 1208 m, *Rhabdammina abyssorum*; 532 m, *Ammolagena
clavata*; 550 m, *Trifarina angulosa*; 803 m, *Uvigerina mediterranea*; 945 m, *Cyclammina cancellata*; 966 m, *Crithionina hispida*. Factor analysis on reduced datasets
gives three main clusters and the MOW assemblage is dominated by epifaunal
taxa (*Cibicides lobatulus, Cibicides refulgens, Cibicidoides* sp., *Discanomalina semi-
punctata, Lepidodeuterammina ochracea*). The latter occur in areas of maximum
current influence in the upper part of the MOW at velocities of 26–50 cm s^{-1}
(Figure 7.4). The greatest abundance is in the most elevated positions above the
sediment substrate where current velocities are higher and presumably food
particles are more abundant. On flat slabs the standing crop is equal to that of
the adjacent sediment (2 individuals 10 cm^{-2}) and large blocks are inhabited
only on the upper surfaces. In distal settings where the current velocity is lower
and there is sandy silt to silty clay, the standing crop is 14–39 individuals
10 cm^{-2}. Common tubular *Rhabdammina abyssorum* provides anchorage for *Crithionina pisum, Trochammina squamata, Saccammina sphaerica* and *Rosalina anomala*.
These forms do not attach higher than 2.8 cm above the sediment surface so that
they remain in the sublayer with its maximal content of detrital particles. With
one exception, substrates elevated >0.5 cm above the sediment surface are
inhabited by epifaunal foraminifera only at depths >550 m.

Further sampling of MOW >150 μm assemblages at 1085–1209 m from the
Gulf of Cadiz and from off Sines, Portugal (Schönfeld, 2001; WA-195) shows
that in the top 1 cm (including the overlying suspended material) there is

dominant *Rhizammina algaeformis*, *Saccorhiza ramosa* and *Uvigerina mediterranea*. The standing crop is low (23–62 individuals 10 cm^{-3}). The sediment was examined down to 10 cm but there are few living foraminifera deeper than 2 cm. There is a high positive correlation between standing crop and pore-water oxygen levels ($r = 0.86$–0.99) and also with chlorophyll pigment equivalents ($r = 0.89$–0.95). All are statistically significant at the 99% confidence level. Schönfeld considers oxygen to be the more important control. Although many species tolerate low oxygen levels, their maximum occurrence is the most reliable guide to their degree of tolerance. *Melonis barleeanum*, *Neolenticulina peregrina* and *Trochammina* spp. tolerate suboxic to dysoxic pore waters but have distinct maxima under oxic conditions close to the sediment surface. Therefore, Schönfeld does not consider them to be indicators of suboxic or dysoxic conditions. However, *Globobulimina affinis* and *Bathysiphon capillare* have maximum abundance under dysoxic conditions.

The slope west of Portugal is bathed in north-flowing MOW between 600 and 1250–1500 m. The velocity is sufficient to erode the sea floor and form ripples in the sediment (Schönfeld, 1997). Two down-slope traverses sampled by Seiler (1975; WA-123) have >63 μm assemblages dominated by various taxa but no single species is widely dominant. Off Mondego there is: 816 and 1268 m, *Trifarina bradyi* and 1010 m, *Marsipella cylindrica*. Off Sines there is: 618 m, *Melonis barleeanum*; 611 m, *Trifarina bradyi*; 996 m, *Globobulimina* spp. The >250 μm assemblages from this area (Schönfeld, 1997; WA-196) vary with depth. The most widespread assemblage is that of *Uvigerina mediterranea* at 246–968 m on muddy sand. At depths greater than 1000 m the sediments are silty clays. From 1103–1300 m there is *Saccorhiza ramosa* with subsidiary *Uvigerina mediterranea*. *Rhabdammina abyssorum* dominates a single assemblage at 820 m. Schönfeld used R-mode cluster analysis and his upper-slope association represents MOW: *Uvigerina mediterranea*, *Marsipella elongata*, *Bigenerina nodosaria*. Apart from foraminifera living on and in the sediment, hard substrates such as shells, other animals and clinker from ships, support a diverse foraminiferal epifauna. These forms are especially common from the shelf edge down to the base of the MOW but some are found also beneath North Atlantic Deep Water (NADW). In comparison with the slope off Portugal, the Gulf of Cadiz with its stronger currents has twice the abundance of epifaunal individuals.

In the Bay of Biscay a site on silty mud at 1012 m lies beneath MOW (Fontanier *et al.*, 2002, WA-121). The >150 μm assemblage in the top 1 cm of sediment has *Uvigerina mediterranea* with *Uvigerina peregrina*, *Nuttallides umboniferus* and *Hoeglundina elegans*. Subsurface samples were taken down to 10 cm. The majority of living forms occur in the top few centimetres with 44% down to 7.5 cm. There is no correlation between presence of oxygen into the sediment and depth of life as the latter extends into the anoxic zone. In this area, the contour current speed is much reduced compared with Cadiz and Portugal.

Assemblages not influenced by currents

Norwegian–Greenland Sea. Away from contour currents there is inferred to be a vertical flux of food. The Northeast Water (NEW) Polynya off northeast Greenland is an area of open water on the ice-covered shelf and is one of the largest examples in the Arctic. It is ice-free from April–May until September. The water column is formed of cold Polar Water (PW) down to 140 m, intermediate depths by Knee Water (salinity 33.0–34.4, T <0° C) and bottom water by Atlantic Water (AW) (salinity 34.86, T 0.5–1.0° C). Only one station sampled for foraminifera was in Knee Water (155 m), the remainder being under AW (Ahrens *et al.*, 1997). Primary productivity in summer 1993 was 763 mg C m^{-2} d^{-1} in open water with a maximum of 2250 mg C m^{-2} d^{-1}. The annual flux of particulate organic carbon is ~1.6 g C m^{-2} y^{-1} with maximum values in late summer to autumn. Over the sampling period it was 0.964 g C m^{-2}. A series of samples taken in the summers of 1992–1993 at more-or-less the same site at 285–330 m gives some idea of temporal changes in the >150 µm assemblages (Newton and Rowe, 1995, WA-197). Only the top 1 cm is considered here as the actual numbers counted were not given. In 1992 the dominant forms were *Buccella frigida* and *Elphidium excavatum* with subsidiary *Nonionellina labradorica*. The dominant forms in 1993 were: May, *Pyrgo* spp.; July, initially *Melonis barleeanum* then *Elphidium excavatum*; in August, *Pyrgo* spp. Agglutinated forms were frequently common. The standing crop peaked in July and August 1993 when the supply of food was greatest. The polynya was smaller and had more ice cover in 1992; standing crops were significantly lower than in 1993 suggesting that the supply of food was less.

In the Ahrens *et al.* (1997, WA-198) study, the foraminifera were counted in size fractions but the published data tables give a summary of the >63 µm assemblages. On average, 65% of the foraminifera are in the 63–125 µm size fraction. Agglutinated forms, including *Cribrostomoides subglobosus*, *Saccorhiza ramosa* and *Hormosinella guttifera*, were treated as a group. There are four main assemblages in the top 1 cm of sediment: *Cassidulina teretis* with *Elphidium excavatum* and *Islandiella islandica* (178–400 m), *Textularia* spp. (187–310 m), *Elphidium excavatum* with various subsidiary species (255–338 m) and *Islandiella islandica* (326–487 m). The *Textularia* spp. assemblage occurs in both Knee Water and AW. Standing crop varies from 97–507 individuals 10 cm^{-3} and biomass from 54–288 mg C m^{-2} for the top 1 cm of sediment. Highest standing crops occur at the shallowest stations (vs. water depth, Spearman rank coefficient $r_S = 0.69$, $N = 23$) but no such correlation exists for biomass due to the abundance of large agglutinated forms which strongly influence values. Some sites were sampled to a depth of several centimetres with 95% living in the top 5 cm. A few species are essentially confined to the top 1 cm: *Cassidulina teretis*, *Islandiella norcrossi* and *Saccorhiza ramosa*. Others show an exponential decline with depth: *Elphidium excavatum*, *Cribrostomoides crassimargo* and *Islandiella island-ica*. A few show a subsurface maximum: *Nonionellina labradorica* and *Melonis barleeanum*. *Cibicides lobatulus* occurs not only in the surface cm but also attached to firm substrates. Foraminiferal biomass for the 0–5 cm interval is roughly twice that of the top 1 cm. Ahrens *et al.* found that there is positive correlation between standing crop and sediment chlorophyll content and sediment ATP content. Many of the foraminifera have green-coloured cell contents suggesting feeding on algae. The high standing crops in June could be due to the sedimentation of ice algae whereas the high values at the end of July may be the response to a phytoplankton pulse. There is possible evidence for predation by isopods causing minor fluctuations in standing crop.

Foraminifera live down to 15 cm below the sediment–water interface at water depths of 261–2790 m in the NEW Polynya. Live *Cibicides lobatulus* (given as *Lobatula* by Mackensen *et al.*, 2000) are found in the top 2 cm and have an average stable carbon isotopic composition 0.4 ± 0.1‰ higher than the $\delta^{13}C$ of bottom-water dissolved inorganic carbon (DIC). This species may grow and calcify its test before the ice-free season at a time when $\delta^{13}C_{DIC}$ is higher. *Melonis barleeanum* and *Melonis zaandamae* have a subsurface maximum at 2.5 cm and extend down to 8 cm. They show a negative offset from bottom-water $\delta^{13}C_{DIC}$ of -1.7 ± 0.6‰ in the surface sediment and -2.2 ± 0.5‰ at 3–4 cm. Dead tests do not show a downward decrease in $\delta^{13}C$ and this may be due to selective dissolution removing the light isotope first.

In the Norwegian Sea, freshly settled phytodetritus stimulates organic activity especially by benthic foraminifera (Graf *et al.*, 1995). Unusually high hydrolytic enzyme activity in surface sediment is associated with an abundance of *Hyperammina* and *Reophax* individuals 1–3 cm long and up to 5 mm in diameter. The hydrolytic enzymes are thought to be secreted into digestion vacuoles and used to digest organic matter. This demonstrates that foraminifera are major degraders of organic detritus in this area (Meyer-Reil and Köster, 1991). Soft-shelled *Crithionina hispida* is abundant in the >250 μm assemblage living infaunally in the top 1 cm of sediment in the Norwegian–Greenland Sea (Thies, 1990). Only around 10% of the test is occupied by protoplasm (Altenbach, 1987).

Along the southern Norwegian slope there are two >125 μm widespread assemblages: *Melonis zaandamae* often with subsidiary *Cassidulina laevigata* from 605–1100 m in areas of rapid mud sedimentation; *Cribrostomoides subglobosus* often with subsidiary *Epistominella exigua* from 1100–2005 m (Mackensen *et al.*, 1985, WA-193). Minor assemblages in the deepest water are dominated by *Fontbotia wuellerstorfi* with *Epistominella exigua* and *Triloculina frigida* at 2683–2940 m on sand-grade sediment rich in planktonic foraminiferal tests. This overlaps with the *Oridorsalis umbonatus–Triloculina frigida* assemblage at 2605–3940 m. The upper limit of these assemblages coincides with the 0.5% TOC contour so they may live in a region of lower food supply. Lutze and Salomon (1987) found only three species >250 μm are common: *Cribrostomoides subglobosus*, *Pyrgo murrhina* and *Fontbotia wuellerstorfi*.

Darwin Mound region. The region 185 km northwest of Scotland comprises roughly 100 sandy mounds, each about 100 m in diameter and 5 m high. Some of these are colonised by the ahermatypic deep-water coral *Lophelia pertusa* which is also widely present along the northwest European slope (Hovland and Mortensen, 1999; Freiwald, 2003). A wide range of foraminifera live in and on the reefs including attached forms such as *Acervulina inhaerens*, *Cibicides* spp., *Rosalina anomala* and *Valvulina conica*. In addition, there are free-living forms such as *Adercotryma* cf. *glomeratum*, *Hoeglundina elegans*, *Pyrgo murrhina*, *Quinqueloculina seminulum*, *Saccammina sphaerica*, *Uvigerina mediterranea* and *Uvigerina pygmaea*. Box cores from depths of 946–958 m between the sandy mounds contain dead *Syringammina fragilissima* (xenophyophore) which provides microhabitats occupied by living benthic foraminifera (Hughes and Gooday, 2004).

Bay of Biscay. Four sites were sampled from the shelf to mid bathyal. All have silty mud substrates with well-oxygenated bottom waters. The site at 553 m is under NACW, that at 1012 m under MOW (discussed above), while those at 1264 and 1993 m are under NADW. There are very subtle salinity differences between sites but temperature falls from 11 to 4 °C with increasing depth. On the assumption that primary production is uniform at 150 g C m^{-2} y^{-1} throughout the area, the flux of organic carbon to the sea floor decreases from 9.2 to 3.2 g C m^{-2} y^{-1} and the labile component from 6.6 to 1.8 g C m^{-2} y^{-1} with increasing water depth (Fontanier *et al.*, 2002, WA-121). Each site has a distinctive >150 μm assemblage in the top 1 cm of sediment: 553 m, *Uvigerina mediterranea* with *Uvigerina peregrina*; 1264 m, *Uvigerina peregrina* with *Hoeglundina elegans*; 1993 m, *Hoeglundina elegans* with *Uvigerina peregrina*. The 553 m, site also has abundant arborescent agglutinated forms but these are not included in the counts. The majority of living forms occur in the top 1 cm of sediment: 553 m, 77%; 1012 m, 44%; 1264 m, 62%; and 1993 m, 70%. Fontanier *et al.* calculated the average living depth (ALD) for all species. None of those dominant in the surface 1 cm has an ALD as great as 1 cm and almost all species have ALD of 0.5–0.8 cm so it can be concluded that the 0–1 cm sample is representative of the area beneath it. However, there is one exception. The only species that is under-represented is *Globobulimina affinis* with an ALD of 4.2 cm. It forms only a tiny proportion of the combined 0–10 cm live assemblages at all stations except that at 1012 m where it makes up 29%. There is no correlation between presence of oxygen in the sediment and depth of life as the latter extends into the anoxic zone. The standing crop for the whole cores (0–10 cm) is 187, 17 and 25 individuals 10 cm^{-2} with increasing water depth. Because the flux of labile organic material and water depth are highly negatively correlated ($r = -0.78$), the standing crop has high correlation with both factors (with flux, $r = 0.89$; with depth, $r = -0.90$). The concentration of live individuals in the surface layer shows that they are responding to labile organic matter. The authors place the stations on the trophic conditions and oxygen concentrations (TROX) diagram (see Section 8.5) of Jorrisen *et al.* (1995) with the 553 m site in the eutrophic field and the other sites in the meso-trophic field.

The 550 m station sampled ten times (with replicates on five occasions) between October 1997 and April 2000 has adequate oxygen in the bottom water (205–221 µM) but seasonal variations in surface primary production. There are data for >63 µm assemblages from 0.0–0.5 cm, >150 µm assemblages from 0–1 cm, and from the entire 0–10 cm core (Fontanier *et al.*, 2003; WA-199). *Uvigerina mediterranea* is the main >63 µm dominant species from 0.0–0.5 cm but others are *Reophax scorpiurus*, *Cassidulina carinata*, *Epistominella exigua* and *Globobulimina affinis*. The 0–1 cm, >150 µm assemblages are dominated by *Uvigerina mediterranea* in all replicates and in each month except for July 1998 when *Melonis barleeanum* took over with subsidiary *Uvigerina elongatastriata* and *Uvigerina mediterranea*. Surprisingly, *Melonis barleeanum* is otherwise uncommon being subsidiary only in June 1998. Subsidiary species at several months include *Cribrostomoides subglobosus*, *Reophax scorpiurus* and *Uvigerina peregrina*. Taking the whole 0–10 cm interval, the >150 µm assemblages are consistently dominated by *Uvigerina mediterranea* with subsidiary *Cribrostomoides subglobosus* and *Melonis barleeanum*. Most of the living forms are in the oxygenated surface layer of sediment and most species have ALDs in the top 1–2 cm. The ALD for *Uvigerina mediterranea* is mainly within the top 1 cm although it increases to a maximum of 1.6 cm in December 1998. The ALD of *Globobulimina affinis* shows the biggest range (0.9–4.1 cm) and it is consistently found in the upper part of the anoxic zone. The authors conclude that overall the 63–150 µm size fraction shows greater spatial variability and the >150 µm size fraction shows greater temporal variability. They consider that the peaks of *Epistominella exigua* relate to phytodetritus events prior to the sampling periods. They put forward a model of successive responses to a phytodetritus influx: within two weeks by *Epistominella exigua* and *Hormosinella guttifer*, three to five weeks *Uvigerina peregrina* and *Uvigerina mediterranea*, and five to eight weeks *Melonis barleeanum* and *Globobulimina affinis*. The latter two species live slightly deeper in the sediment. Fontanier *et al.* suggest that there are several phytoplankton events involving different phytoplankton groups. If that is the case, it is surprising that the overwhelming response is primarily by one species, *Uvigerina mediterranea*. If the type of food is important other species might perhaps have become dominant. The fact that two different species can be dominant in replicates from the same time and the standing crops can also differ significantly suggests that the correlation between food and fauna is not too strong unless patchiness of phytodetritus is equivalent. This study shows the influence of choice of size fraction on perceived temporal and spatial variability.

Portugal slope. This is bathed in MOW between 600 and 1250–1500 m. Above 600 m there is NACW and below 1500 m there is NADW (Schönfeld, 1997). Two down-slope traverses sampled by Seiler (1975; WA-123) have >63 μm assemblages dominated by various taxa but no single species is widely dominant. Off Mondego there is: 211 m, *Trifarina fornasini* with *Uvigerina peregrina*; 287 m, *Bolivina tongi*; 377 m, *Brizalina spathulata*; and 1010 m, *Marsipella cylindrica*. Off Sines there is: 455 and 618 m, *Melonis barleeanum*; 500 m, *Brizalina spathulata*; 611 m, *Trifarina bradyi*; 996 m, *Globobulimina* spp.; 1709 m, *Rhizammina* spp.; and 3095 m, *Epistominella vitrea* (possibly = *Epistominella exigua*). There is a marked decrease in standing crop with increasing water depth, from values generally >80 individuals 10 cm^{-3} at depths <1000 m to <25 individuals 10 cm^{-3} at depths >1000 m. The >250 μm assemblages from this area (Schönfeld, 1997; WA-196) vary with depth: 1699 m, *Hoeglundina elegans*; 1871 m, *Globotrochamminopsis bellingshauseni*; 1985 m, *Cribrostomoides subglobosus*; and 2020–2331 m, *Uvigerina peregrina*. Schönfeld used R-mode cluster analysis to recognise associations. Shelf-edge association: *Trifarina fornasini, Textularia sagittula, Uvigerina elongatastriata, Globobulimina* sp.; upper-slope association: MOW discussed above; lower-slope association: *Uvigerina peregrina, Saccorhiza ramosa, Hoeglundina elegans, Cribrostomoides subglobosus, Martinotiella bacillaris, Cibicidoides robersonianus, Cribrostomoides jeffreysii, Globotrochamminopsis bellingshauseni, Cibicidoides kullenbergi*; boundary-layer association: *Reophax bilocularis, Saccammina sphaerica, Cibicidoides pseudoungerianus, Ammolagena clavata, Jaculella obtusa, Ammoscalaria tenuissima*. However, these associations show considerable bathymetric overlap. The standing crops are low (7–69 individuals 10 cm^{-2}).

Northwest Africa.   The aim of the early studies was to investigate depth distributions. Lutze (1980, WA-132) sampled three traverses at approximately 26, 23 and 15° N. The shelf samples are discussed in Section 5.2.7. The bathyal samples range from 202 to 3314 m. The >63 and >250 µm assemblages are quite different in dominant species but they are similar in that few species are dominant in more than one sample. In the >63 µm assemblages, *Epistominella exigua* and *Epistominella levicula* are intermittently dominant from 741–3314 m and *Rectuvigerina arquatensis* from 342–514 m. Species dominant in two samples include *Saccorhiza ramosa* and *Cancris auriculus*. Taxa subsidiary in several samples include *Reophax* cf. *fusiformis*, *Bolivina tongi*, *Cassidulina laevigata*, *Globobulimina* spp., *Melonis zaandamae*, *Trifarina bradyi* and *Uvigerina peregrina*. In the >250 µm assemblages, forms commonly subsidiary and occasionally dominant include *Cribrostomoides subglobosus*, *Cribrostomoides* cf. *triangularis*, *Hormosina globulifera*, *Liebusella? goesi*, *Marsipella elongata*, *Saccorhiza ramosa* and *Trifarina fornasini*. Forms dominant in several samples include *Cancris auriculus*, 201–514 m; *Hoeglundina elegans*, 1070–2896 m; and *Uvigerina* spp. (*elongatastriata, peregrina* and variants) intermittently at 324–2822 m. The standing crop shows big differences: at 26 and 23° N 15–115 individuals 10 cm^{-3} with only two >100; at 15° N, with one exception, they are >100 and range from 40–410 individuals 10 cm^{-3}. These higher values are because the area is one of seasonal upwelling. Haake (1980b, WA-131) sampled the area at 14 and 12.5° N also in the area of upwelling. He found high standing crops with an exceptional maximum of 1470 individuals 10 cm^{-3} at 300 m but most values between 300 and 1000 m are in the range 300–500 individuals 10 cm^{-3}. His key species are essentially the same as those of Lutze. Lutze and Coulbourn (1984) re-analysed Lutze's data and added further samples giving a latitudinal spread from 15–40° N. The most common forms fall into five bathyal groups based on relative abundance: *Cancris auriculus*, upper slope; *Globobulimina* spp., upper slope; *Liebusella goesi*, upper slope; *Uvigerina peregrina finisterrensis*, northern mid slope; *Cibicidoides kullenbergi*, lower slope.

The upwelling region off Cap Blanc from 19–27° N has two areas of high primary production: year-round upwelling at the shelf break with values up to 325 g C m^{-2} y^{-1}; summer upwelling several hundred kilometres offshore with values up to 250 g C m^{-2} y^{-1}, most of which occurs in July and August. In spite of the high organic flux to the sea floor, the bottom waters are well oxygenated. The estimated organic fluxes of labile material are 4.19–1.88 g C m^{-2} y^{-1}. The >150 µm foraminiferal assemblages of the top 1 cm of sediment are dominated by different taxa at different depths and distances from shore (Jorissen *et al.*, 1998; Jorissen and Wittling, 1999; WA-200): 1200 m, *Cibicidoides kullenbergi* with *Uvigerina mediterranea*; 1525 m, *Bulimina inflata* with *Ammoscalaria* spp.; 2002 m,

Fontbotia wuellerstorfi with *Saccammina* sp.; 2530 m, *Ammoglobigerina* sp. and *Cribrostomoides subglobosus*; 3010 m, *Uvigerina peregrina* with *Cribrostomoides subglobosus*. The standing crops are high, with 848 individuals 10 cm^{-3} at 1200 m, and 272–449 individuals 10 cm^{-3} at greater depths. The cores were sampled to a depth of 10 cm and the proportion of the entire standing crop (0–10 cm) in the top 1 cm ranges from 39% at 3010 m to 69% at 2002 m. The 0–10 cm assemblages differ from those of the top 1 cm. At 1200 m, *Bulimina marginata* with *Cibicidoides kullenbergi* and *Uvigerina mediterranea*; 1525 m, *Bulimina marginata*; 2002 m, *Fontbotia wuellerstorfi*; 2530 m, *Melonis barleeanum* with *Ammoglobigerina* sp. and *Cribrostomoides subglobosus*; 3010 m, *Cribrostomoides subglobosus* with *Haplophragmoides canariensis* and *Uvigerina peregrina*. The standing crop in the 0–10 cm interval ranges from 1297 at 1200 m to 359 at 3010 m and a minimum of 272 individuals 10 cm^{-3} at 2002 m. Fragments of *Rhizammina* (not included in the counts) are abundant at 1200 m.

Gulf of Guinea. This has complex oceanography but the key features are the existence of intermediate waters at 500–800 m with oxygen levels sometimes below 2 ml l^{-1}, NADW mixed with up to 20% AABW occurs down to 4500 m, and there is local seasonal upwelling. Data on the slope come from Schiebel (1992), Timm (1992) and Altenbach *et al.* (2003) (WA-133). The very diverse 63–250 μm assemblages are dominated by relatively few species. On the upper slope there is *Rectuvigerina cylindrica* at 240–297 m; *Eggerelloides arcticus*, 310–691 m; *Brizalina albatrossi*, 445–749 m; *Epistominella levicula* (given as *Epistominella pusillus* in Schiebel, 1992, and Timm, 1992), 457–4658 m; *Epistominella exigua*, 1500–3911 m; *Reophax bilocularis*, 1500–4933 m; *Reophax fusiformis*, 1619–4449 m; and *Reophax longicollis*, 2818–4970 m. Altenbach *et al.* (2003) used factor analysis to define seven major groups, five of which occur mainly on the slope rather than the shelf. For the >63 μm assemblages the following are the main species: 250–740 m, *Eggerelloides arcticus*, *Brizalina albatrossi*, *Epistominella levicula*, *Reophax scorpiurus*; 674–2007 m, *Epistominella levicula*, *Brizalina albatrossi*, *Bolivina pacifica*; 1002–4658 m, *Epistominella levicula*, *Epistominella exigua*, *Reophax calcareus*; 1475–4970 m, *Reophax fusiformis*, *Reophax bilocularis*, *Epistominella levicula*. There is a clear relationship with flux of organic material but, as the authors point out, flux is correlated with water depth, sediment grain size, benthic oxygen respiration and chloroplast pigment equivalents. Thus correlation between foraminiferal distributions and an individual parameter is impossible.

Eastern tropical Atlantic. There are several systems of permanent and seasonal upwelling cells that control primary productivity. Two high productivity centres along the continental margin are the Congo plume and the region south of the Congo estuary. Four of the six cores studied by Licari *et al.* (2003; WA-201) are from areas unaffected by upwelling and two are from the influence of the Congo plume. Five are from 1200–1300 m and are at the boundary between Antarctic Intermediate Water (AAIW) and NADW. One, from 2186 m, is under NADW on the Niger Fan. The >125 μm assemblages were studied in 1 cm slices down to depths of 4–8 cm. The organic flux to the sea floor was calculated from satellite and literature data; the former values were preferred by the authors. The Niger Fan 0–1 cm sample from 2186 m has a *Robertina chapmani* assemblage with subsidiary *Reophax scorpiurus* and *Hoeglundina elegans* but in the whole 5 cm core the rank order of the first two species is reversed. Species diversity is low (Fisher alpha 5.6) as is the standing crop of 7 and 13 individuals 10 cm^{-3} for the surface sample and the whole core respectively. The organic flux is 3.6 g C m^{-2} y^{-1}. Three stations span the equator. The northernmost has a *Gavelinopsis translucens* assemblage with *Robertina chapmani* both at the surface and for the whole 8 cm core. The next has *Uvigerina peregrina* with *Robertina chapmani* and *Reophax scorpiurus* at 0–1 cm but *Reophax scorpiurus* with *Lagenammina difflugiformis* and *Uvigerina peregrina* for the whole core. The southernmost sample has dominant *Reophax bilocularis* with *Reophax scorpiurus* and *Uvigerina peregrina* both at the surface and for the whole core. The standing crop is 33–63 individuals 10 cm^{-3} and species diversity is high. The organic flux is 7.1–7.4 g C m^{-2} y^{-1}. Of the two samples from beneath the Congo plume, that on the Congo Fan has *Eratidus foliaceus* with *Uvigerina peregrina* at 0–1 cm while the whole core is co-dominated by *Ammoscalaria* sp., *Cribrostomoides jeffreysii* and *Eratidus foliaceus* with *Uvigerina peregrina*. The standing crop is low (10 individuals 10 cm^{-3}) yet the organic flux is 10.4 g C m^{-2} y^{-1}. Licari *et al.* interpret this as possibly due to fractionation of the organic matter in the water column (due to intense outflow from the River Congo) with smaller and lighter particles having a longer settling time and therefore greater degradation before arrival at the sea floor. Also, there may be an input of terrestrial organic matter and sediment reworking. The second plume sample is dominated by *Bulimina mexicana* with *Uvigerina peregrina* at the surface and throughout the core. The standing crop is very high at 186 individuals 10 cm^{-3} and the organic flux is 9.3 g C m^{-2} y^{-1}.

Southwest Africa. A regional study of the continental margin of southwest Africa, the Walvis Ridge, Angola and Cape basins spans 23° of latitude from 17–30° S (Schmiedl *et al.*, 1997, WA-202). The >125 μm assemblages, with the exclusion of most agglutinated species, were subject to Q-mode principal component analysis to give seven principal components that account for 65.4% of the variation. The authors used multiple linear regression analysis to determine relationships with environmental parameters. There is a strong correlation with water depth, temperature, TOC and dissolved oxygen. In the oligotrophic regions of the Walvis Ridge and the deep basins the assemblages reflect the low fluxes of organic matter and the physical and chemical attributes of the bottom-water masses. The *Rectuvigerina cylindrica* assemblage (PC7) with *Ammoscalaria pseudospiralis* and *Cancris oblongus* is widespread on the shelf and slope down to 700 m. It is related to high organic flux with *Rectuvigerina cylindrica* also correlating with lower oxygen. The *Uvigerina peregrina* assemblage (PC4) is patchily distributed on the outer shelf and upper slope and also at greater depths with a distinct maximum at 2500–3500 m. Although it occurs in areas with lower oxygen values it avoids areas of serious depletion. The *Uvigerina auberiana* assemblage (PC2) includes *Bulimina costata, Bulimina mexicana* and *Reophax bilocularis* and occurs at 700–2000 m on the continental slope with a maximum between 17 and 24° S. The oxygen minimum zone extends from 400–800 m but oxygen never falls below 45 μM. *Reophax bilocularis* with *Epistominella exigua* (PC5) has an upper limit of 1000–3000 m off Africa and a lower limit of >5000 m in the Angola Basin and 4000 m in the Cape Basin. It also occurs on the Walvis Ridge at 2000–3000 m. These assemblages are in high productivity areas and the standing crops exceed 100 individuals 10 cm^{-2}. *Rhizammina* sp. (PC1) occurs on the central Walvis Ridge at 2000–3000 m and on the continental slope south of Walvis Ridge at 1000–2200 m. The *Psammosphaera* spp. assemblage (PC6) with *Cribrostomoides sphaeriloculus, Cribrostomoides subglobosus* and *Oridorsalis umbonatus* occurs mainly between 2000 and 3500 m on Walvis Ridge. The *Nuttallides umboniferus* assemblage (PC3) has common *Adercotryma glomeratum, Ammobaculites agglutinans, Cribrostomoides subglobosus* and *Epistominella exigua* and occurs in the deepest parts of the Cape and Angola basins. These assemblages are correlated with oligotrophic conditions and the standing crop is as low as 20 individuals 10 cm^{-2}. However, *Rhizammina, Reophax, Psammosphaera* and *Lagenammina* are also related to sandy substrates. *Bulimina alazanensis, Cibicidoides kullenbergi, Melonis pompilioides* and *Osangularia culter* living in oligotrophic conditions occur in the oxygen-rich saline core of NADW at 1600–3700 m.

South Atlantic. Between latitudes 35–57° S by far the most abundant >125 μm live assemblage is dominated by *Reophax bilocularis*, sometimes with subsidiary *Rhabdammina* and *Rhizammina* spp. (Mackensen *et al.*, 1993a, WA-203). The next most abundant assemblages are those of *Bulimina aculeata* and *Nuttallides umboniferus*. Taxa dominating a few or single assemblages include *Adercotryma glomeratum, Aschemonella ramuliformis, Cribrostomoides subglobosus,*

Cribrostomoides weddellensis, Hoeglundina elegans and *Trifarina angulosa*. The latter is regarded as a 'strong bottom-current fauna' by Mackensen *et al.* because of its association with coarse sediments. The *Bulimina aculeata* assemblage is considered as a 'northern high-productivity fauna' because it lies south of the Polar Front and north of the average winter ice limit, with a bottom-water temperature of $>0\ °C$ in upper Circumpolar Deep Water (CPDW) in an area of high surface primary productivity. The *Reophax bilocularis* assemblage is widely developed under a variety of water masses ranging from cold AABW to warm NADW. It forms the 'northern-component deep-water fauna' of Mackensen, *et al.* (1993a) because of the presence of NADW. The *Nuttallides umboniferus* assemblage is present at only two stations.

Stable isotopes. Site-specific $\delta^{13}C$ values of epifaunal foraminifera are influenced by: the $\delta^{13}C_{DIC}$ of the bottom water; the saturation state of bottom water with respect to calcite, which may alter fractionation during calcification; and seasonal inputs of food. In addition to these, infaunal taxa are influenced by the $\delta^{13}C_{DIC}$ of the pore water (Mackensen and Licari, 2003) and always have more negative values than epifaunal taxa. Stable isotope analysis of $>125\ \mu m$ live foraminifera from the sediment surface down to several centimetres in the eastern South Atlantic and Southern Ocean show that the $\delta^{13}C$ values of all species on average vary $\pm\ 0.09\%_0$ around a species-specific and site-dependent mean value. Thus, at each station, live individuals of each species have essentially the same carbon isotopic composition regardless of the depth at which they live in the sediment (Mackensen and Licari, 2003). This is in agreement with the previous studies of Mackensen and Douglas, 1989; McCorkle *et al.*, 1990; McCorkle and Keigwin, 1994; Rathburn *et al.*, 1996; Mackensen *et al.*, 2000. It suggests that species moving vertically in the sediment preserve an average isotope signal (Mackensen *et al.*, 2000). Epifaunal taxa such as *Fontbotia wuellerstorfi* and *Cibicides lobatulus* differ consistently and significantly from $\delta^{13}C$ values of infaunal taxa by as much as $1.0\text{--}1.5\%_0$. Furthermore, *Fontbotia wuellerstorfi* gives an almost perfect record of bottom-water $\delta^{13}C_{DIC}$ whereas *Cibicides lobatulus* and *Cibicidoides pachyderma* deviate by about $-0.4\%_0$ and $-0.6\%_0$, respectively. However, where there are ephemeral inputs of phytodetritus this can cause *Fontbotia wuellerstorfi* to be offset from $\delta^{13}C_{DIC}$ by up to $-0.6\%_0$, for example, close to the Antarctic Polar Front (Mackensen *et al.*, 1993b; Mackensen *et al.*, 2000).

7.2.4   Atlantic Ocean: the Mediterranean

This is quite unlike other oceans because it has relatively warm bottom waters. Also, there is net evaporation so that surface salinities increase from west to east. Overall, the area is low in nutrients and is mainly oligo-trophic. De Rijk *et al.* (1999, 2000) use total assemblages to interpret regional distribution patterns. They conclude that present distributions are partly influenced by periods of deep-water anoxia that led to the formation of sapropels, especially the one that ended at 6 ka BP because a new fauna had to recolonise the area once oxygen returned. The distributions are also influenced by the interdependence of water depth, substrate and food supply. Overall there is a decrease in food supply from west to east, both of primary produc-tion and supply of labile organic matter to the sea floor.

In the Gulf of Lions off southern France samples were taken down to 10 cm below the sediment surface (Schmiedl *et al.*, 2000; WA-204). In February, the canyon axis site at 920 m has a 0–1 cm >125 μm assemblage dominated by *Uvigerina mediterranea* with subsidiary *Saccorhiza ramosa* and *Uvigerina peregrina*. The combined 0–10 cm assemblage is dominated by *Chilostomella oolina*, a deep infaunal species, with subsidiary *Uvigerina mediterranea* and *Uvigerina peregrina*. The flux of organic matter is estimated at 19.7 g C m^{-2} y^{-1}. In August the faunas are slightly different. In both 0–1 cm and 0–10 cm samples the domi-nant form is *Uvigerina peregrina* with subsidiary *Uvigerina mediterranea*. *Saccorhiza ramosa* is also subsidiary in 0–1 cm. At the open slope site at 800 m, both the 0–1 and 0–10 cm February assemblages have *Saccorhiza ramosa* with *Uvigerina mediterranea* while the August assemblage has dominant *Uvigerina mediterranea*. The flux of organic matter here is lower (12.8 g C m^{-2} y^{-1}). This is reflected in the standing crop size. In the canyon the values for the 0–1 and 0–10 cm intervals are respectively 60 and 82, and 40 and 106 individuals beneath 10 cm^2. On the slope the comparable figures are respectively 12 and 41, and 43 and 61 individuals 10 cm^{-3}. The proportion of live forms in the top 1 cm ranges from 30 to 72%. Most of the taxa have their ALD in the top 2 cm. The only common form that has a deep ALD is *Chilostomella oolina* (6 cm in February, 4 cm in August). The depth of the oxygenated layer is around 5 cm so oxygen is not a limiting factor for most taxa.

The only other detailed studies of living foraminifera from bathyal depths in the Mediterranean are those of de Stigter *et al.* (1998, 1999) on the southern Adriatic Sea. This area has well-oxygenated bottom waters and primary pro-duction in the surface waters is estimated to be 20–50 g C m^{-2} y^{-1}. For the east Mediterranean, de Rijk *et al.* (2000) estimated the supply of labile carbon to the sea floor as 3 g m^{-2} y^{-1} at 350 m water depth to 1 g m^{-2} y^{-1} at 1000 m water depth. The sediments in box cores from 398–1200 m water depth were sampled by de Stigter *et al.* from 0–5 cm. In all cores the majority of the live forms are in the top 0.5 cm although some forms live down to 5 cm and occasional indivi-duals are found to 10 cm. The 0–1 cm interval in all cores is dominated by fragments of astrorhizids. Here, the relative abundances for each assemblage have been recalculated to exclude all astrorhizids as many fragments may come from a single individual (WA-127). At 398 m in the 0–1 cm assemblage no species reaches 10% abundance but *Adercotryma glomeratum* (8%) is dominant. At 487 m *Saccammina atlantica* is dominant with subsidiary *Glomospira* sp. and *Trochammina* sp. (given as *Trochammina inflata*, a marsh species and therefore likely to be incorrect). At 578 m *Glomospira* sp. and *Anomalinoides minimus* are co-dominant. At 664 m *Adercotryma glomeratum* is dominant with subsidiary *Glomospira charoides* and *Anomalinoides minimus* while at 898 m *Glomospira charoides* is dominant with the other two species as subsidiaries. These results are broadly consistent with the conclusions of de Rijk *et al.* (2000). Standing crop for 0–1 cm is 142 and 141 individuals $10\,cm^{-3}$ at 398 and 487 m, but deeper than this values fall (578 m, 78; 664 m, 18; 794 m, 56; 898 m, 49; 1200 m, 9 individuals $10\,cm^{-3}$). Forms >150 μm in size make up >20% of the assemblages down to 487 m and <15% from 578–1200 m.

7.2.5 Atlantic Ocean: the Americas

Virtually the whole of the North American seaboard is affected by contour currents. The north-flowing Gulf Stream bathes the upper slope from Florida to Cape Hatteras and then it moves out into the open ocean. North of Cape Hatteras, the continental slope and rise are covered by cold water derived from the north – northeast Atlantic Deep Water (NEADW) and NADW. The main feature of the continental-rise oceanography is the southward-flowing Western Boundary Undercurrent at a depth of 2300–3000 m.

Nova Scotia. On the slope north of Nova Scotia, replicate box cores were taken at depths of 512, 1532 and 2694 m (Schafer and Cole, 1982). At 512 m, under the Labrador Current water mass, the dominant species is *Adercotryma glomeratum* with subsidiary *Spiroplectammina biformis* in both replicates. Other replicates have common *Nonionellina labradorica*. At 1532 m, under NADW, either *Nonionella atlantica* or *Elphidium excavatum* is dominant. At 2694 m, under Norwegian Sea Overflow Water in the Western Boundary Undercurrent, the dominant species is *Oridorsalis umbonatus* with subsidiary *Cibicides lobatulus* in one replicate and *Epistominella vitrea* in the other. The standing crop increases from 16 to 564 to 1369 individuals $10\,cm^{-3}$ from 512 to 1532 to 2694 m, respectively. Two further cores from 2938 and 3210 m have standing crops of 2399 and 2949 individuals $10\,cm^{-3}$, respectively. The high values are on coarse sediments beneath the Western Boundary Undercurrent. This is attributed to reduced predation by macrofauna coupled with a food supply brought in by the current.

A transect of box cores taken from the margin off Nova Scotia and the Gulf of Maine to the abyssal plain shows that >150 μm foraminifera live infaunally down to 3 cm at 202 m water depth, 11–13 cm at 2225–3000 m and 4 cm at 4800 m. Corliss and Emerson (1990) consider that the flux of organic matter is the principal control. In shallow water the flux is high and generates low-oxygen conditions at a shallow depth in the sediment. At bathyal depths the flux of organic matter is low so the depth of the redox boundary is deeper. However, this area is under the influence of the Western Boundary Under-current which advects material laterally so estimates of flux based on vertical settling are unreliable. The maximum depth of life in the sediment is under this current. Agglutinated forms were not identified although in some cases they make up >50% of the assemblage (Corliss, 1991; WA-205). The >150 μm assemblages in the 0–1 cm interval are dominated by: 1075 m, undifferentiated agglutinated forms; 1575 m, *Uvigerina peregrina*; 2225 m, *Elphidium excavatum* and *Uvigerina peregrina*; 2975 m, *Hoeglundina elegans*.

Northern USA.   Corliss (1985) first reported >150 μm infaunal for-aminifera living down to 15 cm in a box core collected in September from 3000 m on the New Jersey slope, USA. He found that 73% of the fauna occur deeper than 2 cm and different species have preferred depth distributions. A core taken from the same general area in February by Kaminski *et al.* (1997) has a different distribution of taxa which they link to availability of food. Corliss's core was collected after the summer phytoplankton bloom and has epifaunal *Hoeglundina elegans*, *Oridorsalis tener* and *Fontbotia wuellerstorfi* in the top 2 cm. The Kaminski core was collected during the winter at a time of limited food supply. Most living forms were in the 6–10 cm interval and comprised *Globulimina* spp. and *Chilostomella oolina*. *Melonis barleeanum*, which was deeper infaunal in autumn, became shallow infaunal in winter. These observations clearly show that depth of life in the sediment depends on local circumstances which may change through time.

Cores from canyons off New Jersey, USA, have the following >63 μm assemblages: 1567 m, *Nonionella iridea*; 1832 m, *Globobulimina affinis* and *Bulimina exilis*; 1914 m, *Uvigerina peregrina* and *Bulimina exilis*; 1915 m, *Nonionella iridea*, *Uvigerina peregrina*, *Trochammina quadriloba* and *Reophax scorpiurus* (Jorissen *et al.*, 1994). In the two shallower cores, living forms are confined to the top 1 cm whereas in the cores from 1914–1915 m they extend down to 5 cm. The standing crop of the top 1 cm is 10, 17, 129 and 107 individuals 10 cm^{-3} with increasing water depth. The low standing crops and absence of deep infaunal taxa is attributed by the authors to the relative instability of the sediment in the axial regions of canyons. In a separate study, Swallow and Culver (1999,

WA-206) use cluster analysis to recognise two main groups of stations: 1473–2066 m and 2468–2500 m. Within these major groups, subgroups are recognised based partly on the occurrence of species unique to the group. The main dominant species are *Nonionella* cf. *iridea* (1473–1912 m) and *Reophax bilocularis* (1508–2066 m). Other species occasionally dominant include *Ammoglobigerina globigeriniformis*, *Lagenammina* cf. *laguncula*, *Reophax scorpiurus*, *Thurammina faerleensis*, *Bulimina exilis* and *Uvigerina peregrina*. There is no correlation between the fauna and setting within the canyons. At 1884–2500 m the main forms are *Rhabdammina/Hyperammina*. However, these also occur with equal abundance in the upper bathyal samples but they are never dominant.

In a canyon off Virgina, USA, factor-analysis-defined assemblages include: F5 *Globobulimina turgida* from 200–600 m water depth; F4 *Lenticulina lucida*, *Sphaeroidina bulloides*, *Marginulina bacheii*, *Hoeglundina elegans*, *Islandiella norcrossi*, *Cassidulina teretis*, *Fursenkoina compressa* and *Dentalina communis* mainly from <400 but also extending to 1200 m; F2 *Hormosina globulifera*, *Reophax scorpiurus*, *Cyclammina compressa*, *Ammobaculites agglutinans*, *Haplophragmoides emaciatum*, *Gaudryina curta*, *Uvigerina peregrina* var. *bradyana*, *Lenticulina peregrina* and *Pullenia quinqueloba* from 700–1100 m; F1 *Uvigerina peregrina* var. *dirupta* and *Reophax scorpiurus* from 1200 to 2000 m; F3 *Bulimina affinis* with minor *Hoeglundina elegans* and *Reophax scorpiurus* deeper than 1800 m (Thompson, 1992). The F5 assemblage is related to an oxygen minimum zone (but bottom-water oxygen values are never lower than 3.5 ml l^{-1}), F2 may be related to winnowing by bottom currents, and the upper limit of the F1 assemblage coincides with a sharp decrease in sediment grain size and NADW.

Southern USA. The top centimetre of cores from 337–825 m off Cape Hatteras, USA, have >150 μm assemblages dominated by *Bolivina spissa* often with subsidiary *Globobulimina affinis*, and with subsidiary *Bulimina aculeata* at 577 and 825 m and *Pleurostomella alternans* at 825 m (McCorkle *et al.*, 1997). Live forms extend down 2–4 cm in most cores and in others down to at least 10 cm. Cores from 1470 and 1477 m have *Uvigerina peregrina* assemblages. Most of the species are close to oxygen isotopic equilibrium for bottom water. There are major depletions for foraminiferal carbon isotopes relative to the δ^{13}C of bottom-water DIC. There is first-order correlation between microhabitat depth and intensity of δ^{13}C depletion with top-centimetre taxa being closest to equilibrium and deep infaunal taxa being depleted by 2–4‰. The authors relate this to pore-water characteristics and also point out that calcification of infaunal tests may take place over a short vertical depth.

Three sites at 850 m off capes Hatteras, Lookout and Fear were selected because they show contrasting organic carbon flux (4.7 ± 0.4, 3.8 ± 0.4,

$2.2 \pm 0.2 \, \Sigma CO_2$, respectively). They also differ in bottom conditions: steeply sloped (30–35°) and cut by canyons, slope of 15–20° with biogenic pits and mounds, gently sloping at 2–3° with small burrow openings and phytodetritus aggregates, respectively, from north to south. Only the >300 μm fraction has been studied so the results are not directly comparable with those of other studies. Also, data are given for 0–2 cm and for 0–15 cm. The 0–2 cm assemblage at Cape Hatteras comprises *Nodulina dentaliniformis* (given as *Hormosina* by Gooday *et al.*, 2001a), *Globobulimina auriculata* and *Trochamminella conica*. At Cape Fear the assemblage is dominated by a mudball with stercomata, with *Reophax scorpiurus* in second rank position. The 0–15 cm assemblages are from north to south: *Globobulimina auriculata*; *Hormosina* sp., *Technitella atlantica* and *Reophax* sp.; mudball and *Reophax* spp. Species diversity of the 0–2 cm assemblage varies with location: Fisher alpha, Hatteras, 6.3–7.5; Lookout, 21.2; Fear, 19.7–24.5; *H*, 1.52–2.67; 3.33; 3.19–3.53, respectively. Gooday *et al.* interpret these results to show that calcareous forms are more abundant in areas with high organic flux, and uniserial agglutinated taxa are typical of the oligotrophic and oxic deep sea. However, the upper part of the bathyal zone along the Hatteras–Florida slope is influenced by movements of the Gulf Stream and although it is nutrient-rich with periodic upwelling there must also be considerable advection of organic material. Sen Gupta and Strickert (1982) found seasonal variation in the dominant species which include *Brizalina lowmani*, *Globocassidulina subglobosa*, *Hanzawaia concentrica* and *Planulina exorna*. Overall, *Brizalina lowmani* is most abundant. A single core from the upper bathyal zone of North Florida, where upwelling takes place, is dominated by *Brizalina lowmani* and also has subsidiary *Brizalina subaenariensis* (Sen Gupta *et al.*, 1981).

7.2.6   Atlantic Ocean: Gulf of Mexico

In the northwest, *Uvigerina peregrina* from 311–1050 m appears to favour microhabitats associated with macrofaunal organisms. In areas with low oxygen in the bottom water and a redox boundary within the top 2 cm of sediment, it lives in the near-surface layer. Where the redox boundary is deeper in the sediment, it is able to live deeper (Loubere *et al.*, 1995). Two morphotypes are recognisable: *typica* and *parvula*. Both appear to reproduce in the surface layer but the latter also reproduces down to 3–5 cm. The carbon isotopic signature of the tests does not show a relationship with pore-water values since there is no variation with depth of life. The authors suggest that this species lives in sediments already disturbed by macrofaunal organisms and therefore without stratified pore waters with respect to $\delta^{13}C_{DIC}$.

Hydrocarbon seeps at 543–589 m in Green Canyon in the northern Gulf of Mexico are bordered by mats of the chemolithotrophic, sulphide-oxidising bacterium, *Beggiatoa*. Although the O_2–H_2S boundary is thought to be within 1–2 mm of the sediment–water interface, rose-Bengal-stained foraminifera are present down to a depth of 2–3 cm in sediment which is black and has sulphidic pore water (Sen Gupta *et al.*, 1997; Sen Gupta and Aharon, 1994, WA-207). However, most of these forms have degraded protoplasm as viewed in transmission electron microscopy (TEM) so they were probably dead at the time of sampling. The authors speculate that some were recently dead and some samples may not have been as well fixed as necessary for TEM analysis. Nevertheless, they conclude that foraminifera do indeed live in the sediment. Three of four assemblages have dominant *Brizalina ordinaria* and the other *Gavelinopsis translucens*. Subsidiary species include *Cassidulina neocarinata*, *Brizalina albatrossi* and *Bulimina alazanensis*. There is a complete absence of agglutinated and porcelaneous forms. Diversity is low (Fisher alpha 4.4–4.8) and so is standing crop (30–56 individuals 10 cm^{-3}). *Brizalina albatrossi*, *Brizalina ordinaria*, *Cassidulina neocarinata*, *Gavelinopsis translucens*, *Nuttallides rugosus* and *Trifarina bradyi* are considered to be facultative anaerobes. These species are not confined to seeps but are widely distributed in normal marine environments. This is unlike the macrofauna which has evolved species adapted to chemosynthesis (Sibuet and Olu-Le Roy, 2002). *Uvigerina peregrina* shows $\delta^{13}C$ negative anomalies in the vent assemblages. Sen Gupta and Aharon (1994) speculate that this might be due to the utilisation of vent-derived carbon during feeding.

7.2.7 Indian Ocean: Arabian Sea

During the southwestern summer monsoons, coastal and open-ocean upwelling brings nutrient-rich water onto the continental margin of Arabia. It also leads to the development of an intense oxygen minimum zone (OMZ) between 200 and 1500 m due to the microbial decay of the high primary production brought about by upwelling (Alongi, 1998). A single site from 412 m within the OMZ and with a bottom-water oxygen value of 6 μM (0.13 ml l^{-1}) has a standing crop of 15 107 individuals 10 cm^{-3} in the 0–1 cm interval when all foraminifera, both soft and hard shelled, are taken into account (Gooday *et al.*, 2000). The two top-ranked species are *Bolivina seminuda* and *Bolivina inflata*; spiroplectamminaceans, trochamminaceans and a small *Bathysiphon* are common. Although hyaline forms dominate the >63 μm assemblage, they are less abundant in the >125 μm assemblage. Diversity for the >125 μm assemblage is *H* 1.23 and Fisher alpha 9.28. By contrast, a site from 3350 m from below the OMZ with bottom-water oxygen 46 μM (1.072 ml l^{-1}) has a standing crop of 625 individuals 10 cm^{-3} in the 0–1 cm interval when all foraminifera, both soft- and hard-shelled, are taken into account. Calcareous forms are less common and the assemblage is dominated by allogromiids and saccamminids. This has extremely high species diversity (Fisher alpha 71.7) for the combined hard- and soft-bodied >63 μm foraminiferal assemblage from the 0–1 cm interval (Gooday et al., 1998). Soft-bodied forms with no preservation potential dominate the assemblage and taking only the calcareous forms Fisher alpha falls to 9.5. This deeper site has a higher degree of carbonate undersaturation than the 412 m site.

Monsoon-driven upwelling also takes place on the Pakistan margin and leads to high primary productivity (200–400 g C m^{-2} y^{-1}) so even at considerable water depths the fluxes of C_{org} are still high. Two transects off Karachi cross the associated OMZ (200–1000 m) with five cores in transect 1 (500–2000 m) and four in transect 2 (500–1500 m). Living forms are mainly in the top 5 cm of sediment with clear down-slope trends of ALD: below the OMZ most living forms are in the top 1 cm but at depths greater than the OMZ they live deeper, although calcareous forms are still most abundant in the top 2 cm (Jannink et al., 1998). Eight species are dominant in the two transects: *Ammodiscus* sp., *Bolivina dilatata*, *Bulimina aculeata*, *Bulimina exilis*, *Epistominella exigua*, *Melonis barleeanum*, *Rotaliatinopsis semiinvoluta* and *Uvigerina peregrina*. Suboxic to anoxic conditions prevail within the OMZ and the assemblages are dominated by a few species, which reach high abundances. At 495–556 m *Bolivina dilatata*, *Bulimina exilis* and *Ammodiscus* sp. make up 50–70% of the assemblages. *Bolivina dilatata* and *Ammodiscus* sp. occur only at these stations in the upper part of the OMZ and down to 1.5 and 10 cm, respectively. At 556 m they occur in much greater abundance than at 495 m. *Bulimina exilis* and *Uvigerina peregrina* are most abundant within the OMZ. These forms are perhaps adapted to feed on fresh organic matter. Just below the OMZ at 1000–1250 m *Rotaliatinopsis semiinvoluta* is dominant and may possibly be a marker for the base of the OMZ although this still awaits confirmation. The highest abundances of *Melonis barleeanum* and *Epistominella exigua* are at 1000 and 1472 m respectively. These species are perhaps adapted to feed on more degraded organic matter. The two transects were sampled three weeks apart and the differences in standing crop (transect 2 four times greater than transect 1) may be due to patchiness or rapid response of the fauna to the arrival of food. The authors conclude that there is no simple relationship between standing crop and organic flux and there is no evidence that any of the dominant species are limited by low oxygen in this area.

7.2.8   Indian Ocean: Red Sea

The Red Sea has an anti-estuarine circulation with surface water flowing in from the Indian Ocean and deep water flowing out at the bottom. Thus the deep water of the Red Sea is not in direct contact with that of the world ocean and it has higher salinity and temperatures. It is also oligotrophic. Living forms from 0–5 cm sediment depth in six multicores from 366–1782 m water depth show a decrease in standing crop from 108–16 individuals 10 cm^{-3} with increasing depth. At 366 and 579 m the dominant form is *Cibicides* cf. *subhaidingeri* (given as *Heterolepa*) with subsidiary *Discorbinella* spp. and *Ammoscalaria* sp. at 366 m and *Morulaeplecta* sp. and *Adercotryma glomeratum* at 579 m. The oxygen minimum zone is at 400 m with oxygen values of <0.5 ml O$_2$ l^{-1} and an organic carbon flux of ∼3.5 g C m^{-2} y^{-1}. Buliminids (*Bolivina dilatata*, *Bolivina persiensis*, *Bolivinella* cf. *pescicula*, *Brizalina subspathulata*, *Bolivina* sp. and *Bulimina marginata*) make up 20% of the assemblage. The lower margin of the oxygen minimum zone is at ∼600 m with ∼1–2 ml O$_2$ l^{-1} and a flux of 2.5 g C m^{-2} y^{-1}. At 810 m *Nuttallides pusillus* is dominant with subsidiary *Adercotryma glomeratum* and *Glomospira charoides* in a transitional environment with 1.5–2.5 ml O$_2$ l^{-1} and 1.8–2.0 g C m^{-2} y^{-1}. At 929 m and 1782 m the dominant form is *Textularia cushmani*. At 1161 m the assemblage comprises *Astrononion* sp. with subsidiary *Haplophragmoides bradyi*. Here the water is well oxygenated but the organic flux is low (∼1.2 g C m^{-2} y^{-1}, Edelman-Furstenberg *et al.*, 2001).

7.2.9 Pacific Ocean

All the deep water originates in the Southern Ocean and flows northwards at depths >2500 m. It takes roughly 1000 years for the bottom water to reach the North Pacific and during this time oxygen is reduced and phosphate increased. The continental margin of North America is tectonically active and consequently the transition from continental shelf to abyssal plain is more complex than that of passive margins. Off California, between the shelf and the slope, there is the continental borderland of deep basins separated by sills. The faunas of these are considered in Section 5.6.3.

7.2.10 Pacific Ocean: east

California, USA. Cold seeps with associated bivalve mollusc ('clam') beds and carbonate nodules occur off northern California at 500–525 m on the upper slope. The standing crop of >150 μm foraminifera is 55–68 individuals 10 cm^{-3} in the surface 0–1 cm of sediment but taking the top 5 cm it is 220–308 individuals 10 cm^{-3} (Rathburn *et al.*, 2000, WA-208). Unfortunately, agglutinated forms were treated collectively although they form 37–47% of the 0–1 cm assemblages so it is possible that an agglutinated species is dominant even though it is not named as such. However, from the named forms, *Uvigerina peregrina* is dominant and there are no subsidiary species with abundances reaching 10%. Abundance is greatest in the top 2 cm of sediment but some (including *Uvigerina peregrina*, *Loxostomum pseudobeyrichi* and *Nonionella globosa*) live down to 6 cm and occasionally down to 10 cm. The carbon stable isotopic composition of several species is variable both within and between clam beds and presumably related to differences of pore-water chemistry, although these were not measured. Repetitive sampling of both cold seep (clam field, 906 m) and non-seep sites (clam flat, 1003 m) in Monterey Bay, California, were analysed using ATP, rose Bengal and examination of ultrastructure to detect living foraminifera (Bernhard et al., 2001, WA-209). Pore fluids from both sites have high sulphide concentrations (to >5 mM) but methane concentrations in the clam-field are higher than those in the clam-flat area (311 and 11 µM, respectively). Beggiatoacean sulphide bacteria are present only in the clam field. Using the ATP assay, living 'Textularia' sp., Tolypammina sp., Bolivina pacifica?, Buliminella tenuata, Epistominella pacifica, Fursenkoina rotundata, Loxostomum pseudobeyrichi and Uvigerina peregrina occur in clam-field bacterial mat samples. Assemblages defined on rose Bengal staining are dominated by Fursenkoina rotundata and Epistominella pacifica in the clam field. Non-seep assemblages are dominated by Phthanotrochus arcanus, Bolivina spissa?, Epistominella pacifica or Epistominella smithi. Subsidiary species present in either area include 'Textularia' sp., Bolivina pacifica?, Gyroidina altiformis and Uvigerina peregrina. The authors note that although there are stained Globobulimina spp. none gave a positive ATP result. Otherwise, the rose Bengal and ATP results seem comparable. None of these species is confined to seep sites for they occur elsewhere in normal habitats. The standing crop of two seep assemblages (125 and 213 individuals 10 cm^{-3}) is lower than that of the five non-seep assemblages (570–1022 individuals 10 cm^{-3}) but these values are within the range to be expected in the upper-bathyal zone. Species diversity is mainly moderate (Fisher alpha 5.4–9.4 with one at 22.6; H 1.85–2.37). Hyaline walls predominate and porcelaneous walls are totally absent.

The top 1 cm of cores from 786–3705 m off Point Sur, California, USA, have >150 µm assemblages dominated by: 786 m, Globobulimina pacifica with Uvigerina peregrina, Bolivina spissa and Nonionella spp.; 998 m, Uvigerina peregrina with Bolivina spissa; 3705 m, Globobulimina pacifica with Uvigerina spp. As in the Atlantic Ocean, there are major depletions for foraminiferal carbon isotopes relative to the δ^{13}C of bottom-water dissolved inorganic carbon but the depletion is less than that observed off Cape Hatteras in the Atlantic Ocean (McCorkle et al., 1997). This area has bottom waters depleted in dissolved oxygen (12–20 µM at depths <1000 m; 129 µM at 3705 m) and a low flux of organic carbon to the sea floor. These may partly account for the differences between Point Sur and Cape Hatteras. Bernhard (1992) examined >63 µm assemblages from 624–3278 m and determined live forms using rose Bengal and ATP assays. The standing crop measured by ATP is 0–118 individuals

$10\,cm^{-3}$ and the maximum is at 998 m in the OMZ. The figures using rose Bengal are 9–343 individuals $10\,cm^{-3}$. The former method may underestimate and the latter overestimate the true numbers of living forms.

Gulf of California. Data from 174–3027 m from the Gulf and adjacent Pacific Ocean (Phleger, 1964, WA-161) were re-interpreted using Q-mode factor analysis (Streeter, 1972). With the exception of *Bolivina* spp. (97–732 m, oxygen minimum zone) and *Reophax longicollis* with *Textularia* sp. and *Trochammina vesicularis* (2767–3027 m in the open Pacific Ocean), all other assemblages are represented by a single sample: *Bolivina pacifica* with *Cancris panamensis* and *Cassidulina* sp. (174 m), *Bolivina subadvena* with *Bolivina* spp. (351 m, oxygen minimum zone), *Fursenkoina spinosa* with *Bolivina pacifica* and *Buliminella elegantissima* (816 m) and *Globobulimina pacifica* with *Chilostomella ovoidea* and *Eponides leviculus* (2012 m). Standing crop is generally low except at 174 m off the Colorado River delta where the sediment may be rich in organic matter and off the Rio Fuerte at 732 m (304 and 702 individuals $10\,cm^{-3}$ respectively). Species diversity is low (Fisher alpha 3.6–5.8; H 0.34–0.43) and walls are agglutinated or hyaline. In the Guaymas Basin, the sediments are affected by hydrothermal fluids because of their proximity to a spreading centre (Molina-Cruz and Ayala-López, 1988; Ayala-López and Molina-Cruz, 1994). In the hydrothermal-vent microenvironment there are living *Oridorsalis umbonatus*, *Fursenkoina cornuta* and *Fursenkoina rotundata*. Away from hydrothermal influence there are common *Bulimina spinosa*, *Bolivina seminuda* and *Cibicides* sp.

El Salvador. Three >150 µm assemblages from the slope at 435–885 m on volcaniclastic silty mud are dominated by *Reophax* cf. *comprima*, *Bolivina seminuda* and *Chilostomella cushmani* with subsidiary *Nodulina dentaliniformis* (Smith, 1964, WA-162). This is an area of dysoxic bottom water (0.30–0.54 ml l^{-1}) associated with the oxygen minimum zone. Species diversity is moderate (Fisher alpha 6.8–12.0, H 1.99–2.78) and hyaline walls dominate (80–97%). Analysis of bolivinids morphology (Smith, 1963) shows that some species vary with depth: *Bolivina seminuda* (has clear areas on all chambers in shallow water but only on the last chambers in deep water), *Bolivina subadvena* (variable in width, periphery and amount of sutural crenulation), *Loxostomum pseudobeyrichi* (shallow forms have a narrow keel while in those from deep water it is wide and fragile). Overall, bolivinids on the shelf are small while those on the slope are large. The wall is thick and the pores are large in *Bolivina subadvena* whereas the wall is thin and the pores tiny in *Loxostomum pseudobeyrichi*. It is perhaps for this reason that the latter species is rare in the

OMZ. Others show no depth variation: *Bolivina pacifica, Bolivina semiperforata, Bolivina minuta, Suggrunda eckisi.*

Peru. There are phosphorite hardgrounds encrusted with agglutinated foraminifera on their upper surfaces (Resig and Glenn, 1997). The phosphorites form under dysoxic conditions where bottom currents winnow away loose sediment. The surface waters have high productivity due to upwelling and there are benthic microbial communities, both of which provide a good food resource. The standing crops reach up to 97 individuals 10 cm^{-2} but these figures probably include some dead tests. There is dominant *Ammodiscellites prolixus* at 465–471 m and *Placopsilina bradyi* with *Tholosina bulla* at 538–620 m.

7.2.11   Pacific Ocean: west

Japan. The Kuroshio Current is a western boundary current carrying warm water northwards along the Pacific margin of southern Japan. A branch runs into the southern part of the Japan Sea and flows north as the warm Tsushima Current. The Oyashiro Current flows southwards from the Bering Sea and brings cold water to the Pacific coast of northern Japan. The Tsushima Current flows northwards along the Japan Sea margin of Honshu but Hokkaido is bathed in a cold current flowing south. Beneath the surface water at depths >200–300 m the water is very uniform: salinity 34.0–34.1, *T* 0.0–0.5 °C and well oxygenated. Three >76 μm assemblages from the upper slope at 190–290 m off Hokkaido have *Cribrostomoides* cf. *subglobosus* and *Cassidulina yabei* co-dominant with subsidiary *Involutina* sp. at 190 m, *Adercotryma glomeratum* with *Haplophragmoides bradyi, Trochammina torquata, Trochammina charlottensis* and *Trochammina quadriloba* at 210 m, and *Eggerella advena* with *Haplophragmoides bradyi* and *Trochammina quadriloba* at 290 m. At 190 m calcareous tests are dominant but at 210 and 290 m the assemblages are 87–95% agglutinated (Ikeya, 1970, WA-163). The water here is not under the influence of the surface current. Two samples from the silty mud on the slope off the Oga peninsula, Honshu, have a *Bolivina pacifica* assemblage with *Eilohedra nipponica* and *Thalmannammina parkerae.* This area is beneath cold (0.5 °C) deep water with salinity 34.05 and oxygen 5–6 ml l^{-1} (Matoba, 1976a, WA-164).

The gulf between Hokkaido to the north and Honshu to the south is the meeting area of the warm Tsushima Current flowing out of the Japan Sea and the cold Oyashio Current flowing southwards along the Pacific coast of Hokkaido. The upper slope is under intermediate water. The sediment is mainly sandy mud down to 660 m and mud from 680 m (Ikeya, 1971, WA-166). There are two main >76 μm benthic assemblages: *Nonionellina labradorica,* 430–695 m and *Bolivina* sp., 818–985 m. In addition there is a *Nonionella globosa* assemblage at 320 and 840 m and single occurrences of *Trifarina kokozuraensis* at 240 m, *Uvigerina* cf. *akitaensis* at 300 m and *Epistominella takayanagii* at 505 m. Subsidiary forms include *Eggerella advena*, *Fursenkoina* cf. *apertura* and *Islandiella* cf. *norcrossi*.

Sagami Bay near Tokyo is eutrophic with seasonal input of phytodetritus. Time-series sampling was undertaken at the same site at 1430 m water depth from 1991–7 although it was only during the final year that the input of phytodetritus was monitored (Kitazato and Ohga, 1995; Ohga and Kitazato, 1997; Kitazato *et al.*, 2000a). The sediment is grey mud with biogenic and volcaniclastic grains and many polychaete tubes project above the surface. The thickness of the oxidised layer varies seasonally. A time series of assemblages >32 μm for the top 10 cm of sediment is available for 20 different sampling events from 1993–97 (Kitazato *et al.*, 2000a, WA-210). Most foraminifera live in the top few centimetres but there is some seasonal vertical migration with the majority being concentrated nearest the sediment surface in spring. From 1991 until February 1997 (with the exception of May 1996) the dominant species was *Textularia kattegatensis* commonly with *Bolivina pacifica*, *Chilostomella ovoidea* and *Stainforthia apertura*. During a further seven sampling events in 1997, *Chilostomella ovoidea* was dominant over four events including three consecutive months from October to December. *Textularia kattegatensis*, *Globobulimina affinis* and *Stainforthia apertura* were each dominant at one event. *Textularia kattegatensis* and *Bolivina pacifica* were present as juveniles in spring and as adults at other times of the year. The standing crop fluctuated in a cyclic manner with low values in June–August and high values in November–May/June in 1991–1994. Phytoplankton blooms occurred in February to May 1997 (360–2100 mg C m^{-2} d^{-1}) with a transit time of phytodetritus to the sea floor of around 10 days. Foraminiferal standing crops increased prior to the spring bloom and remained high until October (2017–2962 individuals 10 cm^{-3}) then fell in November and December (1736 and 862 individuals 10 cm^{-3}, respectively). Sediment trap data and sea-floor photographs show that the supply of phytodetritus is high throughout the year suggesting that resuspension of organic matter as a nepheloid layer is important at the sampling site. Three species responded rapidly to the appearance of phytodetritus: *Textularia kattegatensis*, *Bolivina pacifica* and *Stainforthia apertura*. They moved into the fluffy phytodetritus layer and reproduced within one month. Infaunal species showed a different pattern. *Globobulimina affinis* reproduced in May so was perhaps responding to phytodetritus; *Chilostomella ovoidea* reproduced throughout the year so presumably was not.

Based on box cores from the Pacific off Japan and the Solomon Rise, it appears that although the flux of food is an important control, species abundance is also influenced by the available sediment pore space (Shirayama, 1984a). With respect to depth of life in the sediment for meiofauna in general, Shirayama (1984b) concludes that if oxygen availability is a control then this demonstrates that biological interaction operates because the depth of penetration of oxygen is controlled by bioturbation.

South China Sea.   The surface primary productivity is 90–160 g C m^{-2} y^{-1}. There is a strong seasonal contrast with monsoon-driven circulation resulting in summer upwelling off Vietnam and winter upwelling off the northwest of Luzon in the south and these areas have similar primary productivity. The carbonate lysocline is from 2900 m to the carbonate compensation depth (CCD) at 3500 m. Living assemblages >150 μm from the 0–1 cm sediment layer are from 329–4307 m water depth (Kuhnt *et al.*, 1999, WA-211). These assemblages have abundant agglutinated forms, mostly *Rhizammina algaeformis*, *Reophax bilocularis*, *Reophax scorpiurus* and *Saccorhiza ramosa*, any of which may be dominant. Because some of these forms are fragile and not likely to be fossilised, the authors calculated a reduced dataset. The abundant forms are then *Reophax bilocularis* and *Reophax scorpiurus* with a few assemblages dominated by *Ammoglobigerina globigeriniformis*, *Eggerella bradyi* or *Uvigerina auberiana*, the latter at depths <1000 m. Correlation between organic flux and standing crop has been recalculated here using the Suess and Sarnthein equations (see WA-211 for values). The standing crop varies from 10–173 individuals 10 cm^{-3} and correlation with either flux estimate gives $r = 0.65$ or 0.67. Thus, although there is a positive correlation it is not very strong (Figure 7.5). However, Hess (1998) carried out correspondence analysis and found that factor 1 has a correlation of $r = 0.82$ with the Suess estimate of carbon flux.

During 1991, ash from the eruption of Mount Pinatubo, Philippines, was deposited across the South China Sea and buried the fauna to variable depths. In a box core from 2503 m taken three years after the event, stained forms occur both at the surface and at depth. The surface assemblages on top of the ash are dominated by *Nodulina dentaliniformis* (given as *Reophax*) with *Reophax bilocularis* and *Brizalina difformis* (given as *Bolivina*) with a standing crop of 19 individuals 10 cm^{-3}. Although *Textularia* sp. is abundant dead it is not recorded living. The ash extends to 7 cm. The buried sediment at 7–9 cm has a few stained forms (2 individuals 10 cm^{-3}) as does the lower part of the ash (4 individuals 10 cm^{-3}). Another core from 2506 m shows a similar pattern with a surface standing crop of 55 and a basal ash value of 12 individuals 10 cm^{-3} (Hess and Kuhnt, 1996). A core from 4226 m, where the ash layer is only 2 cm thick, has a surface assemblage of *Reophax* spp. and *Rhabdammina* spp. and living forms extend down to 10 cm, the limit of sampling. The colonising

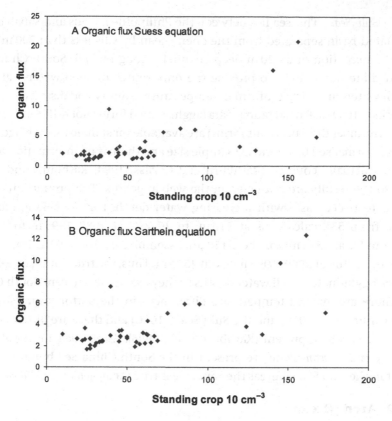

Figure 7.5. Standing crop and organic flux in the South China Sea. Comparisons between the Suess and Sarnthein equations (data from Kuhnt *et al.*, 1999).

species are all infaunal. The authors speculate that *Textularia* sp. is the pioneer coloniser that had been superseded by other forms by the time of sampling. It is possible that the stained forms from the lower part of the ash are still living infaunally. Further sampling revealed the presence of a few epifaunal taxa, such as *Astrorhiza crassatina*, komokiaceans, rhabdamminids and large xeno- phyophores on the surface in summer 1996; they increased in number by December 1996 and were dominant colonisers in 1998 (Hess *et al.*, 2001). Standing crop peaked in December 1996 and the authors suggest that the decrease thereafter may be due to increased macrofaunal bioturbation and predation. After the ash fall, there was a diversity gradient from low to high from proximal to distal but by 1996 the gradient was reduced as the proximal values increased. In the long term there is no reason to expect a pre-ash fauna to be re-established until the ash layer is buried by hemipelagic sediment like that prior to the ash fall.

Sulu Sea.   This sea lies between the Philippine Islands and Sabah and is an isolated basin separated from the open ocean by sills less than 200 m deep with the exception of a 420 m deep channel through which South China Sea intermediate water flows to provide the only source of deep water. Sulu Sea bottom waters are fairly uniform in temperature, salinity and dissolved oxygen at depths >1000 m. Partial faunal data (agglutinated forms not differentiated yet they outnumber the calcareous forms) are available for stained assemblages. The data are normalised to a standard sample size and there is no information on the numbers actually counted (Rathburn and Corliss, 1994; Rathburn and Miao, 1995) so the results are not given in the web appendix. The proportion of calcareous tests decreases with increasing water depth. For the >63 μm assemblages, from 55% calcareous at 510 m, to 26–29% at 1980–1995 m to 14% at 3995 m to 7% at 4515 m; for the >150 μm assemblages, from 39% calcareous at 510 m, to 12–16% at 1980–1995 m to 0 at 4515 m. Thus, the true dominant species must be agglutinated at all water depths as they are in the adjacent South China Sea. There are marked temperature differences in the bottom waters of the South China Sea (<10 °C) and the Sulu Sea (>10 °C) and there are differences in the calcareous taxa present (Rathburn *et al.*, 1996). *Astrononion, Ceratobulimina, Gavelinopsis* and *Osangularia* are present in the South China Sea but are rare to absent in the Sulu Sea whereas the converse is true for *Siphonina* and *Valvulineria*.

7.2.12 Arctic Ocean

The Arctic Ocean receives a large amount of freshwater from melting ice at the surface and the deep water is supplied by dense, saline water from the Atlantic. The waters are well stratified and the deep waters have a long residence time. Over the Eurasian Basin below 200–400 down to 900 m there is Atlantic water which has a residence time of around 25 years and below 900 m there is Eurasian Basin deep water and Eurasian bottom water with residence times of ~75 to ~290 years, respectively (Stein and Macdonald, 2004). The surface water overlying the deep basins has an ice cover 3–5 m thick. During the summer, breaks in the ice allow local primary production if they persist for a week or more but such breaks are ephemeral and they move about under the influence of the wind and currents. Primary production may be ~15 g C m^{-2}y^{-1} and the Arctic Ocean is the least productive of all the oceans. Not surprisingly, therefore, in the central part, standing crops are very low at 1.5–3, rarely 6–8 individuals 10 cm^{-2} (Belyaeva and Khusid, 1990). The meiofauna is dominated by foraminifera, mainly allogromiids (Schewe, 2001).

The Eurasian Basin is subdivided by ridges into the Nansen and Amundsen basins with mean depths of 3800–4000 and 4300–4500 m. The Arctic deep water has a temperature of − 1 to 0 °C and salinity 34.9. Primary production is limited by the seasonal availability of light, the low angle of incidence, and the low nutrient content of the Polar surface water. Export production in the permanently ice-covered areas is lower than that in oligotrophic oceans lacking ice cover and is estimated to be 0.1–5 g C m^{-2}y^{-1}. The most detailed studies of living foraminifera are those of Wollenburg (1992, 1995) and Wollenburg and Mackensen (1998a,b) who give partial data (for those species having an occurrence >5% in at least one sample) for two size fractions, 63–125 μm and >125 μm. However, they did not present data for the >63 μm assemblages so here these have been calculated from the size data. Wollenburg and Mackensen (1998b) recognise eight 63–125 μm and seven >125 μm varimax PCA-factor-defined groups. However, some of the named species never dominate the assemblages and principal components 5–8 account for only a small proportion of the data (usually single samples). Their 63–125 μm factor assemblages are not discussed here as it is not normal to use such a restricted size range. There are numerous >63 μm assemblages (WA-212) but only three are widespread: *Aschemonella* sp., *Placopsilinella aurantiaca* and *Stetsonia horvathi*, all from beneath permanently ice-covered areas. In addition, there is an *Ioanella tumidula* assemblage on Lomonsonov Ridge and Morris Yesup Rise and *Adercotryma glomeratum* occurs in both seasonally ice-free and permanently ice-covered areas. The >125 μm assemblages are dominated primarily by *Aschemonella* sp. or *Placopsilinella aurantiaca*. These results show how the assemblages are controlled by the size fractions examined as the small taxa dominant at >63 μm are lost from the >125 μm assemblages.

Other species associated with seasonally ice-free areas and occurring in abundances >10% include *Hormosinella guttifer* (given as *Reophax*), *Textularia torquata*, *Astrononion gallowayi*, *Cassidulina teretis*, *Cibicides lobatulus* (given as *Lobatula*), *Epistominella arctica* and *Nuttallides pusillus*. Species associated with permanent ice cover include *Reophax fusiformis*, *Hippocrepina flexibilis* and *Oridorsalis tener*. The highest standing crops occur in the seasonally ice-free areas: >63 μm, 87–2593 individuals 10 cm^{-2} compared with 11–666 individuals 10 cm^{-2} for the permanently ice-covered areas. Strong currents depress the values on Morris Yesup Rise. Throughout the area, most living forms are confined to the top 2 cm and a very high proportion is present in the top 1 cm of sediment (Wollenburg and Mackensen, 1998a). This is attributed to the oligotrophic conditions. Overall, the standing crop is positively correlated with the estimated carbon flux ($r = 0.68$, Wollenburg and Kuhnt, 2000). Highest diversities are in the seasonally ice-free areas (with one exceptionally high value) and minimum values are found in current-affected areas. There is a positive correlation of H with organic flux levels <7 g C m^{-2} y^{-1} but a negative correlation at higher values (Wollenburg and Kuhnt, 2000).

Small individuals are typical of Arctic assemblages. Nevertheless, the large miliolid, *Cornuspiroides striatolus*, (maximum diameter 2.71 cm) occurs off Svalbard at depths of 178–702 m on muddy substrates and under the influence of different water masses. The live assemblages are dominated by *Nonionellina labradorica* or *Reophax bilocularis* with *Elphidium excavatum* or *Islandiella norcrossi*. Schmiedl and Mackensen (1993) suggest that the mode of life might be epifaunal with the older part of the test buried in the sediment and the younger part placed above the sediment surface in order to catch suspended food with its pseudopodia. This contrasts with the similar form *Cornuspiroides rotundus* found around Antarctica.

7.2.13 Southern Ocean

The principal hydrographic feature of the Southern Ocean is the ACC in which CPDW flows clockwise around the continent. As it circulates it loses water both at the surface and the bottom and this is replenished by intermediate waters derived from the North Atlantic (NADW) (Kennett, 1982). On the north side of the ACC there is a downwelling zone known as the Polar Front where the ACC meets the subantarctic surface water. The coldest and densest of the oceanic bottom water masses, AABW, forms in the Weddell Sea and flows north into the ocean basins. The Scotia Sea is separated from the South Atlantic by the island arc (Scotia Ridge) comprising South Georgia, South Sandwich and South Orkney islands.

Southwest Atlantic.   A regional study including the Argentine Basin and the Scotia Sea, spans a range of bottom-water masses and productivity regimes (Harloff and Mackensen, 1997, WA-213). They recognise six >125 μm associations based on Q-mode principal component analysis. In addition to the forms named in the associations below, there are other species which dominate individual samples (*Adercotryma glomeratum*, *Earlandammina* spp., *Lagenammina tubulata*, *Marsipella elongata*, *Nodulina dentaliniformis*, *Reophax scorpiurus*, *Bulimina aculeata*, *Cassidulina teretis* and *Nonionella bradyi*).

The *Trifarina angulosa* association occurs along the upper slope from the Falkland Plateau into the Argentine Basin mainly at depths of 400–1400 m. Along the Argentine upper slope there is the north-flowing Falkland or Malvinas Current composed of AIW. This current prevents organic particles from settling onto the sea floor and keeps them in suspension. Nevertheless, this association coincides with an area of high standing crop with up to 600 individuals 10 cm^{-3}. The *Uvigerina peregrina* association includes *Bulimina aculeata*, *Cibicidoides bradyi*, *Uvigerina hispida* and *Uvigerina proboscidea*. It occurs at 2400–3300 m on the Argentine Basin continental slope with high organic flux and in the area of NADW. The *Epistominella exigua* association with *Eggerella bradyi*, *Oridorsalis umbonatus*, *Nuttallides umboniferus* and *Globocassidulina subglobosa* occurs mainly under upper circumpolar deep water but also under NADW on varied sediments with a low organic content.

The *Lagenammina difflugiformis* association with *Reophax bilocularis* and *Reophax micaceus* is present on muddy sediments beneath AABW or CPDW in the Scotia Sea and in the South Atlantic. The *Psammosphaera fusca* association occurs on sandy sediments low in organic carbon under AABW or circumpolar deep water in the Scotia Sea and in the South Atlantic. The *Cribrostomoides sub-globosus–Ammobaculites agglutinans* association with *Reophax bilocularis* and *Reophax spiculifer* covers the floor of the Argentine Basin under AABW and along the South Sandwich Trench under Weddell Deep Water. Harloff and Mackensen (1997) emphasise the interdependence of parameters and point out the roles of organic flux, bottom currents, sediment grain size and water masses as controlling factors.

Three samples studied by Echols (1971, WA-214) come from the Scotia Sea (south of South Georgia) while the rest come from the Weddell Sea side of the ridge. A >63 μm *Fursenkoina earlandi* assemblage is present in the Scotia Sea off South Georgia at 1032 m and in the Weddell Sea off South Orkney at 560–1279 m and South Sandwich at 558 m. A *Haplophragmoides quadratus* assemblage with subsidiary *Adercotryma glomeratum* and a *Rudigaudryina inepta* assemblage with subsidiary *Alabaminella weddellensis* are also present in the Scotia Sea. In the Weddell Sea, an *Alabaminella weddellensis* assemblage occurs from 490 and 1589 m off South Orkney, off South Sandwich at 2345 m together with sub-sidiary *Adercotryma glomeratum* and *Eponides tumidulus* and in the open sea at 2968 m. There are isolated occurrences of a *Pseudobolivina antarctica* assemblage with subsidiary *Adercotryma glomeratum*, an *Alabaminella weddellensis* and an *Epistominella exigua* assemblage with subsidiary *Fursenkoina earlandi*. The stand-ing crops range from 71–298 individuals 10 cm^{-3} compared with 6–140 for the South Sandwich area and 58–480 for the area towards the Falkland Islands (Basov, 1974). Echols related the distributions to water depth and the effects of carbonate dissolution. Mikhalevich (2004) also considers water masses to be an important control. In the Weddell Sea the upper slope is influenced by the ACC. Below 1500–2300 m the number of species is reduced. Both the shelf and bathyal zone have a high proportion of endemic species but many species have circum-Antarctic distributions.

Weddell Sea. Box cores have yielded three main >125 μm assem-blages: *Trifarina angulosa* from 310–1521 m along the shelf break and upper continental slope on sand and gravel substrates influenced by bottom currents;

Bulimina aculeata from 1499–1948 m on fine-grained, slightly more organic-rich, sediments beneath the core of the relatively warm Weddell Deep Water (>0 °C); *Cribrostomoides subglobosus* from 1866–4541 m on a continental terrace. The *Bulimina aculeata* assemblage is regarded as indicating high surface-water productivity in areas where the bottom is bathed in CPDW or its derivative, both in the Weddell Sea (Mackensen *et al.*, 1990, WA-215) and the South Atlantic (Mackensen *et al.*, 1993a). Multivariate analysis of environmental and faunal data does not reveal any correlations between the two and there is no direct correlation with water masses. In addition to these main assemblages, there are several with few or single occurrences: *Reophax pilulifer, Hormosina robusta, Nodulina dentaliniformis, Reophax fusiformis, Reophax bilocularis, Globocassidulina crassa, Nonionella iridea*. These may reflect local blooms of taxa at the time of sampling. The standing crop ranges from 11–238 but is mostly <70 individuals 10 cm^{-3}. Two box cores from 1795 and 1985 m have >125 μm *Bulimina aculeata* assemblages with subsidiary *Cribrostomoides subglobosus, Reophax bilocularis* and *Reophax fusiformis* (Mackensen and Douglas, 1989, WA-216). The standing crops are 34 and 43 individuals 10 cm^{-3}. Living forms occur down to 5.5 cm; at 1795 m only 36% of the live forms are present in the 0–1 cm interval and 59% in the 0–2 cm interval but at 1985 m 74% are present in the top 1.5 cm. In both cases *Bulimina aculeata* remains as the dominant species if the whole core data are considered. In the Weddell Sea, there is dissolution of calcareous tests soon after death and some loss of fragile agglutinated tests (see Section 9.3.1).

In a further study in the Weddell Sea, the entire range of >63 μm foraminifera, including soft-shelled taxa, from 1100, 2080, 3050, 4060 and 4975 m, show the assemblages to be diverse (Cornelius and Gooday, 2004). Many of the soft-shelled forms are undescribed and unnamed. Among the hard-shelled taxa, calcareous forms make up 65% at 1108–1120 m, 52% at 2074–2084 m, 31% at 3038–3065 m and <15% deeper than this probably due to increasing dissolution. The single most abundant species is *Epistominella exigua*. In phytodetritus aggregates, this dominates with subsidiary *Alabaminella weddellensis* and *Tinogullmia reimanni*. There is a decrease in standing crop with increasing water depth, from a maximum of 576 individuals 10 cm^{-2} at 2080 m to a minimum of 240 individuals 10 cm^{-2} at 4975 m. However, there is considerable variation in replicates from the same station (304, 333, 1090 individuals 10 cm^{-2} at 2100 m). The proportion of agglutinated tests increases from 35% at ~1100 m to 84% at 4975 m. The local CCD is considered to be ~4000 m (Mackensen *et al.*, 1990) but live forms extend deeper than this although they may not be preserved once dead. Although Cornelius and Gooday argue that the decrease in hyaline forms with increasing depth is due to decreased availability of food, because food, depth and increased corrosivity of the water are co-variant, their explanation cannot be proved. A recently described monothalamous agglutinated form from the Weddell Sea occurs over a bathymetric range of 1080–6330 m yet there are no genetic differences between individuals from different water depths suggesting gene flow over this great range (Gooday et al., 2004).

In the South Atlantic, cores taken near Bouvet Island at water depths of 486–1897 m have live Bulimina aculeata mainly in the top 1 cm in an area where the average organic flux is approximately 83 g C m^{-2} y^{-1}. The δ^{13}C value does not vary with depth of life in the sediment and reflects the pore water $\delta^{13}C_{DIC}$ values (Mackensen et al., 2000). This species is active and presumably calcification takes place mainly in the surface layer. With the exception of one station, it shows a negative offset from bottom water $\delta^{13}C_{DIC}$ by $- 0.6 \pm 0.1$‰. The exception is where there is a higher organic flux and the offset is $- 1.5 \pm 0.2$‰. The stable isotopic composition of high-salinity shelf water and ice shelf water overlying the Weddell Sea shelf has only small differences in $\Delta\delta^{13}$C and δ^{18}O and beyond the resolution of analysis of foraminiferal shell calcite (Mackensen, 2001).

Ross Sea. In McMurdo Sound, Ward et al. (1987, WA-217) distinguish three live >63 μm assemblages: harbour, shallow water (<620 m), deep water (>620 m); the 620 m isobath is the depth of the CCD. According to the authors, agglutinated foraminifera predominate in the harbour and deep-water areas while agglutinated and hyaline forms are almost equally abundant in the shallow-water assemblages (but the figures cannot be calculated because the data table lists only the dominant species). *Reophax subdentaliniformis* is dominant in all areas, *Reophax pilulifer* is dominant only in deep water, *Portatrochammina antarctica* and *Fursenkoina earlandi* are dominant in both harbour and deep-water areas, *Textularia antarctica* is dominant in the harbour and deep-water areas while *Trifarina earlandi* (which is very similar to *Trifarina angulosa*) is dominant only in shallow water. Thus the differences between the areas are subtle. In the western Ross Sea two out of six samples from 450–1100 m water depth are dominated by *Trifarina angulosa*. Each of the rest has a different dominant species: *Portatrochammina antarctica*, *Reophax fusiformis*, *Nonionella bradii*, miliolids (Asioli, 1995, WA-218). Although Asioli links *Trifarina angulosa* to bottom currents and *Nonionella bradii* to high organic input, she does not provide any direct ecological data and these assumptions are not well supported. For instance, one of the two occurrences of *Trifarina angulosa* is on a bank (which may have currents) while the other is in a basin (which almost certainly does not). In Terra Nova Bay only the sample from 220 m has calcareous foraminifera and the assemblage is dominated by *Cibicides* (*lobatulus, refulgens*, sp.). The remaining assemblages are mainly *Portatrochammina antarctica* with subsidiary *Pseudobolivina antarctica*. Species which are dominant or subsidiary in one or a few assemblages are *Miliammina earlandi*, *Nodulina dentaliniformis*, and *Pseudobolivina antarctica* (Violanti, 1996, WA-219). Morphogroup analysis shows that the assemblages are dominated by small planoconvex trochospiral trochamminids with subsidiary flattened tapered forms (Violanti, 2000). Like the Weddell Sea, the Ross Sea is strongly affected by the shallow CCD, hence the presence of agglutinated assemblages deeper than a few hundred metres. In Prydz Bay, east Antarctica, glaciomarine and diatom oozes from 410–987 m have living dominant *Miliammina arenacea* with *Recurvoides contortus*, *Reophax subfusiformis* and *Psammosphaera fusca* (Schröder-Adams, 1990).

7.2.14 Abyssal plains and trenches

Beyond the continental rise there are abyssal plains and hadal trenches. These have depths >4000 m and are generally deeper than the CCD. Therefore, the fauna is primarily organic-walled or agglutinated. In some areas xenophyophores dominate the macrofauna visible on the sea floor (Tendal and Gooday, 1981). They are often associated with high levels of organic-matter input (Levin and Gooday, 1992) and some feed on phytodetritus (Richardson, 2001) but their role in carbon cycling remains enigmatic (Levin and Gooday, 1992). The xenophyophores of the Atlantic Ocean are generally smaller than those of the Pacific (Gooday and Tendal, 1988).

Northeast Atlantic. The mudball komokiacean *Edgertonia floccula* is the most abundant macrofaunal organism >500 μm on the continental slope and abyssal plain in the northeast Atlantic (Shires *et al.*, 1994a). *Arborammina hilaryae* is a 1.5–2.0 cm high arborescent agglutinated form that has no chamber to contain protoplasm. Instead, it is thought to be contained in the tests of planktonic foraminifera incorporated into the wall (Shires *et al.*, 1994b). It may be a suspension feeder or osmotroph living in oligotrophic conditions. Xenophyophores are scarce on the Porcupine and Cape Verde abyssal plains with densities of 5 and <2 individuals m^{-2}, respectively (Gooday, 1996b). Episodic growth of *Reticulammina labyrinthica* took place over a period of two to three days interspersed with inactive intervals of 51–70 days and led to a three to ten-fold increase in volume over eight months (Gooday *et al.*, 1993). The estimated length of life is one to two years. The xenophyophores provide microhabitats for other organisms which congregate around, beneath and within their tests.

Nova Scotia. On the rise, depths >4000 m are beneath the Western Boundary Current. Although such depths are often considered to experience only low velocity currents such as ~3 cm s^{-1}, in this area there are periodic 'benthic storms' when velocities at 10 m above the sea floor are 15–23 cm s^{-1}. Such events occur at around ~21 d intervals and last for 7 ± 5.8 d. Samples taken in July 1982 at 4187 m had a mean standing crop of 118 ± 45 while those taken in April 1983 had a mean of 3.6 foraminifera 23 cm^{-3} for the top 0.5 cm of sediment (Aller, 1989). For the 0.5–1.0 cm layer, the values were 10 ± 9 and 5.3 ± 7 individuals 23 cm^{-3} respectively. In July, the level of current disturbance was low and standing crop was high in the surface 0.5 cm. In April, there was an intermediate level of current disturbance which led to erosion of the sediment surface and standing crops were low. The 0.5–1.0 cm level was less affected by these events.

Sargasso Sea. Slope assemblages of robust suspension-feeding tubular rhabdamminids are partly influenced by the Western Boundary Undercurrent (Kuhnt and Collins, 1995a). A sample from the southern part of the Gulf Stream close to the area of formation of cold core rings (which move at a speed of 5 cm s^{-1} for westward-moving and 25–75 cm s^{-1} for eastward-moving rings) has a tiny branched komokiacean with thicker and coarser walls living infaunally in the top 2 cm of sediment. The core rings have trophic characteristics of slope water being enriched in nutrients, phytoplankton and planktonic foraminifera; this increases the supply of food for detritus and suspension-feeding benthic foraminifera. Finely agglutinated, small forms occur beneath the oligotrophic water mass of the central Sargasso Sea. The size of tubular forms such as *Rhabdammina* and *Rhizammina* is generally related to the availability of food, with small forms in oligotrophic conditions and large forms where more food is available (Kaminski and Kuhnt, 1995). Such tubes provide a firm substrate for the attachment of smaller foraminifera; inside *Bathysiphon rusticus*: *Crithionina mamilla*, *Glomospira gordialis*, *Placopsilina bradyi*, *Saccodendron* sp., *Thurammina papillata*, *Tolypammina fragilis*, *Tolypammina* aff. *schaudinni* and *Tumidotubus albus*. The orientation of the aperture suggests that these forms graze on bacteria (Gooday and Haynes, 1983).

Pacific. A single sample of biosiliceous ooze from 5289 m and below the CCD in the North Pacific is dominated by soft-shelled, monothalamous, mainly saccamminids, typically <120 μm in size (Gooday *et al.*, 2001b). The standing crop for the 0.0–0.5 cm layer is 229 individuals 10 cm^{-2} for >63 μm and 427 individuals 10 cm^{-2} for >32 μm. The numbers are lower for the 0.5–1.0 cm interval: 61 and 117 individuals 10 cm^{-2}, respectively. Taking the top

1 cm the values are 290 and 545 individuals 10 cm^{-2}, respectively. These are high because of the inclusion of soft-shelled taxa which make up 31% of the assemblage. If these are excluded, the standing crop of the 0–1 cm >63 μm fauna is 202 individuals 10 cm^{-2}. Soft-shelled allogromiids and saccamminids are represented by 20 species plus *Lagenammina* sp. in the >20 μm assemblage from the 0–6 cm layer at 7800 m in the Peru–Chile Trench off South America (Sabbatini *et al.*, 2002). In the Marianna Trench, Challenger Deep, at 10 897 m, the greatest depth in the ocean, the red pelagic clay contains four rose Bengal-stained *Lagenammina difflugiformis* with tests composed of anorthite grains preferentially selected from the sediment (Akimoto *et al.*, 2001). Subsequent sampling using a remotely operated vehicle has yielded a diverse assemblage of soft-shelled taxa such as *Chitonosiphon*, *Nodellum* and *Resigella* together with *Leptohalysis* and *Reophax* and with a standing crop of 449 individuals cm^{-2} (Todo *et al.*, 2005). The xenophyophore *Ocultammina profunda* is abundant at 8260 m in the Ogasawara Trench off Japan (Shirayama, 1984a).

7.2.15 Encrusting forms on polymetallic structures

Encrusting epifaunal foraminifera, including large agglutinated forms, are found on manganese nodules on abyssal plains, crusts in the bathyal zone and associated with hard substrates near hydrothermal vents in various settings. Some of the agglutinated forms have iron and manganese incorporated in the test: *Saccorhiza ramosa*, *Tolypammina vagans* (Dudley, 1976; Dugolinsky *et al.*, 1977) and Riemann (1983) suggests that foraminiferal stercomata may play a role in the formation of the nodules. However, Thiel and Schneider (1988) consider this to be unlikely or negligible. The data on faunal occurrences are not always confirmed through staining but at least the attached epifaunal dead forms are *in situ* and not transported in from elsewhere. Most of the published comments about mode of life of the foraminifera, their feeding strategies and their relationship with nodule growth are speculative rather than based on observation or experiment.

Large agglutinated foraminifera commonly encrust manganese nodules in the equatorial north Pacific and central Pacific (Mullineaux, 1987; 1988a; 1989). The number of epifaunal foraminifera and percent of area covered are higher in the equatorial than central area. They include allogromiids, ammodiscaceans, lituolaceans and calcareous forms such as *Pyrgo*, *Quinqueloculina*, *Bulimina*, *Cibicides* and *Patellina*. Mat- and tunnel-forming morphologies are more common than tubular and chambered forms. There is vertical distribution of taxa with some forms preferring smooth substrates such as those found at the summit and others the rough texture found near the base of nodules. Numbers are higher at the summit. Flow conditions are stronger at the summit and particle contact is greater near the base. It might be expected that suspension-feeding forms are more common at the summit. In a separate study, Dugolinsky *et al.* (1977) suggest that forms living on the rough surfaces close to the sediment–water interface may influence the physical growth of the nodules.

Agglutinated foraminifera (*Thurammina papillata*, *Ammodiscus* sp., *Tolypammina* sp., *Haplophragmoides* sp. and a saccamminid) also live in crevices in nodules (Maybury, 1996). It is possible that this strategy incidentally prevents the nodule from breaking up as the foraminifera cement the sides of the crack together although this has been questioned by Thiel and Schneider (1988). They may also be responsible for fusing small nodules into large polynodules (von Stackelberg, 1984). Broken tests on the nodule surface may be evidence of predation and this may be a reason why the organisms mainly settle in hollows or crevices.

On the European margin off Portugal, there are polymetallic crusts at bathyal depths that have not only agglutinated forms (*Saccorhiza*, *Sorosphaera*, *Trochammina*, *Eggerella*) but also hyaline *Carpenteria proteus* (Schaaf et al., 1977; Bignot and Lamboy, 1980). Watch-glass-shaped *Abyssotherma pacifica* lives attached to hard substrates in hydrothermal fields (Brönnimann et al. 1989b; Lee et al., 1991a). It feeds at least partly on bacteria (Lee et al., 1991b). Indurated sediments from a hydrothermal venting area on the Juan de Fuca Ridge, northeast Pacific, provide hard substrates on which agglutinated foraminifera attach. Many have tubular, winding tests and they overgrow one another: *Tolypammina vagans*, *Tumidotubus albus*, *Crithionina*? sp., *Placopsilina bradyi* and *Ropostrum amuletum* (Jonasson and Schröder-Adams, 1996). Also present are otherwise free-living forms which in this case are attached: *Subreophax aduncus*, *Saccodendron heronalleni*, *Reticulum reticulatum*, *Lana spissa* and *Trochammina globulosa*.

Experiments on the East Pacific Rise to investigate recruitment to hard surfaces show that foraminifera are part of a diverse epifauna that develops (van Dover et al., 1988). In another experiment on Cross Seamount, although simple agglutinated foraminifera were the main colonisers, forms found free elsewhere also formed part of the fauna on the hard substrates: *Eggerella* sp., *Discanomalina* sp. and *Globocassidulina* sp. (Bertram and Cowen, 1994).

7.3   Summary and conclusions

Progress in understanding some aspects of the ecology of the deep sea over the past two decades has come from technological developments such as satellite imagery of ocean colour (as a proxy for primary production), ability to measure primary production *in situ*, and improved bottom sampling (especially the development of the multicorer, box corer and unmanned vehicles). However, in relation to the fact that oceans cover a high percentage of the Earth's surface, very little is known. Indeed, large areas of the Indian and Pacific oceans have never been sampled. Most of the data are from continental margins and especially from the eastern Atlantic.

There seem to be five main controls that structure deep-sea foraminiferal assemblages: supply of food; sediment geochemistry; corrosivity of bottom waters with respect to $CaCO_3$, lysocline and CCD; bottom currents; hardgrounds such as polymetallic nodules and crusts.

Food. The amount, kind and quality of food influence numbers and biomass. Some authors believe that apart from controlling numbers of individuals, organic flux also controls the species composition of assemblages (Lutze and Coulbourn, 1984; Jorissen *et al.*, 1995; 1998; Altenbach *et al.*, 1999; Gooday *et al.*, 2001a). One of the unanswered questions is whether the flux of organic carbon is a true measure of food availability. The organic matter reaching the sea floor is composed of labile material from recently dead plankton, refractory material derived from the plankton, refractory material of terrestrial origin and faecal material from feeding especially by zooplankton. It is clear that deep-sea foraminifera have resource partitioning because some species respond to seasonal inputs of phytodetritus (so-called phytodetritus species such as *Alabamminella weddellensis* and *Epistominella exigua*) whereas the majority do not. Even among the phytodetritus species there may be preference for one type of alga rather than another although this remains to be evaluated. During the periods of the year when no phytodetritus is contributed to the sea floor, the phytodetritus species live in low abundance in the surface sediment. Therefore their presence does not require a continuous supply of phytodetritus. Probably because the flux of organic carbon to the sea floor is such a crude measure of real food, individual species show a broad range of tolerance to flux levels. Peak abundance of some taxa can be related to narrower ranges of flux with lower abundance associated with the extremes of organic flux (Figure 7.3). There is some evidence that with respect to food calcareous taxa are more opportunistic than agglutinated forms (Jorissen *et al.*, 1998; Gooday, 2002; Gooday *et al.*, 2001a). High-flux taxa are infaunal and tolerate varying degrees of oxygen depletion. The low-flux taxa listed are mainly epifaunal and are not subject to oxygen depletion. There may be a threshold at 2–3 g C m^{-1} y^{-1} with species adapted to conditions higher or lower than this (Altenbach *et al.*, 1999). Organic flux may not be such an important controlling factor at depths <1000 m. At such depths, there is a greater likelihood of lateral advection by contour currents so flux estimates based on vertical settling will be unreliable. Gooday (2003) suggests that continental-margin faunas will be affected by sediment types, bottom currents, oxygen depletion and organic flux whereas abyssal-plain fauna will be influenced primarily by organic flux.

Sediment geochemistry. Infaunal taxa live at various depths in the sediment and although some species are consistently found deeper than others (e.g., *Chilostomella oolina, Globobulimina* spp., *Melonis barleeanum*), the precise depth at which they live is controlled by a combination of pore-water geo-chemistry (oxygen, nitrate/nitrite, ammonia, H_2S, etc.) and the associated bacteria (archaea and sulphate-reducing bacteria, Hinrichs and Boetius, 2002; unknown microbes responsible for recycling N, Mn, Fe and Si, Boetius *et al.*, 2002). Infaunal foraminifera may feed on refractory organic matter and/or the bacteria associated with its decomposition. They are able to move through the sediment tracking their preferred habitat and this is shown by the $\delta^{13}C$ values of their tests. Whereas it is to be expected that these will be negative with respect to ambient bottom-water values, the values should also become more negative with increasing depth in the sediment. However, that is not always the case and the foraminiferal values tend to be an average of the pore-water values. This suggests that the tests have been secreted at different depths in the sediment during the active life of the individual. Overall, it appears that most forms live in the top 1–2 cm and that forms living deeper than this do not do so in a consistent manner from one area to another.

Throughout most of the ocean there is abundant oxygen in the bottom waters due to the flushing by deep water formed at high latitudes (NADW, AABW). Because the sediments are generally low in organic carbon, dysoxic conditions are rarely found close to the sediment surface. Most abyssal plains and lower bathyal zones fall in the oligotrophic side of the TROX model (see Section 8.5) and oxygen is not a limiting parameter. However, beneath areas of upwelling, there are oxygen minimum zones where the bottom waters may become dysoxic. Furthermore, the bottom sediments have a high organic-carbon content so anoxic pore waters occur close to the sediment surface (the eutrophic side of the TROX model) and oxygen may become limiting. Although no species is confined exclusively to low-oxygen conditions, certain species, especially of *Bolivina* and *Brizalina*, are able to survive there (although they also live in low numbers under more oxygenated conditions elsewhere). Some continental margins are mesotrophic and occupy the central part of the TROX diagram (e.g., Bay of Biscay, France).

Corrosivity of bottom waters. Although the concept that bottom-water mass characteristics influence foraminiferal assemblages is less in vogue than it was two decades ago, it clearly plays a role. Apart from subtle differences of temperature and salinity, there are major differences in the amount of corrosion of $CaCO_3$. With increasing water depth and increasing dissolved CO_2, there is increased dissolution. The lysocline for aragonite is shallower than that for calcite. In the deep sea, *Nuttallides umboniferus* is able to live between the lysocline and the CCD (Mackensen *et al.*, 1993a). The CCD varies in depth, being deeper in the Atlantic than the Pacific and it shallows around Antarctica (Kennett, 1982). At depths deeper than the CCD, only agglutinated and organic-walled foraminifera survive. Thus, the faunas of the abyssal plains and trenches lack calcareous taxa. In the Southern Ocean, where the CCD extends into shallower water, calcareous forms are sparsely present in living assemblages and are dissolved soon after death.

Bottom currents.   The circulation of deep oceanic water masses impacts on continental margins and the flanks of mid-ocean ridges where contour currents touch the sea floor. Under such conditions, there is lateral advection of organic material (food), and disturbance of the sediment surface (erosion, formation of ripples, sediment redeposition, etc.). In such settings, epifaunal foraminifera attach themselves to firm substrates (shells, bigger animals, rocks, etc.) that project above the sediment into the lower part of the current in order to benefit from suspended food particles, which they trap with their outstretched pseudopodia. Thus, there are distinctive faunas of *Fontbotia wuellerstorfi*, *Cibicidoides* spp., *Cibicides lobatulus*, *Cibicides refulgens* such as are found on the European margin beneath MOW. In other areas, breaking internal waves and eddies from major currents may cause 'benthic storms', ephemeral bursts of energy that affect the sea floor periodically. In the HEBBLE (High-Energy Benthic Boundary-Layer Experiment) area near the Western Boundary Undercurrent, the affected areas have large tubular agglutinated *Rhabdammina* and *Rhizammina* whereas the quieter areas unaffected by the 'storms' lack these forms.

Polymetallic nodules.   On the abyssal plains, especially in the Pacific Ocean, but also at bathyal depths along continental margins where there are currents, polymetallic nodules (so-called manganese nodules) and crusts develop. These hard substrates provide anchorage for a variety of encrusting foraminifera, many of which are simple, tubular agglutinated forms. As with other areas affected by currents, the hard polymetallic substrates project above sediment surface allowing the foraminifera to benefit from suspended food particles carried by the currents. Many of the taxa appear to occur only on the hard substrates so this must be a primary control on their occurrence regardless of any other factors that may also exert an influence.

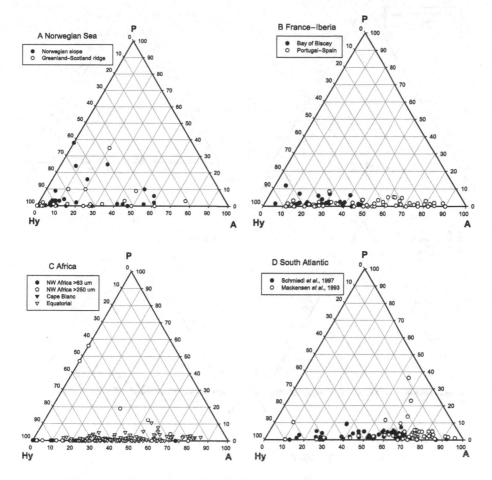

Figure 7.6. Ternary plots of walls for the deep sea: A = agglutinated, P = porcelaneous, Hy = hyaline.

Wall structure.   There is some variation but overall deep-sea assemblages are composed of mixtures of hyaline and agglutinated walls (Figures 7.6, 7.7). Porcelaneous tests generally make up <5% of the total (93% of the points are 0–5%; Figure 7.8). However, there are a few with higher proportions of porcelaneous walls and these come from >250 μm assemblages from off northwest Africa (Figure 7.6C), >125 μm assemblages from the South Atlantic (Figure 7.6D) and from >125 μm in the Norwegian Sea (Figure 7.6A). Thus, with a few exceptions, the deep sea is not a favourable environment for porcelaneous taxa. The main factor controlling the predominance of agglutinated walls is corrosivity of the bottom water. Overall there is an increase in agglutinated tests with increasing water depth, but around Antarctica the CCD

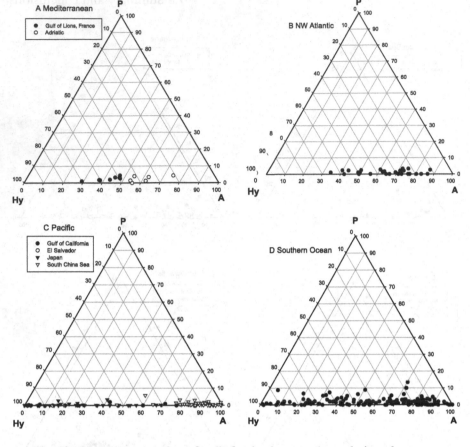

Figure 7.7. Ternary plots of walls for the deep sea: A = agglutinated, P = porcelaneous, Hy = hyaline.

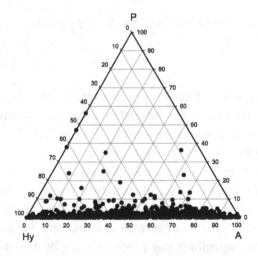

Figure 7.8. Summary ternary plot of walls for all deep-sea assemblages: A = agglutinated, P = porcelaneous, Hy = hyaline.

rises to shallow depths. All hadal assemblages are entirely agglutinated or organic-walled.

Standing crop. Most information is from the continental slope. Although the maximum values reach an exceptionally high value of 6661 individuals $10 \, \text{cm}^{-3}$ off Japan, for the Atlantic it is 1616, for the Southern Ocean 572 and for the Arctic Ocean 2593 individuals $10 \, \text{cm}^{-3}$. Off Japan, typical values are 1000–3000 but elsewhere they are generally 20–200 individuals $10 \, \text{cm}^{-3}$. Overall, in any one area there is a trend towards lower values with increasing water depth.

Species diversity. Ecologists are interested in local and regional species diversity. Although diversity patterns in the deep sea have attracted much attention and theorising, such patterns are merely part of the global patterns of diversity spanning all types of environment. Thoughtful reviews of the problem are those of Gooday (1999) and Levin *et al.* (2001) although the model given by Levin *et al.* is too simplistic to be valid. The influences on species diversity vary according to the scale being examined and many factors are interdependent, i.e., factors do not operate in isolation from one another. At a local level, there are patches relating to topography, nutrients or biogenic features; patch dynamics involves competition, facilitation, predation, small scale disturbance and recruitment. At a regional level, there is spatial hetero-geneity of habitats, metapopulation dynamics, dispersal and environmental gradients of production, water flow, oxygen, disturbance and sediment het-erogeneity. There are also global influences but for foraminifera our knowl-edge of these is not yet adequate for consideration. In addition there are geological influences that shape the ocean basins.

High-diversity assemblages have a few common species and a long tail of rare species (Figure 7.9). In Section 2.11 it is shown there is a linear relation-ship between Fisher alpha diversity and H_{max} (Figure 2.3). For H_{max}, all species in a sample must be equally abundant and this is never the case so H is always smaller than H_{max}. Nevertheless, for the Atlantic Ocean ($n = 340$) there is a linear relationship between Fisher alpha and H and the correlation is high $r = 0.87$ (Figure 7.10) even though the ranges of both are large (Table 7.1). Also, the dataset includes assemblages of a wide range of sizes, from >63 to >250 µm. Thus size fraction does not influence the relationship between the two measures of diversity. For analysis of the results in greater detail, it is more convenient (for the reader) to use plots with arithmetic scales so that values can easily be read off. The Norwegian Sea and the Greenland–Scotland ridge data plot with relatively low diversity values (Figure 7.11A). The Bay of Biscay

Table 7.1. Statistics for diversity in the Atlantic Ocean

	Fisher alpha	H
Minimum	2.61	1.09
Maximum	35.93	3.83
Median	15.50	3.10
Mean	16.03	2.94
Standard deviation	6.83	0.53

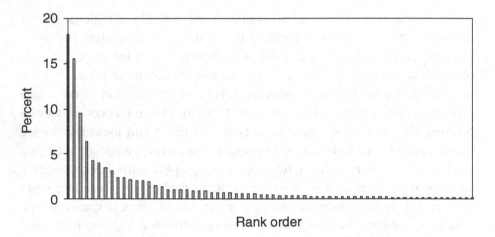

Figure 7.9. The rank order of species in a high diversity assemblage.

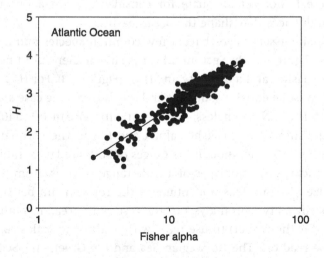

Figure 7.10. Species diversity in Atlantic assemblages.

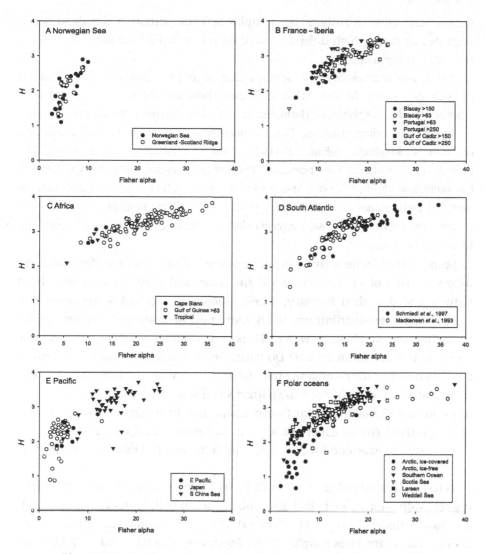

Figure 7.11. Species diversity for all Atlantic Ocean assemblages ($n = 342$).

to the Gulf of Cadiz has few values with Fisher alpha <10 (Figure 7.11B). For the region from Cape Blanc to the South Atlantic, relatively few values are lower than Fisher alpha 10 and the maximum is 37 (Figure 7.11C, D). The Mediterranean plots in the same area as the Bay of Biscay and several other areas. Data for the Pacific Ocean are sparse in relation to the size of the ocean. Values are low around Japan and in the east Pacific (near methane vents; off El Salvador). Higher values prevail in the South China Sea (Figure 7.11E). Polar seas span a wide range from low to high diversity (Figure 7.11F). The ice-covered Arctic Ocean has lower diversity than the ice-free areas. The Southern

Ocean has values mainly >Fisher alpha 9. Thus, although the deep sea is regarded as having high diversity, the data summarised here show that this is not universally true.

There are remarkably few species that regularly appear as dominant or subsidiary (>10%). In the Atlantic Ocean these are *Adercotryma glomeratum, Cribrostomoides subglobosus, Hormosinella guttifera, Reophax bilocularis, Reophax scorpiurus, Saccorhiza ramosa, Cancris auriculus, Cassidulina teretis, Cassidulina carinata, Chilostomella oolina, Cibicides lobatulus, Cibicides refulgens, Cibicidoides kullenbergi, Cibicidoides robertsonianus, Epistominella exigua, Fontbotia wuellerstorfi, Globobulimina affinis, Globobulimina auriculata, Hoeglundina elegans, Melonis barleeanum, Melonis zaandamae, Nuttallides pusillus, Nuttallides umboniferus, Oridorsalis umbonatus, Trifarina angulosa, Uvigerina elongatastriata, Uvigerina mediterranea* and *Uvigerina peregrina.*

Some authors believe that a proper understanding of foraminiferal deep-sea diversity will not be achieved until the many soft-shelled forms have been named and described (Gooday, 1999). Gooday also posed some questions: are cosmopolitan distributions really more prevalent among deep-sea species than they are in shallow water? Does a latitudinal gradient exist among modern deep-sea foraminifera? Do bathymetric trends in foraminiferal diversity down slope differ fundamentally from the parabolic patterns exhibited by metazoa; and, if so, why? Are the patterns and scales of foraminiferal diversity (both local and global) similar to those reported for metazoans or to those of other protists? The considerable amount of foraminiferal data reviewed here can be used to examine local diversity patterns and to answer some of these questions.

Is there a latitudinal gradient in diversity? There is no obvious gradient. The highest values for hard-shelled assemblages are in high latitudes (Figure 7.11F) and low latitudes (Figure 7.11c). The extremes for assemblages including soft-shelled forms are Fisher alpha 55–65 (northeast Atlantic) and 71.7 (Arabian Sea). Overall, there is great variability at all latitudes and this suggests that local rather than regional factors are important. Such factors include flux of food, corrosivity of bottom water, bottom currents and substrate type.

Do bathymetric trends in foraminiferal diversity down slope differ fundamentally from the parabolic patterns exhibited by metazoa and, if so, why? The extensive dataset for the Gulf of Guinea ranges from 201–4970 m and was collected by two workers using the same taxonomic scheme (Schiebel, 1992; Timm, 1992) so is ideal for testing trends with water depth. There is a weak negative correlation with Fisher alpha, $r = -0.43$ and with H, $r = -0.42$. There is a great range in values at any given depth and no diversity maximum at any one depth so the pattern is not parabolic. Again, this suggests that local factors are more important than the regional depth change (with its probably flux-related co-variance).

It might be expected that the Pacific Ocean, being larger than other oceans, would have higher diversity. Unfortunately, the data are insufficient to test this. At the moment, it can be said that the range of diversity is comparable between the Atlantic Ocean and the Arctic Ocean. On the limited data available, the Southern Ocean is less diverse than the Arctic and Atlantic oceans and this may be because the ACC isolates the faunas from those in the wider ocean.

The future.   The ecology of deep-sea foraminifera remains at an early stage of investigation because of the difficulties in gathering data. In the years ahead, with continued technical developments in sampling; remote measurement of sea floor and sediment water properties, fluxes, and rates of change; and *in situ* experiments, there will undoubtedly be exciting new developments and better understanding of processes controlling deep-sea foraminiferal ecology.

8

Summary of living distributions

8.1　Introduction

In general, no attempt has been made in Chapters **4–7** to account for species distributions except in a very general way. The aim of this chapter is to summarise important spatial and temporal patterns of species and assemblage distributions based on field surveys and to comment on possible ecological controls.

8.2　Major new findings

There is increasing awareness that, unlike physical or chemical systems, biological systems are not well ordered. This is undoubtedly due to their complexity and to the multiplicity of interactions that affect living organisms. From the niche concept we can understand that at any one locality it is the parameter (or parameters) close to the limit of tolerance that limit the local distribution of a species; it follows that *for any one species different parameters limit its distribution at different times and in different places.* Benthic foraminifera have been shown to be more mobile than we previously thought and even supposedly attached epifaunal forms can move around (see Section 3.2.8). Infaunal taxa do not show consistent depth stratification either spatially or temporally because they respond to local geochemical conditions in the sediment. Foraminifera are more tolerant of oxygen depletion than most macrofauna and other meiofauna and some hard-shelled taxa can tolerate periods of anoxia of at least two months. There is a general correlation between the availability of food (flux of C_{org}) and species abundance, at least at water depths greater than about 1000 m. However, it has also been shown that the type and quality of food is important and that taxa show resource partitioning (i.e., different species living together feed on different components of the food supply). This means that food supply is not a single factor but a complex of factors (type, quality, quantity). Seasonal inputs of food (phytodetritus) into the deep ocean are followed by an increase in opportunistic taxa and thus the deep sea shows seasonality, which was an unexpected discovery. The role of foraminifera in recycling carbon has been neglected in the past but is now being determined. Molecular genetics is impacting ecological interpretations by demonstrating the coherence of some morphospecies and the existence of cryptic species in others.

8.3   Response of foraminifera to environmental factors

The response of individuals to specific environmental factors may be determined through experiment and, for species and assemblages, inferred from field studies. Although some progress has been made, in reality very little is known in detail as the response to one parameter may also be influenced by another. The selection of those environmental parameters thought to be major controls on distributions is subject to fashion. Water depth, temperature, salinity, food supply and availability of oxygen have all been in vogue. Yet, in most studies only some of these are measured and others are estimated based on uncertain data. Measurements made only at the time of sampling do not provide information on diurnal or longer timescales. There has been a tendency to assume that foraminifera respond immediately to a change in an environmental parameter. For survival it is much more likely that there is both a lag in response and wide tolerance of variability unless the environmental change is catastrophic and fatal. For life-threatening parameters, it is only when a value approaches or exceeds the tolerance limit for a particular species that it has major impact. Field studies allow correlation between distributions of species and environmental parameters but this is always *circumstantial* even if backed up with statistical analysis, i.e., there can never be *proof* of a correlation. For these reasons, field studies from different areas sometimes provide conflicting interpretations of species – parameter relationships. Therefore, the notes below are brief (see also Section 3.3.3).

8.3.1   Abiotic factors

Salinity. Salinity influences organisms primarily through osmosis. However, the response to a change in salinity may depend on the temperature since osmoregulation, water density, viscosity and ability to dissolve oxygen are all influenced by both factors. Below a salinity of 10–12 a cell may cease to function unless the organism maintains a higher internal ionic composition through osmoregulation (McLusky and Elliott, 2004). There is sometimes a disparity between experimental results and field observations. For instance, in experiments, feeding by *Elphidium crispum* slowed down at brackish salinities but returned to normal when the individuals were placed in seawater; and survival at salinity 15 was aided by low temperature. Although *Elphidium crispum* can withstand salinity 15–40 under experimental conditions (Myers, 1943; Murray, 1963) it never occurs in estuaries and appears to be stenohaline. The range of salinity tolerated by foraminifera is from 0 to ~70. The majority of species and highest species diversity are found at normal salinities (32–37). There are progressively fewer species and lower species diversity towards the extreme salinity values.

Temperature.   The major distribution patterns of marginal marine and inner-shelf foraminifera are clearly controlled by temperature. Some species have minimum thresholds for reproduction (especially larger foraminifera) and all physiological processes are to some degree influenced by temperature. Nevertheless, some species are able to tolerate extreme cold and are adversely affected by higher temperatures. The maximum temperature for survival is ~45 °C. The lowest temperatures under aqueous conditions are found in high-latitude deep ocean (−1.9 °C). However, brackish marsh foraminifera at high latitudes must experience severe freezing during the winter (down to at least −10 °C).

Oxygen.   The solubility of oxygen in sea water is low and varies with temperature and salinity. At atmospheric pressure, the solubility at 20 °C and salinity 31 is 241 μM whereas for freshwater the value is 291 μM and 456 μM at 0 °C (Fenchel and Finlay, 1995). For *Ammonia* group, oxygen <54 μM slows the metabolic rate as the organism struggles to cope with the adverse conditions (Bradshaw, 1968). Mesocosm experiments show that foraminifera placed in very low oxygen conditions migrate to higher values (Alve and Bernhard, 1995; Langezaal *et al.*, 2003). Some hard-shelled forms tolerated anoxia for two months under experimental conditions but the authors speculate that prolonged anoxia, around six months, would be fatal (Moodley *et al.*, 1997). On the assumption that the response of a species to an environmental gradient is a bell-shaped curve, a mathematical model has been devised that attempts to quantify this (Sjoerdsma and van der Zwaan, 1992). The model has been tested on the Parker (1954) total assemblage dataset on the Gulf of Mexico for oxygen and organic flux but still awaits a test on data on living forms. Pores in hyaline walls are traversed by an organic membrane so they do not give open communication between the interior and exterior of the test. From experiments using labelled CO_2, it was determined that the membrane is pervious to dissolved gases and that the pores are conduits for gas exchange between the cell-body cytoplasm and the ambient environment (Leutenegger and Hansen, 1979). In *Ammonia 'beccarii'* grown under experimental dysaerobic conditions, the porosity of the chamber wall is higher than that of chambers grown under high oxygen conditions (Moodley and Hess, 1992) perhaps providing supporting evidence for the role of pores in gaseous exchange. Infaunal foraminifera commonly live under reduced oxygen conditions. It has been shown experimentally that shelf forms from the North Sea *(Eggerelloides scaber, Quinqueloculina seminulum, Ammonia* group, *Elphidium excavatum)* can survive anoxia for 24 hours and can feed and grow at oxygen concentrations of <12 µM (dysoxic). They can also tolerate high oxygen levels (225 µM; Moodley and Hess, 1992). Under anoxic experimental conditions, although several species survived 21 days exposure to H_2S, prolonged exposure led to death. Therefore H_2S may be a significant environmental control (Moodley *et al.*, 1998). The key aspect of oxygen is that it is limiting only when close to the limit of tolerance for a given species. Once above that level, it is no longer limiting and can be ignored.

Tides and currents. Tides introduce physical energy into environments and mesotidal and macrotidal conditions create powerful tidal currents. In marginal marine environments there are commonly large diurnal tidally induced variations in salinity but because the waters are constantly moving the diurnal temperature range is small. *Haynesina germanica* is a species that seems to thrive under these conditions. Microtidal regimes are much less variable; the diurnal range of salinity is usually small and during the summer the surface-water layer can become much warmer than that below as little vertical mixing takes place. This may favour certain species such as *Ammotium cassis* and *Ammobaculites crassus/ exiguus.* Currents cause physical disturbance and under severe conditions may sweep away all loose sediment. The critical threshold for the disappearance of free living foraminifera is the onset of sediment mobility. The precise current speed at which this will happen depends on the grain size of the sediment. Current-swept areas have distinctive epifaunally attached assemblages.

Substrate. Hard surfaces (rocks, shells and other biological structures) provide a substrate for attachment of epifaunal and epiphytic taxa. In warm oligotrophic environments most benthic organisms living on hard substrates recycle their nutrients by endosymbiosis (Hottinger, 1984, 1990). Some algal substrates are available seasonally and this influences the timing and duration of the life cycle of those organisms that live epiphytically upon them whereas seagrasses tend to be more permanent and do not exert such a time control (Hottinger, 1990). Thus foraminifera living epiphytically on algae are small and have short life cycles. In the English Channel, most of the foraminiferal fauna is epifaunal on shells, hydroids, and other firm substrates that are less affected by the mobility of the sediment (Sturrock and Murray, 1981). Soft sediments provide a three-dimensional habitat for epifaunal and infaunal taxa in a wide range of environments. Although in a general sense there may be more organic matter in fine-grained sediment than in sand this is not necessarily a reliable guide to the availability of food. For instance, sands with low TOC on the Danish slope of the Skagerrak, North Sea, have high standing crops due to high fertility (Alve and Murray, 1997).

8.3.2   Biotic factors

Competition and predation. Little is known about these topics (see Section 3.2.13). Whereas an individual may respond in a positive or negative way to other environmental variables, there can be no positive response to predation. There is no tolerance threshold; an individual is either eaten or it is not. Although predation undoubtedly limits numbers it cannot limit species distributions. It is assumed by some authors (e.g., Jorissen, 1999) that an infaunal mode of life is favourable for foraminifera because it reduces potential predation pressure but this has never been investigated. Since a large number of macro-faunal organisms are infaunal unselective deposit feeders it may be that infaunal foraminifera are exposed to increased rather than reduced predation pressure.

Food supply. Except for those species that feed on live food or have endosymbionts, labile organic matter and the bacteria feeding on it are the primary sources of food. Because of the individual preferences of each species (resource partitioning), each needs its own unique measure of food supply but this has so far never been achieved. In those studies where an attempt is made to quantify food it is commonly expressed in general terms (g C_{org}). In marginal marine and shelf environments correlation between food supply and for-aminiferal abundance is complicated by the covariance of organic carbon with other parameters such as sediment grain size and geochemistry but for the deep sea the situation is clearer as the low supply of food makes it a primary limiting factor (see Section 7.2.2). However, in some shallow-water studies there is a clear correlation between an algal bloom and an increase in standing crop (e.g., in a fjord, Gustafsson and Nordberg, 1999).

8.3.3   Combined abiotic and biotic factors

Disturbance. All environments are dynamic and disturbance is brought about by physical factors such as waves and tidal currents and through the activities of other organisms (locomotion, feeding, burrowing, etc.; see Section 3.2.8). This may generate local microenvironments with distinctive geochemistry (e.g., around burrows). Even in the deep sea disturbance is considered to be an important feature of the environment and may be responsible for increasing diversity.

8.4   Microdistribution patterns

Theoretically there are three main distribution patterns: random, uniform, and aggregated (clumped). In most cases individuals are not uniformly distributed but aggregated into patches and this occurs on a variety of scales (Valiela, 1995). The distributions of taxa that live epifaunally attached to substrates raised above the sea floor are always clumped as they are confined to suitable substrates. In a mesocosm experiment, *Pelosina* cf. *arborescens*, epifaunal and anchored in the sediment, was aggregated except when abundance was low and then it was random (Gamito *et al.*, 1988). Both spatial distribution and the relative and absolute abundances of organisms are affected by this patchiness. Foraminifera show an aggregated pattern because of the effects of microenvironments and reproduction. These microdistribution patterns are three dimensional in space (laterally and vertically in the sediment and also on substrates elevated above the sea floor) and are also variable temporally. Buzas *et al.* (2002) call them 'pulsating patches' and consider them to be asynchronous or aperiodic as reproduction occurs at different times rather than being synchronised. All species show dynamic change both temporally and spatially and in the case of rare species their patches are unlikely to remain in the same place through time. Thus, with reference to a fixed spot on the sea floor, it is not surprising that through time some species will be ephemeral in their occurrence (Murray, 2003a).

With few exceptions (Buzas, 1968b, 1970; Murray and Alve, 2000a) investigations of living assemblages have not considered possible impacts of patchiness because they have not analysed replicate samples. There is an increase in aggregation as the abundance of individuals increases (Buzas, 1968b) and also as the size of the study area increases (Buzas, 1970). In Rehobeth Bay, USA, the three dominant forms, *Ammonia* group, *Elphidium clavatum* and *Haynesina germanica* (as *Elphidium tisburyenis*) all show heterogeneous distributions with no simple pattern of low or high abundance. Yet the assemblage as a whole is homogeneous over an area of $1500\,m^2$ suggesting that the abundance of one species compensates for that of another to give the most efficient use of habitat space (Buzas, 1970). In the intertidal zone of the Hamble estuary, England, patchiness is on a scale of a few centimetres (Murray and Alve,

2000a). The causes of spatial distribution patterns are complex and not fully understood. They include grazing and predation and these may be the main causes of pre-reproductive deaths (Murray and Bowser, 2000). Even birds foraging for food in intertidal sediments may disturb and mix sediment with their feet and their beaks.

Rose Bengal-stained benthic foraminifera have been recorded down to at least 60 cm in marsh environments (Section 4.2.3) perhaps due to open burrows (Goldstein et al., 1995). Indeed, burrows provide particularly favourable microenvironments because they are often richer in food than the surrounding sediment (Thomsen and Altenbach, 1993). Foraminifera are mobile and can migrate through the sediment in search of food or more favourable oxygen conditions (e.g., Barmawidjaja et al., 1992; Jorissen et al., 1992; Linke and Lutze, 1993; Alve and Bernhard, 1995) so the distinction between infaunal and epifaunal is to some extent arbitrary. Certain taxa even live in anoxic sediments and may be facultative anaerobes (e.g., Bernhard, 1989, 1992; Bernhard and Reimers, 1991; Moodley and Hess, 1992; Moodley et al., 1998a).

Typically the highest numbers are found in the surface 0–1 cm layer both in shallow water and the deep sea and in oligotrophic and eutrophic settings (e.g., Antony, 1980; Gooday, 1986a; Mackensen and Douglas, 1989; Bernhard and Reimers, 1991; Barmawidjaja et al., 1992; Murray, 1992; Rathburn and Corliss, 1994; Bernhard and Alve, 1996; Hannah and Rogerson, 1997; Castignetti and Manley, 1998; de Stigter et al., 1998; 1999; Jorissen et al., 1998; Fontanier et al., 2002) and often in the top few millimetres (Murray and Alve, 2000a; Alve and Murray, 2001). There is generally no clear pattern of species distribution with respect to depth in the sediment from one area to another. Several authors have pointed to the control of the redox boundary (availability of oxygen) as the lower limit of life in the sediment (e.g., Richter, 1964a; Shirayama, 1984b; Ohga and Kitazato, 1997; Langer et al., 1989) and there are examples of foraminifera living deeper than the redox boundary. However, Rathburn and Corliss (1994) specifically stated that there is no control on depth distribution in the Sulu Sea and according to Jorissen et al. (1998) 'foraminiferal microhabitats are only indirectly controlled by pore water oxygen concentrations'. Others have pointed to the availability of food as the principal control (Corliss and Emerson, 1990; Jorissen et al., 1992; Hohenegger et al., 1993; Linke and Lutze, 1993; Wollenburg and Mackensen, 1998a).

Thus the interpretations sometimes appear contradictory. There is certainly no global pattern of a specific taxon occupying one particular subsurface zone. The reasons for this are partly explained by several authors as can be seen from these quotations: Buzas et al. (1993): 'Because each core captures a variety of individuals found together in one instant of time, we are simply observing a transient phenomenon of a group of animals with a three dimensional aggregated spatial distribution'; de Stigter *et al.* (1998): 'microhabitats of benthic foraminiferal species are not static, but may be highly variable from place to place, as determined by local environmental conditions'; Jorissen *et al.* (1998): 'The vertical zonation of foraminiferal species in the sediment shows a close correspondence with the depth distribution of oxic respiration, nitrate and sulphate reduction'. However, Jorissen (1999) concludes that although these local patterns can be recognised there is no universal strict vertical stratification. Previously, some authors considered that different morphotypes live at different levels in the sediment (Corliss, 1985, 1991; Corliss and Chen, 1988; Corliss and Emerson, 1990; Corliss and Fois, 1990; Rosoff and Corliss, 1992) but de Stigter *et al.* (1998) point out that the 'so-called epifaunal morphotypes may also be common in subsurface microhabitats' and they caution against over simplistic interpretations. Jorissen (1999) points to the significant exceptions to the proposed correlation between morphotypes and microhabitat and van der Zwaan *et al.* (1999) consider the morphotypes concept to be meaningless on a global scale.

8.5   Interdependent controls on microdistributions

Corliss and Emerson (1990) linked the interplay of food and oxygen as principal controls on infaunal-species microdistributions and Jorissen *et al.* (1995) formalised this into the TROX (trophic conditions and oxygen concentrations) conceptual model. The TROX-2 model was introduced to take into account other geochemical changes and the role of competition for labile organic matter (van der Zwaan *et al.*, 1999) (Figure 8.1A).

The *x*-axis represents three parameters: nutrient levels (oligotrophic, mesotrophic, eutrophic), food supply and oxygen. Nutrients control phytoplankton primary production in the ocean and benthic primary production in areas shallower than the euphotic zone. The flux of food to the sea floor controls the numbers of individuals. The nature of the food controls the presence or absence of species (at least for those that are not omnivores). The *y*-axis represents depth in the sediment and a range of bacterially moderated biogeochemical boundaries (oxic/suboxic/anoxic, nitrate, Fe^{4+} and Mn^{3+} reduction, sulphate reduction and methanogenesis). The essence of the models is that infaunal species are shallow both in oligotrophic environments (due to the limiting control of food) and in eutrophic environments (due to the restricted availability of oxygen). Oligotrophic examples are the deep sea, especially the lower bathyal and abyssal zones, and shallow tropical reefs. In the deep sea, the foraminifera are predominantly epifaunal or very shallow

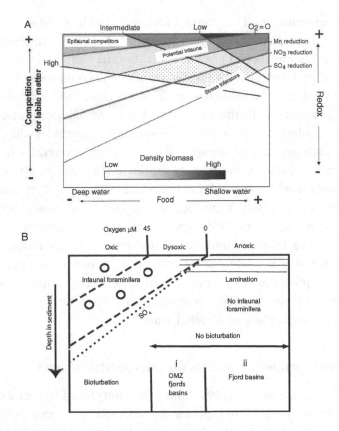

Figure 8.1. A. The TROX-2 model of van der Zwaan *et al.* (1999). B. The TROX upper-right corner. Black stars indicate infaunal foraminifera. Part A is reprinted from *Earth Science Reviews*, 46, Van der Zwaan *et al.*, Benthic foraminifera: proxies or problems? A review of paleoecological concepts. p. 231, Copyright (1999), with permission from Elsevier.

infaunal. Even when exceptional inputs of food arrive (such as phytodetritus falls) epifaunal taxa predominate. In reef environments, most of the foraminifera are epifaunal because they have endosymbionts and therefore require light. In mesotrophic environments the infauna may penetrate deeply into the sediments due to the availability of both food and oxygen and this is typical of most shelf and upper-slope environments.

The most important part of the diagram is the right-hand side (see Figure 8.1B). Oxygen is a controlling factor only where it is absent or in limited supply: anoxic bottom water in contact with the sediment surface has no fauna (i); where there is dysoxic bottom water, the redox boundary is close to the sediment surface, the sediments may have a surface mat of *Beggiatoa* and they may be laminated; only certain species can survive (ii). Where there is oxic bottom water and an oxic surface layer of sediment oxygen is no longer a limiting factor (threshold varies between species but is always a value lower than $1\,\mathrm{ml}\,\mathrm{l}^{-1} = 45$ μM). All the oxic environments are bioturbated by macrofauna. Therefore, instead of there being a regular stratification of geochemical boundaries they are distorted and diffuse. The depth of life in the sediment is controlled by microhabitats created by bioturbation and the availability of suitable food. Foraminifera may respond to any of the geochemical boundaries by feeding on the bacteria or by migrating to a different level in search of more favourable conditions. There is abundant evidence that the depths of the biogeochemical boundaries in the sediment change with time/season. Indeed, in some settings the suboxic/anoxic boundary temporarily migrates out of the sediment into the lower part of the water column and the infaunal foraminifera move with it to become epifaunal on raised structures. There are commonly seasonal changes in nutrients and flux of food and this also affects microdistributions. Low-oxygen foraminiferal assemblages are generally dominated by perforate hyaline taxa although sometimes agglutinated forms are also common (Sen Gupta and Machain-Castillo, 1993; Bernhard and Sen Gupta, 1999). Some species tolerant of low oxygen are also tolerant of hydrogen sulphide (*Bulimina mexicana, Epistominella smithi, Globobulimina* spp., *Gyroidina altiformis, Uvigerina peregrina*; Rathburn *et al.*, 2003). However, experiments indicate that although there may be short-term tolerance (21 days), longer exposure (66 days) leads to death. The presence of sulphide inhibits reproduction and thus controls distribution patterns (Moodley *et al.* 1998). There are no species confined to low-oxygen conditions as they all also occur in oxic environments.

8.6   Time-series datasets and environmental variability

The majority of ecological studies are spatial and based on a single sampling survey but a few are temporal (time series) based on samples collected regularly, often monthly, over a period of at least one year. The advantages of time-series data are that temporal variability can be examined on a variety of scales: month to month, seasonally, one year to the next. Because of patchiness ideally replicates should be taken on each sampling occasion so that error bars may be fitted to the data and a statistically meaningful analysis carried out. However, replicates are rarely taken because gathering time-series data is very time-consuming for the observer. For instance, in the study of two intertidal stations in the Hamble estuary, replicate cores were taken at each and sectioned into seven slices (2.5 mm thick from 0–1 cm, then 1 cm thick down to 3–4 cm). This yielded 28 samples each month (Murray and Alve, 2000a; Alve and Murray, 2001).

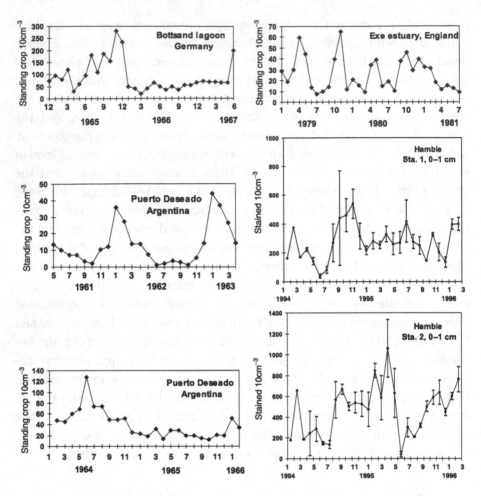

Figure 8.2. Time-series data on standing crop (with source of data). Bottsand lagoon, Germany (Lutze, 1968). Puerto Deseado (Boltovskoy, 1964; Boltovskoy and Lena, 1969a). Exe estuary, England (Murray, 1983). Hamble (Murray and Alve, 2000a).

Because there is both temporal and spatial variability, all time-series records are 'noisy' (see Figure 8.2). Whereas it might be expected that there would be considerable variations from month to month due to seasonal differences, if a repeated annual cycle is present there should be similarity between the annual means (Murray, 2000b). Of the examples studied, only the Exe and Hamble sta. 1 have statistical similarity between consecutive years. Apart from differences in the standing crop of the whole assemblage, individual species show variability that may or may not follow the pattern of the assemblage. In Puerto Deseado, *Elphidium macellum* peaked in the summer in 1962 and 1963 but also in the winter in 1964. At the intertidal Bahrain site, one dominant species (*Ammonia 'beccarii'*) had similar numbers throughout the study while *Elphidium advenum* increased fourfold; two initially minor species, *Brizalina pacifica* and *Nonion* sp., underwent a dramatic increase in the later part of the survey (Basson and Murray, 1995). Is that natural variability or does it represent an environmental change? The answer remains unknown.

Time-series data can also be used to analyse patterns of species diversity. Apart from the dominant species that may be present on every sampling occasion, time-series sample chance encounters (Lazarus occurrences) of rare species continuously present in low abundance, and also arrivals of new species and disappearances. There is a cumulative increase in the number of species through time. Using SHE analysis (see Section 2.11) the time series from the two Hamble stations and that from Bahrain show a log series distribution while that of the Exe estuary does not show any pattern (Murray, 2003a).

8.7   Large-scale distribution patterns

8.7.1   Common species

In this book, assemblages have been named from the dominant species. Considering the large number of modern benthic foraminifera (several thousand species) there are remarkably few species that are dominant in assemblages. By far the majority of modern species occur in low abundance throughout their range. Even for those species that are common locally, they are also rare over a much greater distribution field. In the compilations of data on species ecology (Appendix), the comments apply to the common occurrences. Low abundances of these species occur under less restricted conditions.

8.7.2   Test wall

The functions of the test are discussed in Section 3.2.15. Conventionally, agglutinated walls are regarded as more primitive than porcelaneous or hyaline, the latter being regarded as most advanced especially where canal systems are developed. Since there is an environmental pattern in the distribution of wall types this must be of some ecological significance. All walls contain organic material either as cement (agglutinated) or as a template (calcareous). Only hyaline walls are perforate and although the pores are crossed by an organic membrane this nevertheless allows exchange of gas between the foraminiferal cell and the ambient water. However, mitochondrial activity in pseudopodia may be more important than that in pore plugs (Bernhard and Sen Gupta, 1999) so lack of pores in agglutinated and porcelaneous walls may not be a disadvantage. Organic-cemented agglutinated walls can be deposited in any water regardless of corrosivity with respect to

CaCO$_3$. They therefore inhabit the widest range of environments from high intertidal to the greatest hadal depths and from almost freshwater to hypersaline. Although calcareous walls can be secreted in corrosive water the majority of species live satisfactorily only above the CCD although some live between the lysocline and CCD (*Nuttallides umboniferus*). Porcelaneous walls are best developed in normal marine and hypersaline environments, mainly less than 100 m deep in temperate and tropical climates. However, there are a few deep-sea forms and some that live in subarctic waters.

On ternary plots summarising the results from each environment, marshes occupy the whole triangle whereas all other environments have more restricted distributions. Exclusively brackish marshes plot along the hyaline–agglutinated axis, but most marshes have very variable salinities especially in the summer and this permits the growth of small miliolids.

Marginal marine environments (estuaries, lagoons, fjords, deltas but excluding marshes) plot either along the hyaline–agglutinated axis (brackish) or along the hyaline–porcelaneous axis (normal marine and hypersaline; Figure 8.3A). Shelf seas show a similar pattern except that most examples plot in the lower part of the diagram (Figure 8.3B). Deep-sea assemblages plot along the hyaline–agglutinated axis generally with <10% porcelaneous walls although there are some exceptions (Figure 8.3C). Thus, the addition of new data has somewhat broadened the original fields defined by Murray (1973, 1991).

8.8   Species diversity

For benthic foraminifera, the problems of determining the processes that influence diversity are compounded by the inconsistent use of size fractions, inclusion or exclusion of certain groups (soft-walled forms, xenophyophores, etc.) and the relatively small amount of information, which is mainly from the margins of the Atlantic Ocean. Diversity can be viewed on a local scale (within-patch diversity or alpha diversity; this should not be confused with the Fisher alpha index) and regional (gamma) diversity based on pooling data across a region. In the case of the foraminiferal data examined here, we are concerned primarily with patch-scale diversity.

The basic pattern demonstrated by Murray (1973, 1991) is confirmed by the ~5000 analyses presented here (Table 8.1, Figure 8.4). All marginal marine environments and brackish shelf seas (Baltic) have low diversity with the majority of values of Fisher alpha <4 except in normal marine and hypersaline lagoons (< 6). The majority of shelf and deep-sea assemblages have Fisher alpha 5–15 or 5–20 respectively although there are some examples ranging lower and higher. For the information function, H, there is a similar

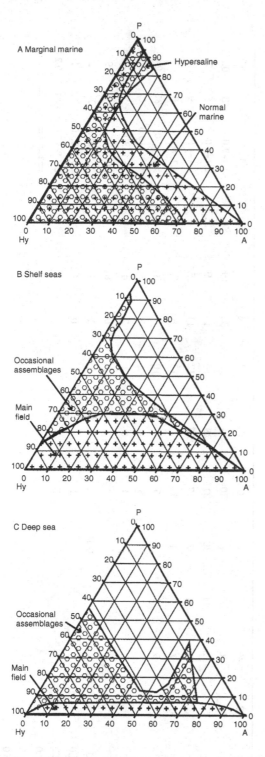

Figure 8.3. Summary of ternary plots of walls (traingle corners represent 100% of the labelled component: A = agglutianted, P = porcelaneous, Hy = hyaline). As marshes occupy the whole field they are not plotted. A. Marginal marine environments (excluding marshes). B. Shelf seas (excluding phytal assemblages in the Baltic). C. Deep sea.

Table 8.1. Summary of species diversity: Fisher alpha, $H \geq 10\%$ shaded.

Fisher alpha	Brackish marsh	Mangal	Normal marine marsh	Brackish estuary lagoon	Normal marine estuary lagoon	Brackish to hypersaline lagoon	Hypersaline lagoon	Fjord	Delta	Brackish shelf
n	788	60	24	783	539	76	111	253	213	187
Mean	1.1	1.5	1.6	1.7	3.9	2.3	2.8	2.4	4.6	1.3
St. dev.	0.7	1.4	1.1	1.0	2.6	1.4	2.1	1.9	3.5	0.8
	%	%	%	%	%	%	%	%	%	%
<1	63	47	42	32	6	13	23	24	8	54
1.1–2.0	29	30	25	40	20	34	28	33	19	26
2.1–3.0	5	12	21	20	21	29	10	16	19	16
3.1–4.0	3	0	8	7	17	16	12	11	15	4
4.1–5.0	0	7	0	1	13	3	10	6	8	1
5.1–6.0	0	5	4	1	6	3	10	4	3	
6.1–7.0				0.1	5	0	2	2	3	
7.1–8.0					4	3	2	2	4	
8.1–9.0					3		3	0	8	
9.1–10.0				0.1	1		1	1	4	
10.1–11.0					1			0	4	
11.1–12.0					1				2	
12.1–13.0					1				3	
13.1–14.0					0.4				0	
14.1–15.0					0.2				1	
15.1–16.0									0.0	
16.1–17.0									0.5	

Fisher alpha	Shelf	Bathyal
n	1272	543
Mean	8.6	12.3
St. dev.	4.9	6.2
	%	%
<1	1	0.2
1.1–2.0	3	0.2
2.1–3.0	6	2
3.1–4.0	9	4
4.1–5.0	8	7
5.1–10.0	37	24
10.1–15.0	27	30
15.1–20.0	7	21
20.1–25.0	1	9
25.1–30.0	0.4	2
30.1–35.0	0.2	1
35.1–40.0	0.1	0.2

Table 8.1. (cont.)

H	Brackish marsh	Mangal	Normal marine marsh	Brackish estuary lagoon	Normal marine estuary lagoon	Brackish to hypersaline lagoon	Hypersaline lagoon	Fjord	Delta	Brackish shelf	Shelf	Bathyal
n	788	60	24	783	539	76	111	253	213	187	1272	543
Mean	1.10	1.17	1.28	1.08	1.66	1.42	1.43	1.39	1.81	0.91	2.37	2.68
St. dev.	0.49	0.62	0.53	0.47	0.59	0.43	0.71	0.58	0.74	0.49	0.63	0.64
	%	%	%	%	%	%	%	%	%	%	%	%
< 0.50	13	12	4	13	1	1	12	7	4	24	0.3	0.2
0.51–1.00	28	35	33	26	12	12	20	19	8	42	3	1
1.01–1.50	36	23	29	42	25	47	22	30	23	18	7	6
1.51–2.00	20	18	25	17	33	34	25	28	25	14	17	8
2.01–2.50	3	12	4	1	18	3	16	14	16	2	26	18
2.51–3.00			4		9	3	5	3	21		33	28
3.01–3.50					0.4				3		13	35
3.51–4.00											1	3
4.01–4.50											0.1	

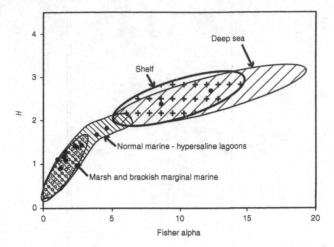

Figure 8.4. Summary of species-diversity data showing the fields for the main values (excluding Radford, 1976a,b, data, due to oversplitting of species). A small number of assemblages plot outside these fields (see Table 8.1).

pattern with the majority of values <2.5 in the marginal marine environments while in shelf and deep seas they range from 1.51–3.50 with a few assemblages ranging even higher. Thus at patch scale the diversity of the deep sea overlaps that of shelf seas and ranges only slightly higher. This means that although regional diversity may be high, at patch level each foraminiferal assemblage includes only a tiny subset of the many rare species (i.e., the subset is different from one sample to another).

8.9   Colonisation/recolonisation

The widespread distribution of benthic foraminifera in environments from freshwater (naked forms only) to high intertidal to deepest hadal indicates that colonisation actively takes place. Also, dispersal is important for maintaining species diversity. The processes of colonisation can be deduced from field observations and experiments.

Field evidence of colonisation can be obtained from environments that experience seasonal variation and also from areas that have undergone environmental change. In seasonally variable environments such as estuaries, although the dominant species may be present throughout the year some additional species are ephemeral and appear for only part of the year. In a time-series study of two intertidal mudflat sites in the Hamble estuary, England, ephemeral species included forms sourced from nearby marshes and also from the nearby marine shelf (Alve and Murray, 2001). Although the number of species ranged from 5 to 22 per sample, a total of 35 species from the two stations was recorded over the 27-month survey period. A progressive environmental change has taken place in the Skagerrak, North Sea, over the past six decades. Since 1937 *Trochamminopsis pusillus* and *Saccammina* spp. have entered the area and become common (Alve and Murray, 1995a). An abrupt environmental change takes place when an ash fall from a volcano is deposited on the sea floor over a very short time. Under such conditions, the pre-existing fauna will either be severely stressed or completely killed off. In this case there are new opportunities for colonisation. The 1967–70 eruptions of Deception Island, Antarctica, killed off much of the fauna. By 1975 a new foraminiferal fauna had become patchily established but this was different from that previously described from the area (Finger and Lipps, 1981). Opportunistic *Stainforthia fusiformis* was an early colonising species even though it had not been recorded there by Earland (1934) prior to the volcanic activity. The 1991 eruption of Mount Pinatubo in the South China Sea buried the fauna to a depth of 2–6 cm. The earliest coloniser was considered to be *Textularia* followed by *Reophax* and *Quinqueloculina* by 1994 (Hess and Kuhnt, 1996). In these two cases, even three to five years after the catastrophic event that caused mass mortality of the foraminifera, the recovery process was still only in its early stages. A freshly deposited turbidite in a submarine canyon was within four months being colonised by *Technitella melo* accompanied by *Stainforthia concava, Bulimina marginata* and *Brizalina subaenariensis* (Anschutz *et al.*, 2002). A reef flat destroyed by a hurricane was recolonised probably in less than 10 years (Collen, 1996).

Experiments to determine colonisation must be carefully designed to be representative. One obvious limitation is the isolation of the experimental sediment substrate from adjacent sediment by the sides and floor of the container in which it is housed. Some studies have used a limited grain size (e.g., sand) while others have used marine sediment that has been treated to kill off the existing fauna (azoic or defaunated) (see review by Alve, 1999). In shallow water, colonisation of azoic sediments occurs within days or weeks (Schafer, 1976; Ellison and Peck, 1983; Buzas *et al.*, 1989). Alve considers that this is due to the easy physical transfer of individuals from the surrounding sediments in these medium-energy environments. She therefore undertook an experiment in a low-energy fjord in which the experimental azoic substrates were isolated from the surrounding sediments. Opportunistic *Stainforthia fusiformis* was an early coloniser and tubular taxa were late immigrants. Even after seven and a half months the fauna as a whole was not comparable with that of the surrounding sediments. The process also operates slowly in the deep sea as shown by experiments carried out at 3900 m (Panama Basin; Kaminski *et al.*, 1988) and 1445 m (Sagami Bay; Japan, Kitazato, 1995). In both cases the colonisation took place over nine to twelve months. Recolonisation of previously anoxic environments (Alve, 1994), severely polluted sediments (Cato *et al.*, 1980) or dredge spoil (Schafer, 1982) likewise may take one or more years. Under natural conditions, a newly deposited volcanic ash lacks nutrients and food suitable for foraminifera so the recolonisation process cannot get properly under way until food is available. In the same way, experiments using pure sand (Buzas, 1993) or glass beads (Kitazato, 1995) have shown a delay in colonization and a slow increase in standing crop levels. If the species are introduced onto substrates that are unoccupied (e.g., newly deposited volcanic ash) the opportunists are the early colonisers and it may take many years before a more diverse fauna is fully established. Alve (1999) suggests that the rate of colonisation depends on the energy of the environment, with faster speeds at higher energy. Surprisingly, sediments containing significant amounts of $CuCl_2$ were colonised with the ambient fauna after only 32 weeks exposure in Oslo Fjord (Alve and Olsgard, 1999). Laboratory-tank experiments to investigate colonisation of sterile substrates indicate that *Ammonia* group, *Buccella frigida*, *Elphidium clavatum* and *Haynesina orbiculare* are active colonisers in water of 6 and 16 m depth. *Quinqueloculina seminulum* is also active in the 6 m experiment and *Saccammina atlantica* in the 16 m experiment based on communities from off Nova Scotia, Canada (Schafer and Young, 1977).

Hard substrates are also affected by colonisation and such substrates tend to be discontinuous and separated by sediment. Artificial substrates simulating seagrass were readily colonised under experimental conditions and the main control was seasonality rather than substrate type although there were minor differences in response to different substrates (Ribes *et al.*, 2000). Biotic substrates such as shells and hydroids are colonised although there seems to be a minimum size for epifaunal organisms to settle successfully. Similarly, it is generally large pebbles and rocks that are affected (e.g., Gulf of Cadiz; Schönfeld, 2002). In a field experiment, manganese nodules at 1240 m were colonised after seven weeks (Mullineaux, 1988a). Basalt plates were colonised (Mullineaux *et al.*, 1988) in hydrothermal-vent communities at 2505 m on the East Pacific Rise with especially high standing crops in the cool area. Indeed, foraminifera are very successful colonisers near hydrothermal vents (van Dover *et al.*, 1988). Experiments at depths of 800–1985 m on a seamount showed rapid colonisation of plates in 1989, at a time of high organic flux, and slow colonisation in 1991 and 1993 under low organic flux (Bertram and Cowen, 1999).

These examples show that the processes of colonisation operate continuously and they affect already populated as well as new, unpopulated, substrates. If the introduced species are able to compete with the existing fauna then they become established and this can lead to an increase in species diversity. If they are unable to compete, or if conditions turn adverse after a short time, they do not become established. Nevertheless, their dead tests are added to the dead assemblages and this partly explains why dead assemblages are commonly more diverse than the living assemblage at the same site. Although colonisation of new substrates may take a few years, it is instantaneous on a geological timescale.

8.10   The species pool

Syntheses of ecological data show that certain associations of species characterise different environments (Murray, 1991) and different biogeographic provinces (e.g., Culver and Buzas, 1999). However, in reality, apart from the dominant species that define an association, the subsidiary species commonly differ from one sample to another within the same area. In an analysis of faunal changes resulting from successive transgressions and regressions on the continental shelf off North America, it was suggested that there is a neritic species pool: 'Immigrants and emigrants shuffled back and forth to the species pool while extinctions and originations continually altered its composition' (Buzas and Culver, 1994, p. 1441). This may be a rather simplistic view. Dispersal of species, perhaps through propagules (Alve and Goldstein, 2002), must lead to continuous potential colonisation of all environments by new species. However, those species that can no longer be successful in a changing environment probably die *in situ* rather than retreating to the species pool because there is no single location housing all species (Murray, 2003a). In modern settings the species pool can be considered to be the sum of the species available to colonise a broad environment. The species pool of an estuary could be regarded as all the species occurring in the various subenvironments. It follows that no one subenvironment would have all the species present in the species pool. From an analysis of the changes through time, it appears that many species behave on an individual basis to changing environmental conditions (Buzas and Culver, 1994). Indeed, it may be that each individual foraminiferan responds independently. On a local scale individual species distributions are pulsating patches of varying size, which show changing degrees of continuity or discontinuity and constantly changing degrees of overlap through time (Buzas *et al.*, 2002). Small-scale immigration and emigration must take place continuously to bring about those changes.

8.11   Biogeography

Biogeography refers to global patterns of species distribution that have developed as a consequence of evolutionary, geological and ecological processes. The term province is used to group faunas having similarity over a connected geographic area. A species is endemic if confined to one province and pandemic (cosmopolitan, ubiquitous) if present in several. A species that is abundant is likely to have a widespread distribution (although in theory it could be restricted to a few localities) but a species that is rare is more likely to have a restricted distribution. However, there are commonly millions of foraminifera living in a few km^2 of sea floor so that even a rare species occurring as one to two individuals in a few cm^3 of sediment may have a density of 10^{10} over a few km^2 (Buzas and Culver, 2001). Also, many rare deep-sea species are cosmopolitan.

It has long been realised that shallow-water faunas (from marginal marine to outer shelf) are primarily controlled by water temperature and this can be seen from the biogeographic maps produced by Cushman (1948) and Boltovskoy and Wright (1976); however, the precise criteria for defining the limits of their provinces were not explained. During the 1970s and 1980s Culver and Buzas systematically explored the biogeography of the benthic foraminifera from the continental margins of Central and North America (Buzas and Culver, 1980; 1986; 1989; 1991; Culver and Buzas, 1981a,b). They defined their provinces on groups of communities associated in space and time based on cluster analysis of species presence/absence using live, dead and total assemblage data (having first established a single taxonomic scheme). In their review, Culver and Buzas (2000) point out that the shallow-water provinces are similar to those recognised for other organisms. They also found depth-related provinces and noted that the differences between shallow (< 200 m) and deep (>200 m) provinces at the same latitude were greater than between adjacent shallow-water provinces (Culver and Buzas 1981a,b; 1982) which is perhaps not surprising. The deep provinces are especially well developed on the wide Atlantic and Gulf of Mexico margins but less so on the narrow Pacific margin. It was suggested that the deep-related provinces might be controlled by water masses (Culver and Buzas, 1982) and even the shelf provinces seem related to separate water masses. Altogether, nine shallow and two deep provinces are recognised around Central and North America (Figure 8.5). The boundaries between the provinces are generally clearly defined except for those between the Arctic and its neighbouring Aleutian and Atlantic Northern inner-shelf provinces. Around most of the margins, a single province spans the width of the continental shelf. However, in addition, outer-shelf–upper-slope provinces occur in the Gulf of Mexico and the Atlantic margin north of Cape Hatteras (Culver and Buzas, 1981a,b; 1983; 2000). The provinces on the eastern seaboard are clearly recognisable on Figure 5.6.

Figure 8.5. Biogeographic provinces (delimited by thick lines) on the continental margins of Central and North America (after Culver and Buzas, 1999); inner shelf except in the Atlantic Northern and Gulf of Mexico where there are inner-shelf and outer-shelf-slope provinces.

The historical development of biogeographic distributions is partly controlled by changes in the shape of ocean basins arising from plate tectonics. It is also controlled by the evolution of new taxa. Some brackish-water agglutinated foraminifera that are common along the Atlantic seaboard of North America (e.g., *Ammotium salsum*) appear to be recent introductions into northwestern Europe (rare there compared with North America). On the other hand, *Ammotium cassis* is abundant on both sides of the ocean (Ellison and Murray, 1987), although it may be a recent introduction into the Baltic. Buzas and Culver (1986; 1989) found that modern foraminifera around North America originated in all latitudes of the global ocean over a time period extending from the Cretaceous. There is no geographic centre of origin from which species disperse and this is consistent with the concept of the species pool. It is common for rare species to be ignored in ecological studies as many authors disregard occurrences of less than 5%, 3%, etc. Yet it is the rare species that hold the key to different species diversities. Indeed, the majority of species diversity is composed of rare species, i.e., those occurring as one or a few individuals in an assemblage. The log series is one of several techniques that can be used to calculate the number of species occurring with just one individual (Buzas *et al.*, 1982; Hayek and Buzas, 1997; Buzas and Culver, 1999). Based on data available in 1991 for Central and North America, this ranges from 139 out of a total of 458 species (Arctic) to 301 out of 1188 (Caribbean) (each with different numbers of localities and occurrences; Buzas and Culver, 1991). Only 5% of the species (112) are pandemic whereas 57% (1326) are endemic. Because species diversity is lowest in the Arctic and highest in the Pacific and Caribbean, more endemic species are present in the latter two provinces. The provinces were grouped into five areas for further analysis. Each area has a log series species distribution and occurrence and therefore can be predicted by the single proportionality constant, α (not to be confused with Fisher alpha). For a database of 2673 localities most of the difference in species richness is accounted for by species occurring $\leqslant$10 times (rare); species occurring once account for 81% of the difference in species richness between the Caribbean and Atlantic (Buzas and Culver, 1999). Most rare species have no fossil record and probably evolved recently. Areas with higher species richness probably have fewer extinctions and/or more originations of species.

Less attention has been paid to bathyal and abyssal distributions. A popular view based on the results of the Challenger expedition (Brady, 1884) is that many deep-sea species are cosmopolitan. More recently, it has been shown that the larger agglutinated forms found in the western North Atlantic and the central North Pacific are very similar (Schröder *et al.*, 1988). A latitudinal diversity gradient exists in the bathyal Atlantic Ocean. Maximum values are found in low latitudes and decrease slightly from 30–70 °S while in the northern hemisphere it is variable from low latitudes until 65 °N where it drops (Culver and Buzas, 2000). The causes are not fully understood. This pattern has broadly existed for the past 36 my; Thomas and Gooday (1996) speculate that it might in part be productivity-related, especially to seasonal input of phytodetritus in high latitudes, but this is hard to understand.

8.12   Standing crop and biomass

The measured values depend on the type of sampler used and the size fractions examined as well as natural variation in these parameters. In all environments there are samples with few or no live individuals. At the other extreme, maximum values of standing crop may reflect a local bloom so should perhaps be treated with caution. The typical values cited here (Table 8.2) are remarkably similar across a wide range of environments – from a few tens to a few hundred individuals 10 cm^{-3}. However, there are some exceptions. Marshes in the UK and Atlantic USA have typical standing crops of up to 1000 and in Japan up to 5000 individuals 10 cm^{-3}. Likewise, in the deep sea off Japan typical values are 1000–3000 individuals 10 cm^{-3}. Some extreme habitats have very low values (<50, e.g., the ice-covered Arctic Ocean).

There is very little information on biomass and not all data refer to the whole assemblage. Most estimates are based on calculated volumes of tests. Altenbach and Struck (2001) consider that such estimates should be corrected

Table 8.2. Summary of standing-crop data (individuals $10\,cm^{-3}$ for the 0-1 cm sediment layer) based on the data in the Web-Appendix tables.

Environment	Maximum	Typical values
Marsh	11 000	50-200
Estuaries	3404	50-300
Deltas	8240	10-500
Fjords	2109	50-500
Shelf seas	4714	50-800
Deep sea	6661	10-200

by multiplying by 0.35 to adjust for the incomplete filling of tests by protoplasm. In the deep sea the flux rate of C_{org} and corrected benthic foraminiferal biomass show a significant positive correlation (Altenbach and Struck, 2001). Uncorrected values expressed as wet weight of organic matter g m^{-2} range from 0 to the following maxima in: estuaries, 7.6; hypersaline lagoon, 4.06; brackish Baltic, 16.31; normal marine shelves, 4.91 (Murray and Alve, 2000b); upper-bathyal deep sea, 8.5 (adjusted to an uncorrected value from Altenbach and Struck, 2001, Table 1). Considering all the uncertainties in the data, the range of maxima is remarkably small. Perhaps this is not surprising as the standing crop data discussed above also show similarities between different environments.

8.13   Role of benthic foraminifera in modern ecosystems

In spite of their abundance in modern sediments, benthic foraminifera are usually ignored by biologists investigating modern ecosystems. Thus, in their review of estuarine ecology McLusky and Elliott (2004) do not cite a single reference to the role of foraminifera in the recycling of C_{org}; yet protozoans in general are important grazers of biofilms in these environments (Patterson et al., 2004). It has been postulated that in the deep sea benthic foraminifera may be major elements in recycling C_{org} due to the fact that they account for ~50% of the eukaryotic biomass (Gooday et al., 1992a) and the same may be true for shallow water too (Moodley et al., 2000). This hypothesis remains to be tested.

8.14   Comments on other topics

Ecophenotypes.   It has long been known that environment influences test shape and that stress may induce deformity (see Section 10.7). In *Elphidium crispum*, the chambers added in spring (abundant algal food) are longer, narrower and thinner than those produced during the autumn (low food supply) and this may cause the test periphery to be notched (Myers, 1943); this was confirmed through feeding experiments (Murray, 1963). In a review of whether changes of ecological parameters influence morphological variation, it was concluded that almost no variables operate independently and that there is no broad pattern (Boltovskoy *et al.*, 1991). However, changes in shape and wall thickness in symbiont-bearing larger foraminifera are related to penetration of light with depth (Reiss and Hottinger, 1984). The ultimate expression of environmentally controlled morphological variation is the concept of ecophenotypes. An ecophenotype is 'a phenotype exhibiting non-genetic adaptation associated with a given habitat or to a given environmental factor' (Haynes, 1992). The concept has been applied to morphologically variable groups such as *Bulimina aculeata, Bulimina marginata, Elphidium* and *Ammonia* from around the North Atlantic (Feyling-Hanssen, 1972; Chang and Kaesler, 1974; Poag, 1978; Poag *et al.*, 1980; Miller *et al.*, 1982; Walton and Sloan, 1990; Collins, 1991). It was stimulated by experiments on *Ammonia*, where cloned individuals gave rise to seven different morphotypes (Schnitker, 1974) although it is doubtful whether the cultures were pure clones (Haynes, 1992). A field survey of morphological variation in relation to salinity showed no correlation between the three morphotypes and environmental variables (Malmgren, 1984). Molecular genetic analysis has shown that morphologically similar forms may be genetically different (Pawlowski *et al.*, 1995; Holzmann and Pawlowski, 1997; Holzmann, 2000; Hayward *et al.*, 2004a). Thus, at present it may be wise to avoid the ecophenotype concept until the morphological and genetic approaches have been harmonised.

Seep/vent faunas. Foraminiferal faunas found in association with hydrocarbon seeps and vents have been described from the Gulf of Mexico (Section 7.2.6) and California (Section 7.2.10). No species are uniquely confined to seeps (Rathburn *et al.*, 2003); all the forms found in seeps also occur where hydrocarbons are lacking. Detailed examination of the protoplasm reveals that there are no prokaryotic endosymbionts that might facilitate foraminiferal colonisation of such habitats (except for one individual of *Uvigerina peregrina* with rod-shaped bacteria in its pores above the pore plug, Bernhard *et al.*, 2001). The main distinctive feature of these faunas is that the carbon stable isotope signature is different. Strongly negative $\delta^{13}C$ values suggest that carbon derived from hydrocarbons is utilised in feeding.

Stable isotopes. Unlike limiting factors, there can be no response from an organism to changes in the isotopic composition of the ambient water.

However, stable isotopes present in the test give a mixed record of both ambient conditions and cell physiology and this makes them useful in palaeoceanography. The basic principles are clearly explained in a review by Rohling and Cooke (1999) and this should be consulted for more detail. There are three stable isotopes of oxygen (^{16}O, ^{17}O, ^{18}O) and normally the ratio $^{18}O/^{16}O$ is studied; for carbon there are two isotopes, namely ^{12}C and ^{13}C. Molecules consisting of light isotopes react more readily than those with heavy isotopes and partitioning of isotopes between substances is called fractionation. In the analytical process, quantitative estimates of isotopes are made by comparing results for a known standard (R_{std}) with those for an unknown sample (R_{sam}) and the differences are defined as

$$\delta_{sam}\text{\textperthousand} = 10^3 \times (R_{sam} - R_{std})/(R_{std}).$$

A positive value of δ indicates enrichment in the heavy isotope. The commonly used standard was PDB (the guard of a belemnite from the Cretaceous Pee Dee Formation) which by definition has $\delta^{18}O = 0$. Now two standards are used, NBS-18 (carbonatite) and NBS-19 (limestone). Disequilibrium refers to isotopic compositions that are not in equilibrium between the mineral phase and ambient water.

The geological $\delta^{18}O$ signal predominantly relates to global ice volumes with temperature playing a secondary role. However, in modern seas the global ice volume is constant so temperature plays a primary role. Deep-sea benthic-foraminiferal calcite secreted under conditions of stable low temperature and absence of photosynthesis (no symbionts) shows $\delta^{18}O$ disequilibrium and this suggests that food supply and microhabitat effects play a role. Larger foraminifera with symbionts have decreasing $\delta^{18}O$ values with increasing growth rates (Ter Kuile and Erez, 1984; Wefer and Berger, 1991). From a re-evaluation of the data on epifaunal *Cibicidoides* and shallow infaunal *Uvigerina* Bemis *et al.* (1998) find that the former is close to isotopic equilibrium with ambient seawater while the latter shows slight enrichment. The aragonitic tests of *Hoeglundina elegans* show enrichment of 0.78 ± 0.19‰ (Grossmann, 1984b).

The global carbon cycle relates to two main reservoirs, organic matter and sedimentary carbonates. The organic carbon cycle in the marine and terrestrial biospheres is based on CO_2 fixation into biomass through photosynthesis: $CO_2 + H_2O +$ energy (sunlight) $\rightarrow CH_2O + O_2$. Respiration using oxygen follows the reverse process. The inorganic-carbon pool of the oceans is controlled by the carbonate reactions. Total dissolved inorganic carbon consists of bicarbonate, the carbonate radical and carbon dioxide. The average $\delta^{13}C$ of the organic carbon reservoir is -25‰ and that of the carbonate reservoir is 0‰. Extremely depleted values are recorded in hydrocarbon seeps (−35 to −80‰). During photosynthesis, ^{12}C is preferentially taken up and marine phytoplankton has organic matter with $\delta^{13}C$ −20 to −23‰ with respect to ambient seawater. Therefore surface waters become relatively enriched in ^{13}C. As organic matter is remineralised in the water column, ^{12}C is returned to the water. Thus, there is a gradient from higher $\delta^{13}C$ in surface waters to lower values in deep water. The relative age of deep water can be determined from lateral gradients in $\delta^{13}C$ as the deep water values reflect the influences of the time of exposure to decay of organic matter, the export production along its pathway (amount of organic matter remineralised) and temperature-controlled rapidity of decay. The young deep water of the North Atlantic has $\delta^{13}C + 1.0$‰ (compared with surface + 1.6‰), whereas old Pacific deep water has $\delta^{13}C − 0.2$‰ (average surface waters + 1.5 to + 2.0‰). Benthic foraminifera commonly show $\delta^{13}C$ disequilibrium due to one or more causes: uptake of metabolic CO_2 during shell secretion, symbiont photosynthesis, growth rate, variation in carbonate-ion concentration in ambient water. Infaunal foraminifera are exposed to pore water which may have a different $\delta^{13}C$ value than the overlying bottom water. There may be depletion of 1‰ per centimetre depth within sediment (Grossmann, 1984a,b; 1987; McCorkle et al., 1985) and Rathburn et al. (2003) suggest that a range of 4‰ in a species is not uncommon. Although McCorkle et al. (1990) found a down-core gradient in $\delta^{13}C$, Mackensen and Licari (2003) found that there was no down-core trend and they attribute this to infaunal taxa being mobile and secreting the test at different sediment depths. The precise controls on the carbon isotopic composition of tests remain uncertain (Rathburn et al., 2003).

8.15   Final conclusions

There is now a large body of data on benthic foraminiferal distributions and a greater understanding of some of the factors that control them. In shallow waters the principal controls appear to be temperature (on a latitudinal scale), salinity, energy levels and substrate, and food availability on a local scale. On the outer shelf and upper slope the principal controls are food supply, energy levels (especially currents) and substrate. The lower-slope, abyssal and hadal areas are limited by very low food supply but are also influenced by carbonate dissolution. Oxygen depletion may be important in unventilated basins (especially fjords, California borderland) and in some oxygen minimum zones. The roles of intra- and interspecific competition remain poorly understood. It is perhaps worth restating that *those factors close to the threshold of tolerance limit the distribution of a given species and different factors will be limiting at different times and in different places.*

9

Taphonomic processes: formation of dead and fossil assemblages

9.1   Introduction

Fossil foraminifera are studied in order to reconstruct palaeoecological (including palaeoceanographic) conditions and for biostratigraphic correlation and dating. Subrecent foraminifera are used to investigate environmental change over the last few hundred years. In order to make these interpretations it is necessary to know whether the fossil record is representative of that originally living. Taphonomic processes are those that change the composition and preservation from living to dead to fossil assemblage. Living assemblages are transient, with temporal changes in absolute and relative abundance of species in relation to seasonal or longer variations in the environment. There are several stages in the transition from living assemblages to the fossil record:

- Life processes that affect the contribution of tests to the sediment.
- Postmortem processes that alter the proportions of species tests in the sediment over a period of decades to a few thousand years (depending on the rate of sediment accumulation) in the taphonomically active zone (TAZ) influenced by macrofaunal bioturbation.
- Diagenetic effects that may further alter the faunal composition from the time they pass into the historical layer, i.e., below the TAZ.
- Weathering effects at outcrop.

The aim of this chapter is to explore the consequences of these effects in order to determine how good a record is preserved. This knowledge is essential for the correct interpretation of past environmental conditions. Past reviews include those of Murray (1973; 1976a; 1991), Martin (Martin, 1999; Martin and

Liddell, 1991) and Scott *et al.* (2001). For a theoretical approach to assemblage formation see Loubere *et al.* (1993).

9.2   Life processes influencing the contribution of tests to the sediment

The production of tests and their rate of contribution to the sediment depend on rates of reproduction and rates and modes of death for each species (see Section 3.2.5). At present there are no reliable data on the absolute number of tests contributed to the sediment by these processes. It is important to emphasise that comparison of *relative* abundances (i.e., percentages) between living and dead assemblages is not a guide to *absolute* differences of production for the reasons explained in Section 2.10. Nevertheless, we are all tempted to make these false comparisons. The only reliable guide to changes of absolute abundance is the *rate of accumulation of tests.*

9.3   Postmortem processes affecting the preservation of foraminifera

The main postmortem processes are destruction of tests, transport and mixing.

9.3.1   Destruction of tests

Loss of agglutinated tests. Agglutinated tests with only a small amount of organic cement are liable to be destroyed through bacterial or chemical decay of the organic matter or through macrofaunal predation (e.g., Kuhnt *et al.*, 2000; Douglas *et al.*, 1980). They may also be destroyed when samples are dried during processing. Down-core reduction in numbers of agglutinated tests has been widely reported (e.g., California borderland, Douglas *et al.*, 1980; Gulf of Mexico, Denne and Sen Gupta, 1989; Red Sea, Edelman-Furstenberg *et al.*, 2001; Weddell Sea, Antarctica, Murray and Pudsey, 2004). In the Red Sea, organic cemented forms are present only in the upper few centimetres of sediment beneath the oxygen minimum zone but away from this they are present to the base of the sampled interval (30 cm; Edelman-Furstenberg *et al.*, 2001). On the Nares abyssal plain the subsurface decrease in agglutinated tests is linked to the transition from an oxidising to a reduced environment (Schröder, 1988). Forms particularly susceptible to loss include komokiaceans and uniserial or unilocular forms with loosely cemented walls (some *Reophax, Rhizammina, Thurammina*). This loss greatly reduces the species diversity: Fisher alpha diversities of live and dead assemblages are 33.0–38.5 and 9.86–11.87, respectively, for a site in Rockall Trough (Gooday and Hughes, 2002). Such a reduction in diversity is much greater than that associated with the loss of calcareous tests discussed below. Because of postmortem loss of agglutinated taxa, Mackensen *et al.* (1990, 1993a) and Harloff and Mackensen (1997) recalculated the deep-sea assemblages to exclude all the fragile agglutinated taxa and termed these 'potential fossil assemblages'. Nevertheless, agglutinated foraminifera are often well preserved in the fossil record particularly in sediments where the rate of accumulation is high (>1 mm y^{-1}, Alve, 1996).

Dissolution of calcareous tests.   Understanding the causes of carbonate dissolution continues to be a problem. To quote Morse and Arvidson (2002): 'A major difficulty in understanding the dissolution kinetics of calcite in seawater is that the saturation state of seawater is generally greater than 0.7 with respect to calcite which is approximately a pH of only 0.2. Thus, an understanding of the dissolution behavior of calcite in the ocean, and influences of factors such as inhibitors and temperature, must be obtained over a pH range of less than 0.2'. They also point out that there is a long way to go to bridge the gap between carefully controlled laboratory experiments (invariably based on mineral rather than biogenically secreted calcite) and field observations in complex natural marine environments.

In general seawater becomes more corrosive with increase in water depth with total dissolution occurring at the calcite compensation depth. The abundance of organic matter is an important factor (Loubere *et al.*, 1993) as the corrosivity of sediment pore water depends on its breakdown. Aerobic decay of organic matter takes place in the presence of free oxygen. Once CO_2 is produced and ammonia is oxidised to nitric acid, carbonate undersaturation occurs at pH $\leqslant$ 8. When the oxygen has been used up, a series of bacterially mediated anaerobic decay processes take over: manganese reduction, nitrate reduction, iron reduction, sulphate reduction, methanogenesis and fermentation (Allison and Briggs, 1991). Nitrate reduction maintains the carbonate saturation level previously present (pH 8). Sulphate reduction leads to bicarbonate production, which in turn increases the alkalinity of the pore water and causes supersaturation of aragonite and calcite. However, in the early stages of sulphate reduction, undersaturation develops. In carbonate sediments lacking iron, H_2S accumulates, pH decreases to $\sim$7, and carbonate undersaturation occurs leading to dissolution. Once 5–7 mM of sulphate has been reduced and alkalinity has reached 10–14 mM, carbonate saturation is re-established. In iron-rich sediments, precipitation of FeS_2 leads to carbonate supersaturation after

0.2–0.3 nM of sulphate have been reduced (Martin, 1999). Bioturbation influences these processes by introducing oxygen into sediments that might otherwise be dysoxic or anoxic. In high latitudes, the formation of sea ice may play a role in dissolution. Sediments with <1% $CaCO_3$ occur beneath areas with seasonal sea-ice cover. During the formation of ice, brine is expelled and this causes an increase in salinity at the same time as the water becomes very cold. Decay of organic matter in the upper water column produces CO_2 that cannot escape to the atmosphere because of the ice cover. This dense water sinks to form a basal layer above the sediment. Also, decay of organic matter at the sediment–water interface is a further source of CO_2. Thus, the dense bottom water may be enriched in CO_2 and therefore corrosive (Hald and Steinsund, 1996). In summary, dissolution of calcareous tests depends mainly on the type of sediment (clastic or carbonate), on the abundance of organic matter, on local pore-water conditions, and intensity of bioturbation.

The best preservation of calcareous tests is in strongly dysoxic to anoxic sediments in the presence of sulphate reduction and absence of bioturbation (e.g., shelves and basins along the Pacific seaboard of North and South America, Phleger and Soutar, 1973). However, Schrader et al. (1983) caution that anoxia leading to the good preservation of calcareous tests is not universally true. In the Gulf of California they found dissolution to be inversely related to primary productivity of the surface waters.

Experimental studies indicate that etching by dilute acid causes transparent hyaline walls to become translucent or opaque due to an increase in surface topography and porcelaneous walls lose their outer layer of crystallites (Murray, 1967b; Murray and Wright, 1970; Peebles and Lewis, 1991). The same applies to dissolution in artificial seawater (Kotler et al., 1992). Etching caused by ingestion of tests during feeding by gastropods produces a similar effect (Walker, 1971). The rate of dissolution depends on wall thickness, wall structure and chemical composition (Corliss and Honjo, 1981; Peebles and Lewis, 1991; Boltovskoy, 1991). There is a good correlation between the proportion of broken tests and undersaturation of calcium carbonate ($r^2 = 0.74$ for Corliss and Honjo data) except in high-energy environments. The pattern of dissolution is etching of the wall surface, breakage of the final chamber, breakage of other chambers and finally total destruction.

Under natural conditions, dissolution is known to take place with varying degrees of intensity in environments ranging from marsh to the deep sea. Based on geochemical studies, Green et al. (1993) consider that calcareous tests in Long Island Sound survive for 86 ± 13 days unless they are moved out of the dissolution zone through bioturbation. Pore-water saturation with respect to carbonate minerals shows cyclicity with saturation in late autumn,

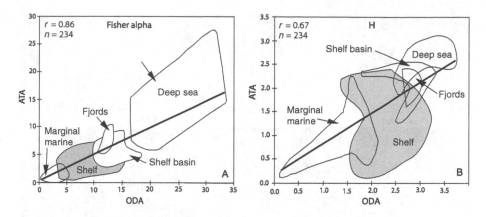

Figure 9.1. Summary diagrams for species diversity plotting ATA vs. ODA.
A. Fisher alpha. B. Information function, H (Adapted from Murray and Alve,
2000b).

undersaturation during winter (when most dissolution takes place) to supersaturation during late spring (Green and Aller, 1998). In Samish Bay,
Washington, USA, dead foraminifera are dissolved within five months (Jones
and Ross, 1979). It is a common misconception that dissolution takes place
only in clastic sedimentary environments. Dissolution is an active process even
in carbonate sedimentary environments due to rapid CO_2 production resulting
from oxidation of organic matter (Peebles and Lewis, 1991). Nevertheless,
dissolution is not usually a major process in carbonate sediments whereas it
may be so in siliciclastic sediments (Kotler *et al.*, 1992).

Evidence of total dissolution of calcareous tests is provided by: the presence
of residual organic linings; tests infilled with glauconite that remain as an
internal mould after the loss of the test (George and Murray, 1977) or fram-
boidal pyrite; or residual secondary organic-cemented agglutinated assemblages. Experimental studies taking an original dead assemblage (ODA) and
treating it with dilute acid to remove the calcareous form to give an acid-
treated assemblage (ATA) simulates the natural process of dissolution. This has
been carried out on material from a wide range of environments (Alve and
Murray, 1994, 1995b; Murray and Alve, 1999a,b; 2000b). Remarkably it appears
that the ATAs preserve much of the species-diversity information of the ODA
(Figure 9.1). This indicates that secondary agglutinated assemblages should not
be ignored as they can be used in palaeoecological analysis.

Bioerosion.   This is the process whereby organisms (bacteria, endolithic
algae and fungi, predators) are responsible for the damage or destruction of

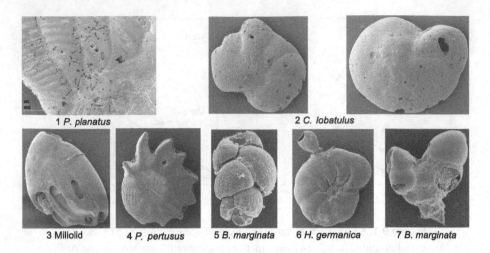

1 *P. planatus*   2 *C. lobatulus*

3 Miliolid   4 *P. pertusus*   5 *B. marginata*   6 *H. germanica*   7 *B. marginata*

Figure 9.2. Scanning electron micrographs of taphonomic changes to tests and other features (longest dimension μm). Microborings: 1. *Peneroplis planatus* (width 380). 2. *Cibicides lobatulus* (330, 350). Abrasion damage: 3. Miliolid (800). 4. *Peneroplis pertusus* (500). Glauconitic chamber infilling: 5. *Bulimina marginata* (240). Other features: 6. *Haynesina germanica* with turbellarian egg case (430). 7. *Bulimina marginata*, twinned test showing abrasion damage (620).

biogenic material. Bacteria are thought to be responsible for the destruction of *Cibicides lobatulus* as they oxidise the organic matrix in the wall and dissolve the calcite microcrystallites (Freiwald, 1995). Endolithic algae and fungi dissolve holes and tunnels in calcareous walls (Golubic *et al.*, 1975; 1984; Kloos, 1982). This type of bioerosion takes place mainly in warm, shallow, carbonate environments but microborings are sometimes seen in tests from deeper temperate and cool areas (Figure 9.2). Martin and Liddell (1991) deployed five tests each of *Archaias angulatus* and *Amphistegina gibbosa* 10 cm above the sediment–water interface in the back-reef environment of Discovery Bay, Jamaica, for three months. *Archaias* was particularly prone to bioerosion whereas *Amphistegina* was more resistant. This fits with field observations in the United Arab Emirates where peneroplids are more affected by bioerosion than rotaliids. Endolithic bioerosion is higher in shallow water because many of the bioeroders require light. It is also higher in carbonate than clastic environments.

9.3.2   Transport

Empty foraminiferal tests are sedimentary grains and therefore subject to transport. However, they differ from other grains in being hollow and in having a great variety of shapes. These parameters affect the traction velocity, i.e., the threshold velocity required to move a grain, and the settling velocity as it returns to the sea floor. The main mechanisms of transport are bed load and suspended load but transport of foraminiferal tests also takes place in mass flow including turbidity currents, via floating plant material, ice and wind. Transport involves loss of tests from the source area and gain of tests in the depositional area. In some cases transport occurs locally in the same environment so the tests are parautochthonous. In other cases, transport moves tests from one environment to another (allochthonous). Transport in suspension is likely to be over the greatest distances (tens or hundreds of kilometres).

   Bed load.   Transport as bed load takes place where the bottom current reaches the traction velocity although bed roughness and the amount by which a grain projects from the bed also play a part in determining whether a grain will be entrained by a current. Flume studies have established that there is a correlation between traction velocities and the mass and shape of foraminifera (Kontrovitz *et al.*, 1978). The medium used was distilled water and the size range of test excluded juveniles so the results are not entirely representative of natural conditions. The results of the Kontrovitz *et al.* study have been used by Snyder *et al.* (1990) as a basis for estimating the traction velocities of common species from the Washington shelf, USA. Three slightly overlapping groups are recognised based on average traction velocities: 4–7 cm s^{-1}, roughly equidimensional; 6–9.9 cm s^{-1}, elongated or coiled, moderately inflated; 9–13 cm s^{-1}, elongated or coiled, highly compressed. Depending on the current velocity, grains may remain in suspension for considerable distances or they may saltate along the seabed. Under the latter conditions, abrasion and test damage may take place (Figure 9.2). Experiments on abrasion of Bahama Bank foraminifera reveal that the tests develop minute scratches and pits, the ornament becomes flattened and, in some cases, they have a polished appearance. Holes form through loss of chamberlet walls in calcareous forms and the plucking out of detrital grains in agglutinated forms (Peebles and Lewis, 1991). In another set of experiments to simulate abrasion in carbonate sediments, equivalent to hundreds of kilometres of transport, damage to tests was insufficient to make them unrecognisable (Kotler *et al.*, 1992). Therefore, Martin and Liddell (1991) suggest that in carbonate environments, even those with high energy, abrasion is not an effective mechanism for seriously damaging tests.

   After transport, grains settle to the sea floor. The settling velocities of tests ~300 μm diameter range from 0.009–0.018 m s^{-1} in distilled water at 20 °C (Haake, 1962; Grabert, 1971). The diameter of trochamminids is three times that of a hydraulic equivalent quartz grain (i.e., they have the same settling velocity). *Ammonia dentata* diameter 380–400 µm has a settling velocity of 0.5–0.8 cm s^{-1} and the hydraulic equivalent quartz grain is 70–90 µm (Seibold and Seibold, 1981). For larger foraminifera, Maiklem (1968) distinguishes between rods (*Alveolinella*), plates (*Marginopora*) and spheres (*Calcarina* and *Baculogypsina* approximate to spheres when the spines are eroded off). Rods settle smoothly with their long axis horizontal, large plates oscillate down and settle on their flat side and spheres settle smoothly. *Marginopora* settles slowly due to its low bulk density caused by the large number of thin-walled chambers. *Calcarina and Baculogypsina* settle 30–50% faster than plates of the same (sieve) size while *Alveolinella* settles fastest of all. Martin and Liddell (1991) carried out flume and settling experiments on warm-water taxa from Jamaica, and found poor negative correlation between traction and settling velocities ($r^2 = -0.22$), due to shape, density and size differences.

Bed-load transport takes place in areas affected by currents or wave disturbance. Tidal currents are greatest in shallow water in meso- and macrotidal regimes. Geostrophic ('contour') currents capable of transporting sedimentary material are present at bathyal depths on continental margins (Iberian margin, Schönfeld, 1997).Wave influence is restricted to depths above storm-wave base so for most areas bed-load transport is confined to shallow waters but in areas of large waves the middle shelf may be affected. Sediments of sand grade and coarser are most likely to undergo bed-load transport and the fine material will be winnowed away. The final result of bed-load transport is to sort foraminiferal tests according to traction and settling velocities (based on size, mass and shape). Forms that have similar traction and settling velocities should occur together in the final depositional area. Their tests are likely to have been abraded and this is shown by chamber breakage and a dull or polished texture. Transport is most active during storms.

The Jutland Current causes long-distance bed-load transport of *Elphidium excavatum* tests from the west side of Denmark >100 km into the Kattegat (Nordberg *et al.*, 1999). *Homotrema rubrum* living on outer Bermuda shoals attached to coral and other debris, is detached by storm waves and currents and through the activities of browsing animals and burrowers. The tests are transported long distances onto the Bermuda slope but not far into the shallow-water lagoons (MacKenzie *et al.*, 1965). In a similar way, peneroplids, soritids and rotaliids from shallow-water carbonate and reef environments are transported from Bahama Bank onto the slope to depths >900 m and are particularly common at 300 m (Martin, 1988). The flat attached tests of *Planorbulina* sp. which during life encrust plants are greatly reduced in the dead assemblages of wave-swept, outer back-reef environments off Key Largo,

Florida, whereas degraded tests of *Archaias angulatus* are more abundant than their living counterparts. This species lives on the inner shelf but dead tests are transported onto the outer shelf. Thus inner- and outer-shelf dead assemblages are similar which means that these environments are not readily separated in the fossil record in this area (Martin, 1986; Martin and Wright, 1988). Li *et al.* (1997; 1998) used foraminifera as tracers of sediment transport during storms in Grand Cayman, Caribbean. Large storm waves throw sediment into suspension in the lagoon and fore-reef. Sand and finer sediment (including foraminifera) are transported shoreward over the reef. Dead tests of *Amphistegina gibbosa*, which lives on the fore-reef but not in the lagoon, show size sorting with a decrease in size from the reef front into the lagoon. As the storm abates, water piled up in the lagoon flows seawards transporting small foraminiferal tests including dead *Archaias angulatus*. This species does not live offshore but small dead tests occur in the sediments there. Thus *Amphistegina gibbosa* is transported into the lagoon and *Archaias angulatus* is transported out of the lagoon. Waves from a hurricane are thought to have been responsible for the more widespread occurrence of dead large foraminiferal tests in Moorea, Polynesia (Vénec-Peyré and Le Calvez, 1986).

Suspended load. The density of protoplasm is probably almost the same as that of seawater so there is no significant difference between a water-filled dead test and a living one with protoplasm. However, living forms are able to seek sanctuary in the sediment or to use their pseudopods to cling to detrital grains to reduce the chance of being transported or taken into suspension. Both living and dead foraminifera have been recorded in plankton samples taken from the water column in various locations around the world (UK: Murray, 1965d, Murray *et al.*, 1982, 1983, John, 1987; Gulf of Mexico: Hueni *et al.*, 1978; California, USA: Lidz, 1966, Loose, 1970; Polynesia: Lefévre, 1984, Vénec-Peyré and Le Calvez, 1986). In these areas there is fairly high energy due to wave disturbance of the sea floor causing fine material to be thrown into suspension. Characteristic features of tests transported in suspension are that they rarely show evidence of damage or abrasion, they are generally thin walled and well sorted, and commonly <200 µm in diameter if rounded or <300 µm in length if elongated. They include naturally small species and the juvenile stages of bigger species. During a storm in the English Channel the average concentration of tests was $3800\,\text{m}^{-3}$ water so large quantities of material are involved. Tests remain in suspension as long as there is sufficient turbulence to keep them there. They are transported backwards and forwards by tidal flow but may show net transport in one direction. Thus some may be redeposited in the area of origin but others will be transported beyond the source area. Off California, foraminifera are transported considerable distances in the foam of breaking waves (Loose, 1970). Suspended-load transport may also take place under very low-energy intertidal conditions. Under calm conditions, where tidal flats dry out in the sun, foraminifera may be picked up by the incoming tide and transported shorewards as observed in an estuary (Siddall, 1878) and in the hypersaline Abu Dhabi lagoon, UAR (Murray, 1973).

In the macrotidally energetic seas of southern England and the southern North Sea, suspended-load foraminifera derived from the inner shelf are transported into estuaries where they accumulate in silty muds in low-energy settings (Murray and Hawkins, 1976; Culver and Banner, 1978; Murray, 1980; 1987b; Wang and Murray, 1983; Reinöhl-Kompa, 1985; Cearreta, 1988, 1989; Gao and Collins, 1995). A small species of *Pararotalia* (<200 μm) is a useful marker of suspended-load transport from the wave- and tide-influenced inner shelf into adjacent coastal lagoons and estuaries in Brazil (Debenay *et al.*, 2001a). In microtidal Cochin lagoon in southern India, small tests (mainly <160 μm) are transported in from the shelf (Seibold and Seibold, 1981). In general there is correlation between the tidal regime of the receiving estuary and the quantity of incoming transported tests. Microtidal estuaries have only a small proportion of allochthonous tests whereas mesotidal and macrotidal estuaries may have more than 50% of the dead assemblage derived from the adjacent shelf (Wang and Murray, 1983; Murray, 1987b). However, the transported component is easily recognised because of its limited size range. Off west Scotland, foraminifera from the surrounding shelf are transported into low-energy deeps (Murray, 2003b). Tests derived from the shelf may also be transported onto the continental slope and into basins (Douglas *et al.*, 1980). Dead assemblages with a high proportion of exotic transported tests have higher species diversity than the associated living assemblages. For example, in the Hamble estuary the Fisher alpha of the live assemblages is 1.5–5.0 whereas that of the dead is 3.0–5.5 (Alve and Murray, 1994).

Turbidity currents and mass flow. Possible evidence for transport by turbidity currents and mass flow of sediment comes from experiment and observation. Sediment trap experiments in Bedford Basin, Canada, showed that the turbid bottom water contained ten species of foraminifera 140–660 μm in size, seven of which had living representatives (Schafer and Prakash, 1968). Shallow-water foraminifera in sediments in Wilmington Canyon, USA, may be in transit down slope, in alternating phases of transport and deposition (Stanley *et al.*, 1986). Prodelta muds give rise to turbidity currents in British Columbian fjords and these carry the benthic foraminifera into the fjord basin (Schafer *et al.*, 1989). At 1650 m on the Rhone deep-sea fan there are shallow-water foraminifera that have been transported down slope (Vénec-Peyré and Le Calvez, 1986) probably in turbidity currents. In Santa Barbara Basin, California, USA, *Brizalina argentea* tests (given as *Bolivina*) are preferentially aligned in a graded unit interpreted as a turbidite (Harman, 1964). An extreme case of turbidite displacement is of calcareous reefal foraminifera from Palau into the adjacent Palau Trench at 8053 m, well below the CCD (Yamomoto *et al.*, 1988).

Ice. Foraminifera may become incorporated in ice where it is in contact with the sediment surface. If the ice then moves elsewhere and melts, the foraminifera are released into a new setting. Examples are known from pack ice in the intertidal zone in Germany (Richter, 1965) and this is an important process around Antarctica (Dieckmann *et al.*, 1987; Ishman and Webb, 2003). There, the process is more complex. Basal freezing of the under surface of grounded ice incorporates sediment into the ice. Further basal freezing adds more material at the bottom. As the ice moves away from the shore, the surface melts and gradually sedimentary material works its way to the surface where some is dispersed by wind. Other sediment is released and settles on the surface sediment below the ice shelf.

Floating plants. In shallow waters there are foraminifera that live epiphytically on seaweeds and seagrasses. During storms, the plants may be uprooted and transported out to sea, along the coast or on to the shore (Bock and Moore, 1968; Boltovskoy and Lena, 1969b). On many shores, piles of uprooted plants, together with their foraminifera, accumulate during winter storms. Some plants living over oceanic waters are pelagic. *Planorbulina acervalis* and *Rosalina globularis* living attached to the pelagic seaweed *Sargassum natans* have been recorded in the Atlantic Ocean more than 5 km from Bermuda (Spindler, 1980). When the foraminifera die they are contributed to the underlying deep-sea sediment.

Wind. Sand dunes adjacent to beaches form from the accumulation of sediment derived from the beach. Onshore winds pick up dry grains and transport them onto subaerial sand dunes. Foraminifera have been recorded from dunes in Ireland (abundant *Cibicides lobatulus*) and the United Arab Emirates (Murray, 1970a; 1991). During the Pleistocene, foraminifera were transported 800 km into the Thar Desert in India (Goudie and Sperling, 1977). It is clearly important to recognise wind transport to avoid a false palaeoecological interpretation.

Predation. The taphonomic consequences are threefold: transport of tests from the position of life to be deposited elsewhere; damage of tests, thus making them more vulnerable to destruction by physical processes; total destruction of tests (Lipps, 1988). In most cases the tests pass through the predator gut without serious damage (Boltovskoy and Zapata, 1980); in some cases there is etching of the calcareous wall and damage to the newer chambers (Herbert, 1991; Langer *et al.*, 1995). The tests are added to the dead assemblage although not always close to the area of life. Some opisthobranch gastropods predating exclusively on calcareous taxa, crush the tests in the gizzard and dissolve them (Cedhagen, 1996).

All these processes potentially introduce bias into the fossil record and blur the boundaries between environments but the importance of this has not yet been quantified for any environment.

9.3.3   Bioturbation and the taphonomically active zone

Epifaunal organisms moving over the sediment surface may physically disturb it. Infaunal organisms moving through the sediment certainly do so. The term bioturbation is used to describe these activities. Some of these organisms also affect sediment chemistry either directly through their secretion of mucus or indirectly through allowing the circulation of pore water. These processes also affect the mechanical properties of the sediment. Some lead to sediment mixing and homogenisation while others lead to unmixing to produce new structures. Among the latter are the 'conveyor-belt' deposit feeders. These feed at depth and transport waste material to the sediment surface where it commonly forms a mound. Also some organisms make burrows for sanctuary rather than feeding. Crabs may move balls of sediment rather than individual grains and these are discharged onto the sediment surface.

The physical effects of bioturbation are advection and diffusion. The use of mathematical diffusion models is based on the assumption that an eddy (particle) biodiffusion coefficient describes the redistribution of particles (see Martin, 1999, for discussion of the relevance of this). However, these models do not take into account the importance of advection or of processes like tube building. Bromley (1996) suggests a tiering model best describes the physical consequences of bioturbation. In the uppermost 2–3 cm layer there is superficial disturbance from epifaunal organisms and mixing by mobile burrowers (including foraminifera, Gross, 2000) with the vast majority of species confined to this layer. Below this are layers with conveyor organisms, open burrows and backfill collectively forming only a small part of the infauna. However, a single model seems inappropriate for the range of environments and sediment types found under natural conditions. The chemical effects include the formation of biogeochemical microenvironments (around burrows, in and around faecal pellets, etc.) instead of an orderly vertical succession of geochemical fronts parallel with the sediment surface. Biological effects also influence the bacterial floras that mediate chemical reactions and open up new habitats for foraminifera by providing shelter and oxygen haloes in dysoxic and anoxic sediments. Macrofaunal bioturbation allows foraminifera to live infaunally down to around 60 cm in certain environments.

The physical, biological and biogeochemical effects of bioturbation extend down to a maximum depth of ~1 m. This zone is sometimes called the TAZ. Below this is the historical layer where material is beyond the influence of short-term postmortem processes, thus being effectively fossilised, and the main processes are diagenetic (Douglas *et al.* 1980). From mathematical modelling of depth of life and sedimentation rate, Loubere (1989) concludes that surface assemblages rarely represent the entire fauna because they do not include deeper infaunal taxa; the most representative dead assemblages are those found at the base of the bioturbated layer. In Bahia la Choya in Mexico there is seasonality in reproduction. Martin *et al.* (1996) suggest that conveyor-belt deposit feeders rapidly move tests into the subsurface historical shell layer where they are preserved. Many of the tests that remain at the surface are dissolved.

9.3.4   Faunal mixing: reworking and relict deposits

There are habitats where foraminifera cling to firm substrates: in carbonate environments phytal foraminifera cling to plants (e.g., miliolids and peneroplids); in shelf environments there are forms clinging to shells, rocks, etc. (M, see Section 5.2.4). When these die they are contributed to the sediment. Thus, the sediment dead assemblages contain indigenous sediment-dwelling forms together with phytal or M forms. Steinker and Rayner (1981) suggest that in such situations although the dead assemblages may be used to interpret the environment of deposition of the sediment, they do not give a good record of the living assemblages due to the effects of postmortem transport and mixing.

In shallow-water environments such as estuaries and lagoons, there are normally channels that migrate laterally through time and as they do so they erode their banks. Subrecent tests of species currently living in the area may be reworked from the older deposits into the contemporary sediment. Unless the older tests are discoloured or damaged, they are indistinguishable from their modern counterparts. This can cause confusion if large numbers of such individuals are picked for AMS (accelerator mass spectrometry) ^{14}C dating as they will give an unreliable date (Cearreta and Murray, 1996).

Because of the postglacial rise in sea level of around 120–130 m during the past 12 000 years, rivers have been adjusting to a higher base level by infilling their estuaries with sediment rather than transporting it onto the shelf. Consequently, many continental shelves are sediment starved. Relict sediments related to earlier sea levels have not yet been buried so the dead foraminiferal faunas are a mixture of modern and subrecent forms (e.g., outer shelf, Bay of Biscay: Andreieff *et al.*, 1971; mainland shelf of California: Douglas *et al.*, 1980). In the Bay of Biscay, outer-shelf bioclastic sand sediments have 80–90% relict tests (broken, infilled with glauconite) and some species indicate water depths 100–120 m shallower than now. Other records are in the Celtic Sea (George and Murray, 1977; Figure 9.2) and off Vancouver (Bornold and Giresse, 1985).

Where older material, e.g., Tertiary or Mesozoic, is reworked from submarine outcrops into modern sediments, the difference in faunas should be easily recognised. Off southern California, USA, Miocene and Pliocene foraminifera are commonly reworked into the modern sediments of the Santa Cruz Basin and may form up to 34% of the dead assemblage (Resig, 1958). They also occur in intertidal assemblages (Cooper, 1961).

9.3.5   Time averaging

An awareness of patchiness in living assemblages is essential for understanding the time-averaged record of the dead assemblages (Murray and Alve, 2000a). If the rate of sediment accumulation is 1 cm in x years, then a 1 cm thick sample will contain the contribution of dead tests derived from a succession of living assemblages over an x-year period. If the sample is from several centimetres below the sediment surface it may also contain deep infaunal taxa that were added much later than the rest of the tests (their age depends on the rate of sediment accumulation). The dead assemblage will also have been modified by a variety of postmortem processes described above. Such a record is said to be time averaged and this applies to all dead assemblages. Except in areas with very high rates of sediment accumulation (>1 mm y^{-1}), the top 1 cm of sediment probably represents a decade to hundreds of years of time-averaged foraminiferal accumulation. Now that it is possible to date individual tests of large foraminifera using ^{14}C AMS dating, it has been shown that both fresh and corroded tests of *Archaias* and *Amphistegina* in surface sediments of a Mexican lagoon range in age up to ~2000 years and the lower limit of temporal resolution of assemblages preserved there may be ~1000 years (Martin *et al.*, 1996).

Mathematical simulation of time averaging has been carried out on marsh assemblages (Martin *et al.*, 2002). Live and dead data were analysed separately and only the most abundant five taxa were used. The seasonal live and dead counts from depths down to 60 cm below the sediment surface were summed to give artificial time averages. Regression analysis was used to determine whether the artificial assemblages could be used to reconstruct past depth of life in the sediment. This worked well for both high and low marsh but less well for intermediate marsh. The authors suggest that on high marsh the absence of bioturbating organisms due to plant root systems is responsible for the excellent results of the regression models. On low marsh, bioturbation opens up habitats for infaunal taxa and these forms bypass the taphonomically active zone. Loubere *et al.* (1993) argue that test production and taphonomy depend on the oxygen content of the bottom water and flux of organic carbon to the sea floor. From mathematical modelling, they conclude that the preserved assemblages depend on the vertical standing crop of the foraminiferal species and their production rates; the rate and vertical distribution of taphonomic processes; and the mode and depth of bioturbation. Their model is based on deep-water environments where transport is not an active postmortem process.

Because of patchiness of living taxa, time averaging may cause the rank order to differ between the living and dead assemblages although the abundant species should be the same in both. Species diversity is generally higher in the dead assemblage because of the presence of a greater number of low-abundance species. This is due to addition of tests from their pulsating patches through time. In areas affected by transport of tests, there may be additional changes in diversity due to the introduction of exotic forms (see Section 9.3.2).

9.3.6 Diagenetic changes

It is beyond the scope of this book to deal with these in detail. Once the time-averaged dead assemblage has passed into the historical layer, it is beyond the influence of organisms and transport processes. The main changes are due to pore-water geochemistry and compaction. There may be further dissolution of calcareous tests or growth of cement on and within tests.

9.3.7 Weathering at outcrop

When strata are exposed at outcrop, they are subject to chemical weathering that may modify or destroy the fossil record. The best way to overcome this problem is always to remove the surface weathered layer and collect fresh material. However, in some situations the thickness of the weathered zone precludes this possibility. The effect of weathering on foraminiferal preservation has never been seriously investigated.

9.3.8   Consequences for interpretation of fossil assemblages

The impression might be gained that taphonomic changes will make a reliable interpretation of a fossil assemblage difficult to carry out but that is certainly not the case. Basically, it is important to recognise taphonomic changes and to take them into account when making an interpretation. The majority of fossil assemblages preserve a good record of the original environment.

9.4   Summary

- Life processes influence the rate of contribution of tests to the sediment.
- Postmortem processes include destruction of tests, transport, bioturbation and time averaging.
- Agglutinated tests with organic cement may be destroyed by bacterial or chemical decay of the cement or through predation particularly in areas of low sedimentation.
- Calcareous tests may be partially or wholly dissolved through corrosive pore water (influenced by bacterial decay of organic matter and bioturbation) and also beneath deep and very cold, dense polar bottom water (lysocline, CCD).
- Transport may be as bed load or suspended load and is caused mainly by tidal currents or waves. Source areas lose tests; depositional areas gain tests.
- Tidally influenced transport is most active in mesotidal to macrotidal marginal marine and inner-shelf environments.
- Wave-influenced transport is most active in inner-shelf seas shallower than storm-wave base.
- Other transport mechanisms are mass flow of sediment, in grounded ice, floating plants and wind.
- Bioturbation mixes sediments in the TAZ, which extends from the sediment surface to the maximum depth of bioturbation to give a time-averaged record.
- Beneath the TAZ is the historical layer in which material is effectively fossilised.
- As long as the effects of taphonomic processes are taken into account, it is possible to make palaeoecological and other interpretations of fossil assemblages.

10

Applications

10.1 Introduction

All applications involving benthic foraminifera must include an understanding of their ecology. The reason for this is that benthic foraminifera are closely linked to their environment of life. Even subtle changes in environmental parameters are matched by faunal change. That is why most of the book is devoted to an ecological synthesis because that provides the modern database with which fossil assemblages may be compared. The aim of this chapter is to describe the rationale and methods used in the various applications and to give case studies as appropriate but no attempt is made to provide a synthesis of all available results. There are five broad groups of applications with considerable overlap between them.

- Sequence stratigraphy – biostratigraphy.
- Interpretation of past environments, such as marsh, brackish lagoon, etc. in the geological record (palaeoecology).
- Interpretation of specific details of environments or processes such as palaeoproductivity, low oxygen, seasonal stratification of the water column, etc. (palaeoceanography).
- A proxy for natural environmental change (e.g., sea level, climate).
- Monitoring of changes induced by the activities of man.

In its broadest sense, palaeoecology might be considered to include palaeoceanography and environmental change but I have chosen to treat them separately as they have become important fields in their own right. Before considering these applications it is appropriate to comment on the use of the modern database as a reference for interpreting the fossil record.

10.2   The present as a key to the past

Interpretations of the fossil record depend largely on comparison with modern ecology (or geochemistry for chemical techniques). Although in a general sense it may be true that environments broadly similar to those we see now have existed through long periods of geological time, it would be wrong to consider that modern and geological examples are exactly analogous. Apart from physical and chemical differences through time, organisms evolve and also exert a profound influence on environments. For instance, prior to the evolution of seagrasses there were no seagrass habitats (providing a substrate for attachment, shelter, and a distinctive epiphytic microflora serving as food) and therefore no seagrass-related foraminiferal assemblages. During the Mesozoic the Lagenina were much more important in habitats now occupied by the Rotaliina. The modern fauna has its origins in the mid Miocene. Furthermore, the niches of modern species are related to the variability of modern environments. If a past environment exceeded the limits of modern examples (e.g., a much higher partial pressure of CO_2) then there is no true modern analogue and any interpretation of such past environments is partially speculative. Also, we cannot be sure that the niche of a given fossil species was exactly the same as its extant representatives. Breeding experiments have shown that representatives of the same morphospecies may not be able to interbreed throughout their geographic range (Kitazato et al., 2000b) and this must indicate subtle differences of niche even within extant species. Additionally, species evolve to adapt to their habitat thus the system is dynamic not static.

Prior to the Quaternary species migrations are generally considered to be more-or-less instantaneous on a geological timescale. In the Quaternary, when a higher-resolution timescale can be established, and when most of the species have modern representatives, there are nevertheless associations of species that do not occur in modern environments – the so called 'non-analogue assemblages'. Do these represent environments beyond the range of modern examples? Or has the duration of the present interglacial not been long enough for species to migrate into most potential habitats? Sea level has risen ~120–130 m in the past 12 ky thus causing most-shallow water environments to migrate shoreward and there has been latitudinal migration of temperature belts as the climate warmed. One view is that species are interdependent so a new combination in a fossil assemblage will be difficult to interpret. Alternatively, if species are considered to be independent of one another, each having its own requirements (niche), then each fossil assemblage can be interpreted in terms of the niches of the component species (subject to the proviso of slightly changing niches). In reality, such problems are of lesser importance in palaeoecological studies where the object is to recognise the broad environment (brackish marsh, upper bathyal, etc.). However, in palaeoceanography where the objective is to detect, and ideally quantify, subtle differences within the same broad environment, these problems become more pressing.

We live in an interglacial and that limits our ecological database. There is no reason to suppose that past interglacials were exactly the same as one another or as that of today. Do high-latitude cold conditions in the present interglacial serve as reliable detailed analogues of past glacial events especially in mid latitudes? Perhaps only partially. In the deep sea, it was only during the late Palaeogene that cold bottom waters developed and during the Mesozoic there is evidence of low-oxygen conditions on a scale quite unknown in modern oceans. Therefore, even broad environments have changed through time and may have no precise modern analogues.

10.3   Sequence stratigraphy – biostratigraphy

Sequence stratigraphy views sedimentary units as products of a dynamic system of erosion, changing supplies of sediment and depositional patterns (systems tracts) in response to a range of factors including uplift/ subsidence, eustatic changes of sea level and climate changes induced by Milankovitch cyclicity. The geometry of depositional units is determined by geophysical methods. The lateral and vertical environmental changes are documented by the fossil record, which provides both a relative timescale (biostratigraphy) and an interpretation of depositional conditions (palaeo-ecology) which aid in determining sequence boundaries in boreholes and at outcrop. In practice, a wide range of microfossil groups are used to provide a synthesis. Benthic foraminifera are the principal microfossils used in recon-structing sea-floor conditions. Because of the importance of relative changes of sea level in the development of sequences, foraminifera are commonly used to estimate palaeo-water depths. Changes of water depth brought about by cli-matic variability are likely to contain signals other than depth (e.g., tempera-ture). Although the ideal may be to determine absolute water depth in reality it is much easier and more practical to determine relative changes (inner shelf, outer shelf, etc.). The ecological data summarised in this volume provide a basis for interpreting past relative water–depth changes. Determination of absolute water depth is sometimes undertaken in an empirical way. The fossil species are categorised into depth groups and the interpretation is based either on some form of averaging (Wakefield, 2003) or other simple mathematical manipulation (Li *et al.*, 2003). Such techniques are probably more reliable for detecting trends than providing accurate palaeo water depths and in this respect they play a useful role in sequence stratigraphy (see the Oligocene example in Li *et al.*, 2003).

A sequence-stratigraphic model for primarily shelf siliciclastic sequences proposed by Leckie and Olson (2003) is based on a database including dead and total assemblages (Table 10.1). The SHE analysis of stratigraphic successions can be used to identify critical bounding surfaces (Wakefield, 2003). Sequence boundaries are marked by a reduction in all measures of species diversity. Transgression leads to new habitats and an increase in species richness, ln(S), and evenness, ln(E). Maximum values of all three SHE components typify maximum flooding surfaces. It is likely that Fisher alpha species diversity

Table 10.1. Sequence–stratigraphic model of siliciclastic foraminiferal-assemblage characteristics (based on data in Leckie and Olson, 2003).

Transgressive surface
- Marked increase in marine foraminifera
- Rapid deepening
- May be a concentrated (lag) horizon

Transgressive systems tract
- Decreased shelf sedimentation (clastic sediment deposited in marginal marine environments)
- Increase in marine microfossils
- Presence of reworked microfossils and/or glauconite

Maximum flooding surface
- Change from deepening upward to shoaling upward trend
- Sediments may be fine grained and there may be dysoxic conditions

Highstand systems tract
- Microfossils diluted by increased rate of sedimentation
- Increased runoff, therefore increased nutrients and productivity
- Down–slope transport of shelf foraminifera
- May be dominance of epifaunal benthic foraminifera

Sequence boundary
- Marked shift to microfossil-poor or barren sands
- Rapid shallowing

would show these trends even more clearly as it is more sensitive than H to environmental change. Nearly all modern environments are part of a transgressive systems tract with maximal sedimentation in marginal marine settings and sediment-starved shelves. Epicontinental seas such as the Adriatic may provide a near analogue of maximum-flooding surface conditions. During eustatic lowstands, shelves are exposed to subaerial erosion but there is no modern analogue.

A distinction is sometimes made between biostratigraphy (ordination and calibration of irreversible bioevents such as speciations and extinctions) and ecostratigraphy (reversible bioevents such as repetition of recurrent associations) (Li *et al.* 2003). However, even speciations and extinctions must be mainly environmentally controlled. It has long been accepted that assemblage biozones based on benthic foraminifera are tied to ecology with most biozone boundaries defined on a faunal change directly linked to a change in environment. Benthic biozonal schemes are usually specific to each basin and some larger basins have more than one scheme for different parts. The need for high-resolution biostratigraphy especially at the scale of reservoirs in single oil fields has led to greater use of ecologically significant faunal changes on a local scale. Such biostratigraphy (1) provides an oil-field-specific stratigraphic framework, (2) tracks facies changes through time and space, and (3) identifies lateral and/or vertical heterogeneity, which are used to determine compartmentalisation in reservoirs. It also plays a fundamental role in 'biosteering' drill sites along near-horizontal target bioevents, sometimes with a resolution of ~30 cm (there are numerous examples in Jones and Simmons, 1999).

For high-quality stratigraphic studies it is necessary to gather high-quality data: faunal counts based on samples from specific levels. Then a variety of interpretational techniques (including multivariate and univariate statistics) can be performed. However, many studies are based on ditch cuttings which do not yield proper assemblages from precise levels. It is clearly more difficult to get reliable interpretations from such material.

10.4 Palaeoecology

Palaeontological data can be used to interpret the conditions under which stratigraphic sequences were laid down. This is done by comparing the composition of the fossil assemblages with data from modern analogues. With passage back in time through the Neogene, Palaeogene, Cretaceous, etc., the fossil faunas are increasingly different from the modern ones. Therefore, the recognition of past environments based on species and genera becomes progressively more tenuous back into the fossil record. However, general attributes of assemblages such as patterns of species diversity and wall structure should be more conservative features (although hyaline walls did not become important until the Triassic). Likewise most postmortem processes of today are likely to be essentially the same as those of the past. Palaeoecological interpretations can be applied not only to the fossil record but also to archaeological investigations.

The procedure outlined by Murray (1991) for making a palaeoecological interpretation of a fossil assemblage is the following.

- Make an assemblage count of the microfauna.
- Check for postmortem effects (transport loss or gain; loss of calcareous and/or agglutinated tests).
- If these effects are minimal, proceed.
- Compare the assemblage species diversity with Figure 8.4 and Table 8.1.
- Compare the wall-structure composition with Figure 8.3.
- Use the Appendix to determine the ecological requirements of the main species (certainly those >10%).

The species-diversity and wall-structure plots will give the range of possible environments. Ecological information for the main species should narrow the possibilities, ideally to a single environment. It may also be necessary to refer to specific chapters to gain additional information. Techniques to refine interpretations in terms of specific variables (oxygen, food supply, etc.) are discussed under palaeoceanography.

Where there is severe postmortem modification it may not be possible to interpret the past environment except in a very general way. However, if dissolution of calcareous tests results in only agglutinated forms being preserved (and there is no evidence of transport) it may still be possible to determine the environment. Experimental dissolution of calcareous taxa from assemblages of known modern environments has shown that the relict agglutinated assemblages retain much useful ecological information both in species distributions and also in species-diversity patterns. Experimental dissolution of original dead assemblages (ODAs) yields acid-treated assemblages (ATAs), which are analogues of fossil assemblages (Murray and Alve, 1999b, Figure 9.1). Although species diversity is invariably lower in the ATA than the ODA of the same sample, nevertheless the pattern of increasing diversity with passage from marginal marine to deep sea is still preserved especially by the Fisher alpha index.

Case study: Rockall Plateau, North Atlantic

There are two reasons for choosing this example: (1) there are faunal counts (essential for using the method outlined above) and (2) there is a major contrast in environments. The western side of Rockall Plateau is a sediment-starved passive margin with a Cenozoic succession extending discontinuously from Palaeocene to Recent. Deep Sea Drilling Project (DSDP) site 553 drilled in a water depth of 2329 m recovered a mixed volcaniclastic and sediment succession from the Palaeogene and biogenic oozes from the Neogene. The Palaeogene successions have undergone some dissolution but the assemblages can still be interpreted from the better-preserved samples (Figure 10.1). Porcelaneous forms are absent and all assemblages are dominated by hyaline tests. With the exception of the inner shelf example, the species diversities fall in the appropriate fields on Figures 8.3, 8.4. During the Palaeogene, the margin of Rockall Plateau was close to sea level. Progressive subsidence was not matched by sediment accumulation as the waters deepened from marginal marine to shelf to bathyal. In addition to identifying the main environments, Murray (1984) also related the assemblages to broad water depths and, for the Neogene, to bottom water masses.

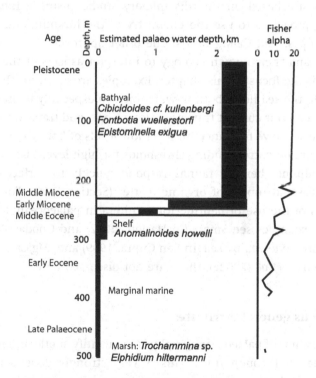

Figure 10.1. Summary of palaeoecological changes in DSDP site 533, Rockall Plateau, North Atlantic (based on data in Murray, 1984). The range of inferred palaeo water depth is indicated by the black shading.

10.5   Palaeoceanography

The aim of palaeoceanography is to reconstruct conditions in past oceans. This means not just recognising the broad environment but attempting to quantify selected parameters including salinity, temperature, dissolved oxygen and surface primary production. Whereas there are several planktonic groups that leave a record of surface oceanic processes (coccolithophorids, diatoms, dinocysts, planktonic foraminifera, radiolaria, silicoflagellates) there is only one important benthic group, namely benthic foraminifera, for recording the history of events on the sea floor in all environments from the intertidal zone to hadal depths. The atmosphere and oceans are closely linked and processes at the ocean surface influence biological activity on the sea floor (benthic–pelagic coupling). The benthic-foraminiferal record can be used to interpret past bottom conditions (bottom-water masses, oxygen minimum zones (OMZs), currents, methane seeps, depth of the lysocline and CCD) and also to give information about surface productivity (flux of organic material from the sea surface). The most robust interpretations of palaeoceanography are based on interdisciplinary studies using a range of techniques. One approach is to use the chemistry of the foraminiferal tests, notably stable isotopes, and Ca/Mg ratios. The other approach is to use the abundance of taxa and their known ecology to interpret aspects of the environment and that is the focus of this chapter. Examples are given for the flux of organic matter to the sea floor, bottom-water oxygen especially in an OMZ, position of former contour current transport along slope, and transport down slope. Although species have tolerance limits at low levels of both oxygen and organic flux, there are no corresponding thresholds for high levels. Because of benthic–pelagic coupling, benthic faunas respond quickly to surface-water changes such as seasonal inputs of organic matter (Section 7.2.2). For comprehensive reviews of the use of benthic foraminifera in palaeoceanography and for additional examples see Smart (in Haslett, 2002) and Gooday (2003). Chemical proxies are reviewed by Lea (in Sen Gupta, 1999) and Mg/Ca by Lear *et al.*, (2002) and Martin *et al.* (2002a); these are not discussed here.

10.5.1 Proxy versus general attributes

The primary aim of palaeoceanography is to quantify subtle differences in selected parameters through time. This is very strongly dependent on the modern ecological database. The concept of a proxy is based on a strong statistical correlation between the abundance of a species and the value of a single environmental factor. However, biological systems are not well ordered; the niche of each species is defined by the limits of tolerance of all the limiting factors and different factors become close to the tolerance limit at different times. These make it difficult to determine statistically robust proxy relationships. However, in environments where the variation is principally in one factor, there is a greater chance of establishing a proxy relationship (Murray, 2001). This may be true for oxygen and organic flux in the deep sea although there may be an oxygen influence superimposed on the flux relationship in dysoxic settings. There is co-variation between oxygen/organic flux and water depth and this explains the absence of isobathyal species (van der Zwaan *et al.*, 1999). Where no reliable proxy relationship exists the alternative is to use the range of ecological attributes of the species present in an assemblage to interpret past conditions. This is less easily quantified and more subjective. It is perhaps worth stressing that virtually all species occur in low abundance in a wider range of environmental conditions than near–optimum conditions of their high abundances. This is clearly shown for organic flux where high abundance occupies a much smaller range than low abundance (Figures 7.2, 7.3). In Chapters 4–7 all the assemblages are defined on dominance and subsidiary species with >10% abundance. It is only these higher abundances that should be used to infer past environments. However, as pointed out by Smart (in Haslett, 2002, p. 56), 'If certain species are always taken to indicate a particular environment then there is only one possible interpretation'. He cautions that 'all palaeoceanographic interpretations are inevitably very subjective'. However, since all techniques, including the use of stable isotopes and other chemical proxies, are based on assumptions, there is no reason to believe that the use of benthic foraminifera in environmental interpretation is less reliable than chemical methods. In combination they both complement and sometimes provide checks on one another.

10.5.2 Transfer functions

Where a proxy relationship exists between foraminiferal abundance and a factor it can be expressed as a transfer function. When data from the fossil record are interpreted using a transfer function the output is a value for the factor. The reliability of the predicted value is dependent on the robustness of the proxy relationship. Almost invariably the modern ('training') dataset is either the dead or total (live plus dead) assemblages. These represent a significant period of time averaging but are nevertheless assumed to be representative of the measured (or inferred) value of the factor. Although all authors state that dead or total data will give a better correlation than live data they do not provide evidence that this is true (i.e., they rarely present the correlation with live data). Transfer functions have been developed for oxygen, productivity, sea level (see below), temperature, salinity and water depth (Sejrup *et al.*, 2004; Debenay and Guillou, 2002; Hayward *et al.*, 2004b). It is true for each of these that the transfer function works best when the interpreted record is from the same area as the training set and, very importantly, *where the variability of the past environment does not exceed that of the training set*. It is also important that interpretations are based on assemblages that have not experienced postmortem alteration. All records are 'noisy'. Ideally, the fossil counts and the parameter values interpreted using the proxy relationships should be accompanied by error bars. Only that way can it be judged whether there are statistically meaningful differences between samples.

10.5.3 Benthic-foraminiferal accumulation rate (BFAR)

The concept that the accumulation rate of benthic foraminifera could be used as an indirect measure of surface-water productivity (i.e., flux of organic matter to the sea floor) was introduced by Herguera and Berger (1991) and has been applied in several studies as discussed below. The BFAR is as a measure of the accumulation of tests per unit area of sea floor per unit interval of time: BFAR cm^{-2} ky^{-1} = tests g sediment × sediment accumulation rate (Herguera, 1992). These BFAR estimates have been used as reliable guides to productivity in general but they are not reliable in areas where carbonate dissolution takes place (Wollenburg *et al.*, 2001) or where there is localised variation in productivity in OMZs (Naidu and Malmgren, 1995).

10.5.4 Organic flux

The importance of organic flux in controlling some deep-sea species especially at water depths greater than 1000 m is now reasonably well established with the recognition of high- and low-flux species and those adapted to seasonal inputs (Section 7.2.2). However, where food supply is very low (as in oligotrophic environments) other factors (temperature, salinity, carbonate corrosivity) may play a more important role in species distributions. The recognition of high flux may be made on the presence or absence of flux-related species including phytodetritus species. Regression of multivariate analysis of dead or total assemblage data and surface productivity has been made especially by Loubere and Qian (1997) in order to establish a transfer function. Limitations to the method are the use of dead or total assemblages that may encapsulate a faunal record different from that of today and the unreliability of modern-ocean productivity data. Gooday (2003) is cautious about their applicability except as a means of recognising relative changes in palaeoproductivity.

Case study: palaeoproductivity in the Arctic Ocean

The Arctic Ocean receives heat from the Atlantic Ocean via the inflow of Atlantic water through the Fram Strait and Barents Sea and this affects the temporal and spatial seasonal ice retreat and also primary production. There is a thirty-fold difference in primary production between ice-free and permanently ice-covered areas but this is not recorded in the TOC content of the sediments as all values are dominated by ice-rafted coal fragments. Therefore, estimation of palaeoproductivity on faunal data is important. The benthic foraminiferal record of two cores from 995 and 2550 m water depth give a 145 ky record of past conditions (Wollenburg *et al.*, 2001). There is taphonomic loss of agglutinated foraminifera in the 995 m core and carbonate dissolution especially in glacial sections in both cores. The modern assemblages are calcareous as the deep water is well ventilated (Section 7.2.12). Only one meaningful canonical factor (CF1) is derived by the correspondence analysis of modern dead assemblages and this has a correlation coefficient $r = 0.9$ with organic-carbon flux. Using this as a 'transfer function' gives reasonable estimates of productivity (within the current range). However, false high palaeoproductivity values are given in part of the core from 2500 m where there are relict agglutinated foraminifera (*Saccammina, Rhabdammina*) because these are absent from the training dataset. In the 995 m core, mean glacial values are roughly a third of modern values but significantly higher than those of the permanently ice-covered areas in the modern Arctic Ocean. This indicates that seasonally enhanced primary production operated during this time. Following meltwater events at 18.1–14 and 11.7–9 calibrated ky, there were rapid shifts to high palaeoproductivity and higher numbers of phytodetritus species. In the 995 m core, stable high productivity at the ice-shelf edge is inferred from the *Melonis zaandamae* principal component (PC) association and highly seasonal fluxes during ice advance or retreat from the site by *Hormosinella guttifer, Cassidulina reniforme* and *Cassidulina teretis*. At the 2550 m site, enhanced productivity is related to the *Fontbotia wuellerstorfi–Globocassidulina subglobosa* PC association. Temperate Atlantic water entered the Arctic Ocean in oxygen isotope stages 6.3, 5.5 and the Holocene and palaeoproductivity maxima indicate extended seasonal ice retreat. During glacial periods there was very low productivity and the poorly ventilated bottom water at the 2550 m site was corrosive but never deficient in oxygen. Thus, here the productivity signal is not confounded by co-variance with low oxygen.

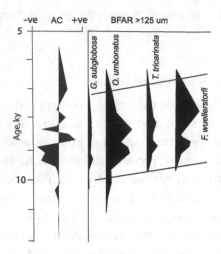

Figure 10.2. Fram Strait core PS 1906. AC = amplitude of change (a ratio of change between adjacent samples). The BFAR peaks during a period of enhanced organic flux. Note the succession of species (based on Nees and Struck, 1999).

Case study: palaeoproductivity in the Fram Strait

The Fram Strait is the gateway between the Norwegian Sea and the Arctic Ocean. It experiences strong seasonality and winter ice cover with a fluctuating ice edge. Warm saline Atlantic water flows north and cold low-salinity Arctic water flows south. The BFAR records spanning the Holocene show an overall peak between ~10 and 7 ky made up of a succession of species (Figure 10.2; Nees and Struck, 1999). Prior to 10 ky the area was ice covered so organic flux was low. The change to higher BFARs is interpreted as indicating breaks in ice cover with consequent increase in productivity. The presence of suspension-feeding, epifaunal *Fontbotia wuellerstorfi* indicates bottom current activity.

10.5.5 Oxygen

Oxygen is limiting only when it approaches the tolerance limit for a species. Once oxygen is above the threshold, it is no longer important (Section 8.3.1). Therefore over large areas of the modern ocean floor oxygen is not limiting because the bottom waters are well oxygenated. This may not always have been the case in the past.

An empirical relationship between calcareous benthic foraminifera (live and dead, >63 μm) and dissolved-oxygen levels in the oceanic bottom water has been established as the BFOI (benthic foraminiferal oxygen index; Kaiho,

Table 10.2. Oxygen conditions and faunal groups (summarised from Kaiho, 1994).

Oxygen conditions	Faunal groups
Oxic	Large planoconvex, biconvex, rounded trochospiral and spherical tests (>350 μm) with thick walls (*Cibicidoides*, *Nuttallides*, *Globocassidulina*) and large miliolids.
Suboxic A	Small individuals of oxic species (<350 μm)
Suboxic B	*Lenticulina*, large *Nodosaria*, *Dentalina*, *Pleurostomella*, large ornamented *Bulimina* and *Stilostomella*, rounded planispiral, flat ovoid, and spherical taxa, small and/or thin-walled, planoconvex and biconvex trochospiral taxa, *Uvigerina*, *Oridorsalis*, *Gyroidina*, *Hoeglundina*
Dysoxic	*Bolivina* spp., *Cassidulina tumida*, *Chilostomella* spp., *Fursenkoina complanata*, *Fursenkoina rotundata*, *Globobulimina* spp., *Rutherfordia* spp, *Suggrunda eckisi*

Table 10.3. Correlation of BFOI with environmental conditions and microfaunas (based on Kaiho, 1994).

Oxygen condition	Oxygen, ml l^{-1}	BFOI	r^2	Foraminiferal indicators
High oxic	3.0–6.0+	50 to 100	0.01	Dysoxic, suboxic, and high ratio of oxic
Low oxic	1.5–3.0	0 to −50	0.32	Dysoxic, suboxic, and low ratio of oxic
Suboxic	0.3–1.5	−40 to 0	0.26	Dysoxic and high ratios of suboxic
Dysoxic	0.1–0.3	−50 to −40	0.48	Dysoxic and low ratios or barren of suboxic
Anoxic	0.0–0.1	−55	0.50	Barren of calcareous forms

1994; 1999). Three faunal groups are recognised: O = oxic, S = suboxic, D = dysoxic (Table 10.2). Where BFOI = {[O/(O + D)] × 100} is greater than zero, the waters are oxic (oxygen > 1.5 ml l^{-1}; Table 10.3). For suboxic and dysoxic environments the calculation is BFOI = {[S/(S + D)] − 1} × 50 giving values ranging from 0 to −50 (although Kaiho recognised a suboxic group C, he excluded them from the calculation). Anoxic environments lacking calcareous benthic foraminifera are given a value of −55. Overall, the correlation between BFOI and overlying bottom-water-dissolved oxygen is $(r)^2 = 0.81$. The values of r^2 for the correlation of BFOI with the ranges of dissolved oxygen (Table 10.3) are not very high even for the most dysoxic state. The correlation between BFOI and both primary productivity and carbon flux to the sea floor is $r^2 = 0.51$ and 0.55, respectively. Kaiho maintains that BFOI is a better indicator of benthic oxygen levels than organic flux but Gooday (2003) considers this problematic. While it is likely that low-oxygen environments can be broadly recognised using this index, the correlation with precise values of oxygen is less certain.

The potentially most successful transfer function for estimating oxygen at the sediment–water interface is that developed by Jannink et al. (in Jannink, 2001) based on the regression O_2 [μM] $= 7.23 + 5.62 \times$ (% living oxyphilic individuals), $r^2 = 0.66$. They tested this on a core from the Adriatic Sea and found that the estimated values parallel the historical records of oxygen measured 2 m above the sea floor (with an offset of 80–100 μM). This requires further testing on other historical records.

Case study: oxygen minimum zone

The northwestern Indian Ocean undergoes seasonal changes in upwelling and primary productivity related to the monsoon. The winter (northeast) monsoon is a period of low productivity. The summer (southwest) monsoon leads to upwelling of nutrient-rich waters, which cause high primary productivity and high fluxes of organic matter to the sea floor. An intense OMZ is developed between 150 and 1200 m with oxygen levels below 2 μM (<0.01 ml l^{-1}). The Murray Ridge rises into the lower part of the OMZ. One core from the OMZ (920 m) and one from beneath it (1470 m; den Dulk et al., 1998) and another from the OMZ on the Pakistan margin (1002 m; den Dulk et al., 2000) provide palaeoceanographic records back to the Quaternary. Although various measures were used for palaeoceanographic reconstruction, only the benthic-foraminiferal data are discussed here. The benthic foraminifera are well preserved. The 150–595 μm species having an abundance $>5\%$ from the Pakistan margin core at 1002 m were analysed using principal component analysis. Only factor 1, accounting for 22% of the variation, is relevant because it separates two ecologically meaningful assemblages. Those species with high positive loadings (>0.5; assemblage 1) include Bulimina striata, Cassidulina carinata, Gavelinopsis lobatulus, Chilostomella oolina, monothalamous species, Sphaeroidina bulloides, Cibicides ungerianus, Hyalinea balthica, Hoeglundina elegans and Quinqueloculina spp. Several of these are intolerant of low oxygen. Although Chilostomella is associated with low oxygen it is deep infaunal so does not reflect bottom-water conditions. Species diversity is high ($H > 3.0$) and dominance is low. Assemblage 1 is interpreted as indicating well-oxygenated conditions and occurs in bioturbated sediment that has macrofauna. It relates to minima in summer-monsoon productivity. Those species with high negative loadings (> -0.36; assemblage 2) include Bulimina exilis, Rotaliatinopsis semiinvoluta, Bolivina alata, Bolivina pygmaea, Fursenkoina bradyi, Globobulimina spp. and Bulimina sp.

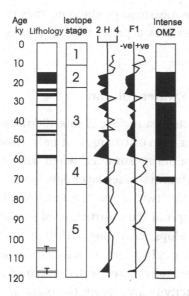

Figure 10.3. Northern Arabian Sea core NIOP455: $H =$ information function; F1 = Factor 1, positive loading, assemblage 1; negative loading, assemblage 2; T = turbidite (adapted from den Dulk *et al.*, 1998).

Assemblage 2 has lower species diversity ($H < 2.0$–3.0) and low equitability due to high dominance. It is interpreted as indicating severe dysoxia and is commonly associated with laminated sediment lacking bioturbation and macrofauna. Assemblage 2 also equates with high organic flux (the cause of the low oxygen) and relates to precession-driven maxima in summer surface-water productivity. Assemblage 1 (positive factor scores) dominates in isotope stages 1, 4 and 5 (Figure 10.3). Assemblage 2 (negative factor scores) dominates in isotope stages 2 and 3 and for brief intervals in stages 4 and 5. The precession cycles are less obvious in stages 2 and 3. Den Dulk *et al.* (1998) suggest that an intensified and colder glacial winter monsoon led to increased winter production. When superimposed on the precession-driven changes in surface-water productivity it led to poorly oxygenated bottom waters during these glacial stages.

The two Murray Ridge cores have assemblages which are divided into five clusters, of which IV and V have low species diversity and taxa indicating oxygen-deficient water while III has higher species diversity and taxa indicative of oxygenated conditions. In particular, den Dulk *et al.* (2000) consider that the abundance of *Quinqueloculina* spp. can be used as a guide to oxygenation as these forms are intolerant of suboxic conditions. However, estimates of surface primary productivity based on BFARs (using the Herguera and Berger, 1991, method) give values twice as great as the measured production. They therefore conclude that this method does not work as a proxy where there is severe dysoxia.

10.5.6 Seasonal stratification of the water column

Heat from the sun received by the sea surface is mixed downward by wind and tide. Strong mixing leads to an even distribution of heat throughout the water column. In temperate regions, during summer, when wind-induced mixing is weak and heat input higher, a warm surface layer develops overlying cooler water. The boundary between the two is the thermocline. Autumn storms lead to increased vertical mixing and the destruction of the layering. Seasonal stratification of this kind develops where $\log_{10}(H/U^3)$ is >2 (where $H=$ water depth and $U=$ the maximum tidal stream velocity at the surface). The boundary between stratified and vertically mixed water is termed a front, is often very sharp, and is a high productivity area. From the point of view of the time-averaged foraminiferal record, because the areas that are stratified in the summer are also vertically mixed in the winter, the resultant assemblages are not entirely distinct. Recognition of the onset of stratification in the examples below is based primarily on the presence of *Stainforthia fusiformis* (as a frontal-zone indicator) together with *Bulimina marginata*.

Case study: UK shelf

Summer stratification develops in the seas to the west and east of Britain. The history of development of stratification during the postglacial rise of sea level is now known from the Celtic Sea (west of UK; Scourse *et al.*, 2002) and North Sea (east of UK; Evans *et al.*, 2002). A 5.68 m core from 118 m in the Celtic Sea spans the last 11 ky. It is divided into three lithological units: an upper (3) and lower (1) mud unit sandwiching a coarse shelly sand and gravel sequence (2); and three faunal zones (F1–F3, Figure 10.4). Zone F1 has sparse abraded tests suggesting transport but the two higher units have rich assemblages. The key parts of the succession are zones F2 and F3. Subzone F2a has a high abundance of *Quinqueloculina seminulum* ($>80\%$) and spans the upward fining sequence from the coarse sand (2) to mud (3). Subzone F2b shows an upward decrease in *Quinqueloculina seminulum* and an increase in *Bulimina marginata*. Zone F3 is dominated by *Bulimina marginata* with some *Stainforthia fusiformis*. Bioturbation has blurred some of the faunal changes but the overall picture is clear. The dominance of *Quinqueloculina seminulum* and its associated fauna is interpreted as indicating vertically mixed waters throughout the year. The incoming of *Stainforthia fusiformis* in zone F2b is taken as representing the

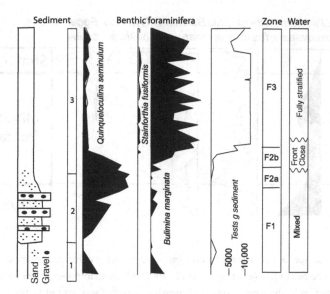

Figure 10.4. The onset of seasonal thermal stratification in the Celtic Sea (adapted from Scourse *et al.*, 2002). Column 1, unshaded indicates muddy sediment.

onset of seasonal stratification (the authors acknowledge that this opportunistic species also occurs in other settings). *Bulimina marginata*, the dominant species of zone F3, also occurs beneath modern stratified waters. The onset of seasonal stratification is dated as ~6.2 ky. The authors suggest that the frontal region was close to the core site from ~8990–8440 cal y BP based on isotopic evidence.

The North Sea core is from 55 m and has a record going back to ~8 ky (Figure 10.5). The sediments range from silt to very fine sand. The common species are *Ammonia batavus* and *Elphidium excavatum*, interpreted as indicators of mixed waters; and *Stainforthia fusiformis*, *Bulimina marginata*, *Hyalinea balthica* and *Eggerelloides scaber*, interpreted as indicators of the onset of seasonal stratification. Fully stratified conditions did not develop until 1158 cal y BP. The time difference in the establishment of stratification in the two areas may reflect the different water depths of the studied cores.

10.5.7 Former contour currents

Where contour currents impact on the sea floor, the vertical flux of organic material is seriously modified by lateral advection. In such settings there are distinctive benthic assemblages.

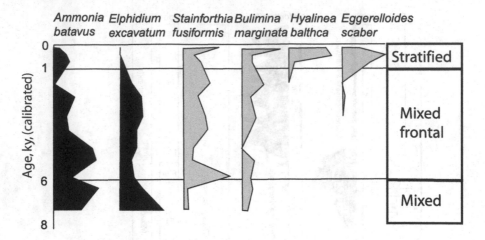

Figure 10.5. Species distributions and interpreted water column conditions in a core from the North Sea (after Evans *et al.*, 2002). Black shading – species indicating mixed water; grey shading – species indicating seasonally stratified water.

Case study: Mediterranean Outflow Water

Although the positions of major currents and water masses may vary with climate change, the basic pattern has been operating for a long period of geological time. The Mediterranean Outflow Water (MOW) emerges from the Mediterranean through the Strait of Gibraltar and the main flow follows a narrow path along the Iberian continental margin at a depth of 600–1500 m. It has an upper- and lower-core layer at 600–950 m and 1000–1500 m, respectively. The lower-core layer meanders west and northwest. The upper core layer flows north with a mean current velocity of ~0.12. m s^{-1} and it has a profound effect on the foraminiferal fauna leading to a distinctive epibenthic assemblage of *Cibicides lobatulus, Discanomalina coronata, Hanzawaia concentrica, Planulina ariminensis* and *Vulvulina pennatula* (Section 7.2.3). Controls on the depth of the current include changes of sea level and associated changes of salinity and density in the Mediterranean. The down-core abundance of the epibenthos group has been used to track the former position of the MOW in the last Glacial and Holocene (Schönfeld and Zahn, 2000; chronostratigraphy-based δ^{18}O of planktonic foraminifera, ^{14}C AMS dating and Heinrich events). Cores from off southern Portugal have a high abundance of the epibenthic group at water depths of 819–1281 m in the Holocene but the abundance decreases with increasing water depth down slope. In cores from the north, only that from 1509 m has a significant content of Holocene epibenthic group. The authors conclude that MOW was ~700 m deeper during the last Glacial probably due to Mediterranean water being colder and more saline. Both layers detached from the continental margin off southern Iberia and there was only low current activity along the margin west of Portugal. The present pattern, with the upper layer flowing along the entire length of the Portuguese slope, was initiated between 7.5 and 5.5 ky ago.

10.5.8 Down-slope transport

The clastic sedimentary particles deposited on the continental slope-rise ultimately are derived from the continents but to reach the ocean they have to pass through the shallow waters of the shelf. Dead benthic for-aminiferal tests are biogenic sedimentary particles that can be transported away from a source area. Because foraminifera live in different environments they behave as 'labelled' particles when transported into a different environment. As discussed in Chapter 9, bed-load transport leads to abraded and damaged tests whereas transport in suspension selectively affects small tests which generally remain undamaged. Mass flow of sediment due to slope failure or local gravity-driven sediment creep may take place on continental slopes. Submarine canyons traversing the slope provide conduits to transport sediment from shallow to deep water.

Case study: Morocco

Inner shelf benthic *Elphidium crispum* and *Amphistegina* sp. are found in turbidites at a present water depth > 4 km. These are thought to be sourced from the Morocco shelf (Wynn et al., 2002). *Amphistegina* is relict from the Pleistocene along the Atlantic margin.

Case study: Bounty Trough, New Zealand

During the late Quaternary, terrigenous sediment eroded from the Southern Alps was transported through Bounty Trough and deposited on the abyssal Bounty Fan at ~4.4 km water depth. The Ocean Drilling Program (ODP) site 1122 was drilled on the levee of Bounty Channel at 4435 m. Both thin sandy turbidites and interturbidite muds have 20–95% displaced outer-shelf and upper-bathyal foraminifera (Hayward et al., 2004a). Turbidity-current activity occurred during interglacials and glacials but was more prevalent during the latter.

10.5.9 Wave climate

The influence of wave disturbance of the continental shelves is controlled by water depth and wave height and the effects are greatest during severe storms. At present wave disturbance of the sea floor on the west shelf of

Scotland causes transport of small foraminiferal tests from the inner shelf into depositional sinks (Murray, 2003b). However, muddy deposits from 4.5–12 ky ago lack transported foraminifera suggesting either that the water was too deep for waves to affect the adjacent sea floor or that the wave climate (direction) was lower (Murray, 2004).

10.5.10 Sea level

The aim of reconstructing past sea levels is primarily to determine vertical fluctuations in the boundary between land and sea. In a general sense, it is possible to recognise this boundary by the respective absence or presence of foraminiferal faunas (although there are some instances where foraminifera are blown from the shore onto land to form part of subaerial sand dunes; Section 9.3.2). The best settings for determining past sea levels are marshes and mangals because they accrete up to the level of the highest tides; they prograde in a seaward direction; they have an inclined surface with different parts of the marsh related to different tide levels and to durations of exposure; and they have distinctive foraminiferal assemblages, which in large part are related to elevation (Section 4.2.3). It is straightforward to relate a succession of assemblages to different parts of the upper intertidal zone in a relative sense. The challenge is to use Quaternary and Holocene fossil foraminiferal assemblages to obtain greater detail regarding past sea level. We might ask: why should there be a close relationship between foraminifera and elevation when there is no reliable relationship between foraminifera and water depth? The first attempt to relate foraminiferal assemblages to marsh elevation, and hence tidal range, was that of Scott and Medioli (1978) using total assemblages from Nova Scotia and California. Since then there have been numerous studies with this aim (see reviews by Scott et al., 2001; Gehrels, in Haslett, 2002). Authors have used either total or dead assemblage baseline (training) datasets. Dead assemblages give a record time averaged over decades. Total assemblages do the same except that they include the living assemblage which, unlike the dead component, has yet to undergo taphonomic change. Several authors have pointed out the importance of recognising faunal changes due to: (1) mixing of surface and subsurface (infaunal components) over time (e.g. Ozarko et al. 1997; Patterson et al., 1999; Hippensteel et al., 2000; 2002) although Patterson et al. (2004) consider the 0–1 cm interval to be the most reliable; and (2) taphonomic changes due to destruction of fragile tests, which may change an assemblage so that it appears to relate to a different sea level.

Although 'sea level' is widely used as a general term in reality the level of the sea is influenced by a range of parameters that cause it to vary continuously. The most obvious of these is tidal but all areas including those with little or no tidal influence experience changes due to barometric pressure and persistent winds. Tides vary in magnitude from microtidal to macrotidal. They also vary according to the phases of the moon, ranging from larger spring tides to smaller neap tides but within each of these there is also variation from one lunar month to the next. A variety of levels are defined, such as highest and lowest astronomical tides, mean high-water spring tide, mean sea level, etc. Modern foraminiferal training sets for transfer functions must be calibrated with appropriate tide levels. At present, authors include all species in their training sets. However, it is known that although some have a positive correlation with elevation, others are indifferent or have a negative correlation. More accurate reconstructions might result by using only those species with a positive correlation with elevation. It is also important to check that the training set is not strongly correlated with salinity rather than elevation (de Rijk, 1995). Some authors favour using local training sets (e.g., Gehrels, 1994). Comparison of local and regional training sets on the reconstruction of sea level in east England led Horton and Edwards (2005) to conclude that the latter is preferable. Local training sets lead to many fossil samples having no analogues whereas regional training sets have few such problems.

Case study: UK marshes

A training set from various locations in the UK has been used to determine past sea level in eastern England (Horton et al., 1999). The 'standardised water-level index' (SWLI) is calculated from:

$$X_{ab} = \{[(A_{ab} - MTL_b)/(MHWST_b - MTL_b) \times 100] + 100\},$$

where X_{ab} is the SWLI of station or tide level a at site b; A_{ab} is the measured elevation of station or tide level a at site b; MTL_b is mean tide level at site b; and $MHWST_b$ is the mean high-water spring tide at site b. All elevations are in m above Ordnance Datum (OD). The addition of the constant (100) ensures that all reconstructed values in the training set are positive. The authors developed a transfer function to relate dead foraminiferal abundance to SWLI. They included species clearly transported in from elsewhere but this is not logical (these species have been omitted from Figure 10.6). They then used a modern analogue technique MAT to test the reliability of the SWLI reconstructions. The tenth percentile of the dissimilarity range calculated between modern samples is used as an approximate threshold between a good analogue ($<$ tenth percentile) or no close analogue ($>$ tenth percentile) (Horton and Edwards, 2005). Clearly accuracy is reduced if there is postmortem loss of tests or transport.

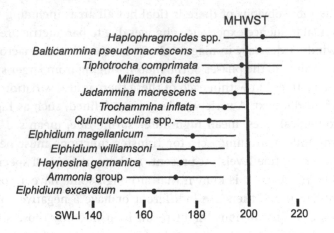

Figure 10.6. United Kingdom marsh training set. Ranges of selected species showing optima (weighted averages) and tolerances (weighted standard deviation) for SWLI (adapted from Horton *et al.*,1999). The range is sometimes termed 'indicative meaning' (Gehrels, in Haslett, 2002).

The transfer function was applied to data from a core from eastern England. Overlying glacial sediment (diamicton) is a peat and olive-grey silty clay. The diamicton and base of the peat are barren but the rest of the succession has a foraminiferal fauna (Figure 10.7). After SWLIs were calculated they were back-transformed relative to OD to give a continuous curve of reference tide level. This has an error range of ± 0.4 m. The mean high-tide level of 3.15 ± 0.41 m in the peat overlaps that of the modern MHWST 3.15 m OD. Changes in the overlying reference water levels may be slightly influenced by selective preservation of calcareous tests or changes in the rate of sediment accumulation. To overcome this problem Edwards and Horton (2004) introduced a new transfer function based on agglutinated foraminifera and test linings. This reduces the number of fossil samples without modern analogues. Only Horton and Edwards (2003) have tested the effect of seasonal changes in dead assemblages on transfer functions. They conclude that the ideal dataset is one that includes samples from all seasons (error = 0.21 m). Otherwise, the next-best alternative is a winter dataset because this has a higher proportion of agglutinated forms (in both live and dead assemblages). A similar approach using a local dataset for Connecticut, USA, predicts former levels with a precision of ± 0.9 m (Edwards *et al.*, 2004).

10.5.11 Storms and tsunamis

Severe storms such as hurricanes and cyclones and seismically induced tsunamis cause temporary flooding of coastal regions and transport

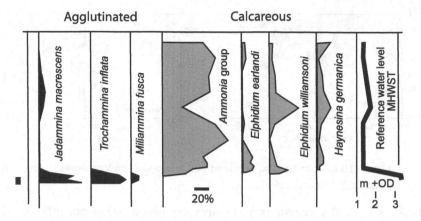

Figure 10.7. Reconstructing Holocene sea level; Lincolnshire, England (adapted from Horton *et al.*, 1999). The black bar on the *y* axis is a dated sample (8205–7977 cal y BP). Agglutinated black, calcareous grey.

shelf foraminifera into terrestrial or marginal marine environments (see review by Scott *et al.*, 2001). The case studies below give examples of transport of foraminifera from a subtidal area onto an adjacent intertidal marsh which is clearly recognisable because of the environmental contrasts. However, at present there are no micropalaeontological criteria to differentiate storm and tsunami deposits but the geological context can be helpful in some cases. Storm deposits intercalated in marsh deposits will be succeeded by marsh having a similar elevation to that preceding the storm. Where coseismic subsidence takes place, the environment above the tsunami deposit represents a lower base level than that beneath the deposit. Therefore, that may be a criterion to recognise a tsunami from a storm event. Nevertheless, tsunamis can travel across oceans and cause severe flooding thousands of kilometres away from the seismic event. In that case, tsunami and storm deposits might be indistinguishable.

Case study: Hurricane Hugo, USA

In 1988 Hurricane Hugo passed over the coast and deposited a 7 cm layer of sand containing subtidal foraminifera (*Ammonia* group and *Elphidium* spp.) in a marsh sequence (in core 1, with *Jadammina macrescens*, *Trochammina inflata* and *Miliammina fusca*) on the landward side of the barrier ridge in South Carolina, USA. A second core, taken from a non-tidal pond 50–75 km north has no lithological evidence of the hurricane but there is a zone of foraminifera intercalated in freshwater deposits with thecamoebians (Collins *et al.*, 1999). The authors draw attention to the need to differentiate between marine

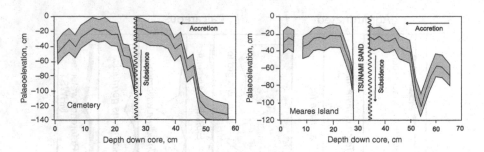

Figure 10.8. Calculated palaeoelevations of two sites on Vancouver Island, Canada (after Guilbault *et al.*, 1996).

incursions in tidal successions and storm deposits in lakes not influenced by tides. In a similar study also in South Carolina, Hippensteel and Martin (1999) use two criteria to recognise storm (washover) deposits in marsh environments: percentage of offshore foraminifera, both modern and Oligo-Miocene; abundance of calcareous foraminifera (the indigenous marsh fauna is mainly agglutinated). They stress that washover intervals are best identified using combined lithological and micropalaeontological analyses.

Case study: Vancouver Island, Canada

Vancouver Island, Canada, lies adjacent to the Cascadia subduction zone between the North American and Juan de Fuca plates. Marsh sequences on the island do not show a simple progressive upward sediment accretion. A core on the landward side of Vancouver Island shows two successive vertical sequences from low to high marsh. Another core has a sequence from low to high marsh, a sandy interval (with few foraminifera) and then middle to high marsh. Guilbault *et al.* (1996) use transfer functions to determine the palaeoelevations of these marsh successions (Figure 10.8). They reach two conclusions from their interpretations of the environments: that the area underwent coseismic subsidence associated with an earthquake (thus causing a repeated marsh sequence); and that the sand represents a tsunami deposit. Following subsidence there was crustal rebound.

10.6   Environmental monitoring

Geologists are aware that the world is constantly changing, albeit slowly. Some changes are cyclic and others unidirectional and, in addition, there are those caused through the activities of man. Major problems in coastal regions include pollution, eutrophication, habitat modification/destruction, aquaculture and introduction of exotic species. Subrecent fossil assemblages of

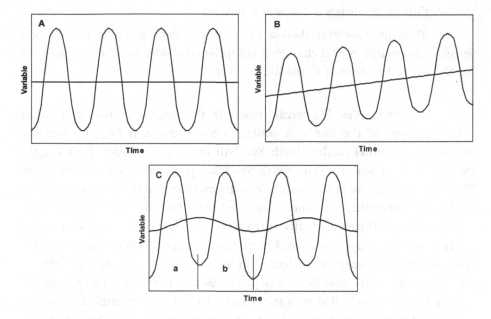

Figure 10.9. Variability and change. A. Variability about a mean with no long-term change. B. Variability with progressive change. C. Variability about a mean that has longer-term variability; segments a and b *appear* to indicate change (after Murray, 2000b).

benthic foraminifera preserve a record of past environments. The aim here is to define the problems and to provide some case studies. No attempt has been made to cover all the available literature. Useful background reading includes Martin (2000) and Haslett (2002).

10.6.1 Natural environmental variation and change

Variability is a fluctuation about a mean and takes place on a variety of time scales from diurnal to seasonal to years/decades or longer (Figure 10.9A). For an environmental change to take place, the mean itself must change (Figure 10.9B) but there may also be cyclic changes on which the variation is superimposed (Figure 10.9C). There is probably a continuum from short-term variability to long-term change and the problem is to know when variability includes a progressive change (Murray, 2000b). Marginal marine environments are especially sensitive to environmental change. In some cases there are long time-series records of hydrographic conditions that make environmental interpretation of past faunas well founded and more reliable.

Case study: changes in southern Scandinavia

This intensively studied area is an excellent example of a region sensitive to environmental change. It comprises the Skagerrak, a branch of the North Sea, and a series of fjords linked to it.

Skagerrak. The Skagerrak basin is the source of the deep-water entering many of the fjords in southern Scandinavia. It lies between the Norwegian Channel in the North Sea (sill depth 270 m) and the Kattegat. Deep–water renewal from the North Sea takes place about every 25 months. The Skagerrak is one of the major depositional areas in the North Sea and is a sink for fine-grained sediments derived from the southern North Sea. It also receives anthropogenically introduced nutrients. An early taxonomic study based on samples collected in 1937 (Höglund, 1947) provides data for comparison with material collected in 1992–93 (Alve and Murray, 1995a). The 1937 sediments now lie at a depth of 9–18 cm beneath the current sea floor and can be sampled in short cores. Alve (1996) determined the foraminiferal assemblages in two 40 cm cores going back to ~1770 (core 74) and 1870 (core 56) based on ^{210}Pb dating. In the lower part of the cores hyaline foraminifera dominate but in the upper part agglutinated forms become dominant. The sediment accumulation rates are 0.15 and 0.23 cm y^{-1} for cores 74 and 56 respectively. Alve used these values to calculate the accumulation rate of foraminiferal tests. In core 74, the accumulation rate of agglutinated tests increased from 20–40 prior to ~1970 to 174 tests 10 cm^{-2} y^{-1} in 1993 (Figure 10.10A) and the same trend but with higher values is present in core 56. This change matches the difference observed between the 1937 and 1992/93 surface samples. Compared with the 1937 data, four agglutinated species have increased in abundance and this is confirmed in the short cores (Alve and Murray, 1995a). *Eggerelloides medius* and *Haplophragmoides bradyi* show an upward trend of increasing accumulation rate. *Trochamminopsis pusillus* was not recorded by Höglund and occurs in low abundance or is absent prior to the mid 1970s but after that has a higher accumulation rate. *Saccammina* spp. are absent prior to ~1970 but increase to 17 and 26% (cores 74 and 56, respectively). Among the hyaline species, *Cassidulina laevigata*, *Nonionella iridea* and *Pullenia osloensis* show an upward decrease in accumulation rate.

Gullmar Fjord, Sweden. This pollution-free fjord (at least since the 1960s) has a sill depth of 42 m and a maximum depth of 119 m. The water is stratified with a low-salinity surface layer, well-oxygenated down to 15 m and

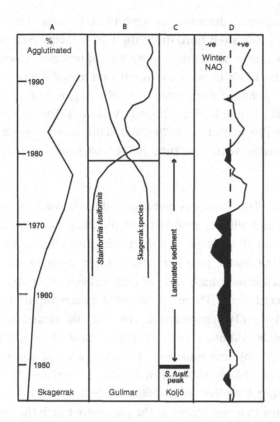

Figure 10.10. Comparison of data for southern Scandinavia. A. Accumulation rate of agglutinated tests in the Skagerrak (from Alve, 1996). B. Generalised faunal change in Gullmar Fjord, Sweden, core G116 (after Filipsson and Nordberg, 2004a). C. Occurrence of *Stainforthia fusiformis* peak in Koljö Fjord, Sweden (after Filipsson and Nordberg, 2004b). D. Oscillations in the winter NAO index (after Nordberg *et al.*, 2000).

intermediate down to a halocline at 50–60 m, and below this normal marine water derived from the Skagerrak and renewed during late winter or early spring. Bottom oxygen records show that dysoxic conditions occurred regularly from the 1970s. Since the 1980s when more observations were made severe dysoxia has occurred on many occasions. There have also been changes of salinity and temperature. Three 50 cm-long sediment cores from 110, 113 and 226 m consist of organic-rich clay. The foraminiferal assemblages fall into two zones: upper, characterised by *Stainforthia fusiformis*; lower, *Liebusella goesi, Hyalinea balthica, Cassidulina laevigata, Textularia earlandi, Nonionellina labradorica* and *Bulimina marginata* (Figure 10.10B; Nordberg *et al.*, 2000; Filipsson and Nordberg, 2004a). The zone boundary dates from winter 1979–1980. Unlike the

Skagerrak, there is an upward decrease in agglutinated tests. The lower zone is a typical normal-marine-shelf fjord-inner-shelf assemblage (Skagerrak–Kattegat fauna). The upper zone is dominated by an opportunist. There is a minor faunal change in the 1980s to early 1990s with increased abundance of *Bolivinellina pseudopunctata*, *Bulimina marginata*, *Nonionella turgida* and *Textularia earlandi*. The major faunal change at zone boundary is coeval with a severe low-oxygen event during the winter of 1979–1980. This was followed by a higher input of phytodetritus which in turn helped to keep bottom-water oxygen levels low.

    Koljö Fjord, Sweden.   This is an open-ended pollution-free fjord with a maximum depth of 56 m and silled connections to the Skagerrak at 8 m and with adjacent fjords at 5 and 10 m. There is a pycnocline/halocline at 15–25 m separating brackish surface water (salinity 20–25) from deep near-marine water (salinity 27–30) which is seasonally dysoxic and occasionally anoxic. This is sourced from an adjacent fjord. There are detailed hydrographic records from 1950. Short cores from 43 m representing the last 200 years are dominated by species of *Elphidium* (*clavatum, incertum, albiumbilicatum, magellanicum*; Filipsson and Nordberg, 2004b), low numbers of individuals and low species diversity. The modern assemblage from 40 m is also dominated by these *Elphidium* species (excluding *Elphidium albiumbilicatum*; WA-104). The authors interpret the record to show that conditions in the past were much the same as now but there are some subtle differences. From 1930 to 1980 the sediment is laminated and conditions were stagnant. However, from 1945 to 1955 there was frequent water exchange (recorded in the hydrographic data) with higher numbers of foraminifera and the incoming of *Stainforthia fusiformis* (Figure 10.10C). From 1980–1996 there were lower salinities and only *Elphidium* was present. From 1996–1999 salinity was high and the bottom water was stagnant leading to laminated sediments with small numbers of *Elphidium* and *Miliammina fusca*. The presence of the latter species is surprising in such deep water.

    Possible climatic control.   Changes in hydrography and faunas in truly unpolluted fjords can be due only to natural causes. Nordberg *et al.* (2000) and Filipsson and Nordberg (2004a) note that in the 1930s the winter NAO (North Atlantic oscillation) index was mostly negative; the faunal changes coincide with a shift to winter positive indices (Figure 10.10D). The latter give mild, rainy winters and westerly winds which lead to reduced deep-water renewal in the fjords. Severe dysoxia can arise through lack of bottom-water exchange (recorded only once in Gullmar Fjord), partial water exchange, renewal water already having low oxygen, or increased oxygen demand due to increased sedimentation of organic material. The authors conclude that the faunal change in Gullmar Fjord from normal marine Skagerrak-Kattegat fauna to opportunistic *Stainforthia fusiformis* is linked to an increase in dysoxia which may, in turn, be due to changes in the winter NAO index. Unfortunately, there are no detailed hydrographic records for the Skagerrak basin but it is highly likely that part, if not all, of the faunal changes recorded there must also be climatically controlled. However, a comparison between the deep-basin amphipod faunas of 1933–7 and 1989–90 shows that there has been an increase in species diversity and change in assemblage composition that is considered to be due to early stages of eutrophication (Miskov-Nodland *et al.*, 1999). Interpretation of continuous plankton records from the North Sea suggest that there are hydrographic changes not directly linked to the NAO index. In the late 1970s there was a cold boreal event with a reduction in inflow of Atlantic Water into the North Sea associated with the 'great salinity anomaly' in the North Atlantic (when both salinity and temperature reached minima). By contrast, in the late 1980s there was a warm-temperate event with an increase in southerly derived (higher temperature and salinity) Atlantic Water into the North Sea. Edwards *et al.* (2002) suggest that enclosed seas such as the North Sea respond to atmospheric variability and temperature changes and that these changes may have been underestimated in the past. Thus, the interpretation of the causes of natural variability and change is not straight-forward.

Case study: Chesapeake Bay, USA

Since the 1930s there has been an increase in oxygen depletion in the bottom waters below the pycnocline in this large microtidal estuary. A long core from the centre of the bay gives a faunal record extending back to AD 1450. *Elphidium selseyense* and *Ammobaculites* alternate in dominance throughout much of the core. *Ammonia* was scarcely present until ~1680 and since then has fluctuated in abundance with particularly high values in the past few decades. Karlsen *et al.* (2000) interpret the alternation of the two main species due to changes in salinity with *Ammobaculites* dominating during periods of high freshwater discharge (salinity <15) and high turbidity and *Elphidium* dominating during dry more-saline phases (salinity >15). They interpret the abundance of *Ammonia* as related to periods of seasonal dysoxia and anoxia commencing ~1970 and brought about by increased pollution (increase in nitrate). Subsequently, Cronin *et al.* (2000) used a linear model to estimate palaeosalinities from the abundance of *Elphidium*. The historical record over the past 500 years shows 14 wet–dry cycles with some severe droughts in the past. Palaeotemperatures have been determined from ostracod and Mg/Ca data (Cronin *et al.*, 2003). They conclude that multidecadal processes typical of the NAO are a feature of the Holocene climate. Superimposed on this are recent changes induced by pollution.

10.6.2 Man-induced change (excluding pollution)

Over the past few centuries marginal marine environments have been destroyed by 'land reclamation', dumping waste, dredging of channels and other civil-engineering works. Marshes have been particularly affected. This destruction results in fragmentation and loss of habitat. There is local extinction of the fauna affecting both common and rare species. It has been argued that because there are many more endemic rare species in the Caribbean than in the Arctic, a localised disturbance in the former would cause greater extinction than in the latter (Culver and Buzas, 1995). Also an increase in water temperature might cause lateral migration of faunal distributions but would probably be less significant than habitat destruction and fragmentation.

Modern coral reefs are affected by human activities including nutrient and other chemical pollution, increased turbidity due to sedimentation, destructive fishing practices and shipping (Hallock, 2000a). In addition, increasing intensities of ultraviolet B (UVB) radiation affect clear tropical waters to a depth of tens of metres (Hallock, 2001). The consequences for the ecosystem are the expulsion of symbionts (bleaching) from host organisms as symbionts are sensitive to slight increases in UVB, and the formation of blooms of UVB-tolerant cyanobacteria. *Amphistegina* from several locations around the world seem to be suffering damage from increasing UVB. In the Florida Keys the maximum loss is in June or early July with partial recovery by late summer. Talge *et al.* (1997) link this with the solar light cycle which peaks at the summer solstice (June) rather than with temperature which peaks in August. Protein-synthesising organelles are rarely seen in individuals having symbiont loss and there may be an increase in shell breakage due to perturbed biomineralisation. *Amphistegina gibbosa* populations have changed their reproductive strategy to successive asexual generations as a possible adaptation to maintaining their numbers (Hallock, 2000a). Increasing nutrient enrichment of coastal waters is a serious problem for larger foraminifera (Ebrahim, 2000) but there will be oceanic refuges that allow survival of at least the deeper-dwelling species. The groups most at risk both from nutrients and UVB are calcarinids and soritids. Hallock (2001) points out that there is not a close link between coral reefs and larger foraminifera. She considers that the loss of reef-building corals through environmental change may not be important for larger foraminifera as in the

geological record the times of expansion of carbonate shelves with abundant
larger foraminifera coincided with periods of reduced reef building.

10.6.3 Monitoring pollution

Comprehensive reviews by Alve (1995b), Martin (2000) and Scott *et al.*
(2001) and a brief review by Murray and Alve in Haslett (2002) provide background references.

The main pollutants are chemical: organic (sewage), nutrients, hydrocarbons,
persistent organic pollutants (POP) and heavy metals; or physical: paper/
wood pulp and thermal. Marginal marine and inner-shelf environments
are particularly affected by pollution as they are closest to source. In many cases
the effluent is discharged from a pipe and this gives a point source. Depending
on the position of a point source discharge of sewage into a lagoon or estuary,
there may be an abiotic zone around the source (due to high oxygen consumption; not always developed), then a zone of high food supply (hypertrophic
zone) beyond which conditions are unaffected (Alve, 1995b). The carrying
medium of sewage is freshwater which is less dense than the seawater
into which it is discharged. Therefore the effluent rises to the surface and
spreads out until it is dispersed and the impact on the sea floor may be lessened.
The main physical pollution is elevated water temperature as around outfalls
from cooling systems. There are data from experiments and two types of
field study: contemporary impact – comparing living distributions in polluted
and unpolluted settings (studies of dead or total assemblages are ignored
here); historical – taking cores of sediment and tracing faunal changes in the
subrecent record from a pre-pollution time in the past to present-day conditions
(termed 'environmental stratigraphy' by Alve, 2000). As noted by Murray
and Alve (in Haslett, 2002, p. 75), 'most impact studies lack quantitative considerations of the faunal responses to the concentrations and bioavailability
of the different kinds of pollutants to which they are/have been exposed
(i.e. dose–response relationships). Future studies should be more interdisciplinary and involve biological (especially physiological) and geochemical as
well as geological approaches'.

Field experiments

LeFurgey and St. Jean (1976) set up two groups (control, treatment)
each of three ponds, the control being 1.3 km from the treatment group. Both
were said to be 'relatively similar' in terms of temperature, salinity and depth
and only the treatment group received sewage. For relative faunal abundance,
two of the treatment ponds show greater similarity with the control ponds

than with the third treatment pond. Species diversity was marginally higher and standing crop was significantly higher in the control ponds. This was an interesting experiment but the authors were aware of design weaknesses. The ponds were artificial and although they were lined with marsh mud each pond had to be colonised by a fauna adapted to a new subtidal habitat (water depth 0.5 m). In the control group, one pond developed a different assemblage from the other two. Also, the ponds were not closed systems as water was introduced from the adjacent creeks. Therefore, although the results appear to show some of the consequences of organic pollution it is difficult to separate out the underlying natural variability.

A colonisation experiment in Oslo Fjord, Norway, involved placing an array of 16 plastic boxes with 10 cm of sediment at a depth of 63 m, for 32 weeks (Alve and Olsgard, 1999). Sediment from the deployment area was sterilised by freezing and for the 12 treatment boxes varying amounts of $CuCl_2$ were added (approximately 150, 300, 950 and 2000 ppm Cu) but not for the four control boxes (Cu 70 ppm). At the end of the experiment, three cores of the deployment area (sea bed) were taken for comparison and cores were taken from each box. All cores had the highest standing crops in the top 1 cm. There was no significant difference between the mean values of the sea bed, control and ~150 ppm-Cu samples: 334, 298 and 277 individuals 10 cm^{-2}, respectively. Mean values of 86 and 81 occurred in the 950 and 2000 ppm-Cu samples. There was a significant negative correlation between standing crop and Cu concentration ($r = -0.57$, probability, $p = 0.022$, $n = 16$). *Stainforthia fusiformis* was dominant in all cores (including the sea bed) except one of the 2000 ppm-Cu replicates. Other species present in decreasing order in the sea bed samples were *Elphidium excavatum*, *Nonionella turgida*, *Bolivinellina pseudopunctata*, *Nonionellina labradorica* and *Bulimina marginata*. In the experimental cores only *Elphidium excavatum* and *Bolivinellina pseudopunctata* were common. Multidimensional ordination (stress 0.09) separates the treatments with the higher Cu values from the rest of the cores. A pairwise, one-way ANOSIM (analysis of similarities) test shows an almost significant difference between the control and 950 ppm-Cu cores, and a significant difference between the control and 2000 ppm-Cu cores ($p = 0.029$) but not with the low Cu samples. Species diversity H of the high Cu (950–2000 ppm) cores approached those of the sea-bed samples (950 ppm: 1.39; 2000 ppm: 1.64; sea bed: 1.92) whereas those with low Cu had lower mean values of H (0.63–0.71). The number of deformed tests was no higher in the Cu-rich sediments than in the control. The experiment clearly shows that the presence of > 900 ppm Cu affected the community structure (species diversity, standing crop) but did not prevent colonisation as all the main sea-bed species colonised the boxes.

Mesocosm experiment

Tri-*n*-butyltin (TBT) is toxic and used in anti-fouling paints. A meso-cosm experiment to test its effects on benthic foraminifera was undertaken using surface sediment (0.08 nmol TBT g dry sediment) from 60 m water depth in Gullmar Fjord, Sweden (Gustaffson *et al.*, 2000). The protocol was: three groups, each comprising four replicate 0.25 m² mesocosms, were set up with the sediment as a substrate; they were continuously flushed with seawater from 35 m in the fjord; and TBT at concentrations of 0.00, 0.02 and 2.00 nmol TBT g dry sediment was added in a sediment slurry and allowed to settle for 48 h before water circulation was resumed. After seven months, the surface 0–1 cm sediment was scraped off, preserved in alcohol and stained with rose Bengal. The overall standing crop and those of the main agglutinated (*Textularia tenuissima* and *Spiroplectammina biformis*) and hyaline species (*Stainforthia fusiformis*) of the 0.02 nmol TBT were higher than that of the control and the 2.0 nmol group (ANOSIM, $p < 0.057$ and $p < 0.029$, respectively). Since an increase in toxin is unlikely to be beneficial to an organism, the authors attribute this increase to reduced predation; i.e., foraminifera are more tolerant of TBT than their predators/competitors. Two less-abundant species (*Nonionella turgida* and *Nonionellina labradorica*) have highest mean standing crop in the 2.0 nmol mesocosms. However, the confidence limits totally overlap those of the control and 0.02 nmol mesocosms. Thus, although they may be more tolerant to this toxin than other species this cannot be proven.

Contemporary-impact field studies

The biggest difficulty with field studies is to separate natural variability (patchy microdistributions, seasonal, annual or longer oscillations in standing crop and diversity) from that induced by pollution. Variability can be assessed only by carrying out time-series studies based on replicate samples. Comparison should be made between several control sites and several polluted (treatment) sites where all environmental factors have exactly the same variability except that the control lacks pollutants. This procedure is difficult to achieve in practice. Commonly only one control site is sampled, of necessity removed from the polluted sites and often no adequate evidence is given that it is otherwise identical to them. There is also need for appropriate statistical analysis of the data. Because of these limitations, many of the published conclusions relating to the effects of pollution are tenuous. Therefore only a brief discussion of results is given here.

In theory organic pollution might provide additional food for foraminifera. The limited amount of experimental data shows that the common estuarine species *Haynesina germanica* does not take up the lipid faecal sterol 5β-coprostanol and therefore does not feed on human sewage (Ward *et al.*, 2003) but this should not be extrapolated to other species. However, an increase in standing crops around outfalls should not be automatically taken as conclusive evidence that foraminifera are directly benefiting from sewage. Macrofauna are often excluded by raised concentrations of suspended sewage so standing crops of foraminifera might be higher through reduced predation (Alve, 1995b). In August 1975 at a Chesapeake Bay, USA, sewage outfall, Bates and Spencer (1979) collected seven samples from within 300 m of the point source and a single 'control' from >2 km away (other sampling sessions when the control site was not sampled are not considered here). The control site was sandy compared with the polluted sites and there is no discussion of how similar or different it was from the polluted sites. Nevertheless, the same species (*Elphidium excavatum*) was dominant at all stations. Both the standing crop of the assemblages (153–534 10 cm^{-3} polluted; 239 control) and the dominant species (81–84% polluted; 82% control) of the control and polluted sites are similar showing no effects of pollution although, based on consideration of total assemblages, the authors concluded that there were limited effects around the outfall.

Aquaculture of fish and bivalves leads to a substantial increase in the loading of organic matter to the sea floor beneath the operation. In Canadian bays, the fauna disappears from part of the affected area during summer and autumn when organic-matter flux is highest and associated with severe bottom-water oxygen depletion (Schafer *et al.*, 1995). As expected, the dominant species show natural seasonal variation superimposed on spatial changes relating to pollution. For instance, in Bliss Harbour, the summer control is dominated by *Leptohalysis scottii* and *Buccella frigida* whereas the cage samples have high dominance of *Elphidium excavatum*. In St. Margarets Bay the rank order of the dominant species (*Eggerella advena*, *Hemisphaerammina bradyi*) changes between control and cage samples. *Eggerella advena* shows negative correlation with organic matter (Pearson correlation coefficient − 0.46 in winter/spring, − 0.45 summer/autumn, p <0.01) but other species show either an inconsistent pattern from one season to another (*Hemisphaerammina bradyi*, *Miliammina fusca*) or only weak correlation. Therefore, the results are inconclusive.

Heavy-metal pollution is commonly associated with organic enrichment so it is difficult to separate the effects (Alve, 1995b). From a comparison of sites ranging from clean (control) to varying degrees of pollution from an oil refinery Sharifi *et al.* (1991) conclude that *Elphidium excavatum* shows a higher tolerance than *Haynesina germanica* to heavy metals (predominantly Cu with some Cr, Ni, Zn, Co, Cd and Pb).

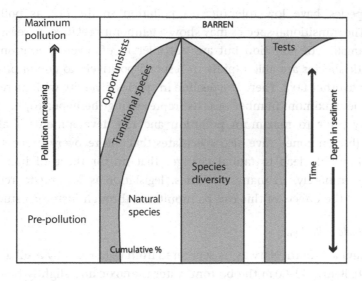

Figure 10.11. Model of the foraminiferal response to the onset and increase in organic pollution (after Alve 1995b).

Historical studies

Under ideal conditions, baseline studies of environments would have been made before a particular environmental impact. However, many impacts started before ecological studies of foraminifera commenced or they affect areas not previously investigated for their modern assemblages. Foraminifera are often the only abundant fossils in otherwise undifferentiated muddy sediments. In areas continuously accumulating sediment at a rate exceeding 1–2 mm y^{-1}, it is possible to reconstruct the recent historical record by interpreting the preserved foraminiferal assemblages. Clearly, attention must be paid to postmortem modification such as mixing through bioturbation or loss of tests (Section 9.3). Sediment may be dated using ^{137}Cs and ^{210}Pb profiles. As ever there is the problem of distinguishing between natural long-term environmental change and that induced by pollution. Nevertheless, the results of such studies are often more convincing than those derived from surveys of contemporary impacts.

Based on case studies of organic pollution, Alve (1995b) put forward a simplistic model of changes that might be expected following the onset of pollution (Figure 10.11). Faunal abundance is relative (per cent) whereas the number of tests is either the number per unit weight or volume of sediment or an accumulation rate calculated from the age of the sediment. Alve divided the fauna into three groups: pre-pollution (her 'natural'); transitional; opportunistic. The 'natural' species have low tolerance to pollution so die out as pollution increases. The transitional species may show a temporary relative increase with a modest amount of pollution but are not tolerant of severe pollution. The opportunistic species are able rapidly to respond positively to conditions that are adverse to other taxa. There is a decline in species diversity with increasing pollution and maximum number of tests (representing the hypertrophic zone) immediately prior to maximum pollution and the development of abiotic conditions (barren zone). Alve also speculates that the recovery process after amelioration would lead to faunal changes that mirror those of increasing pollution. Fortunately, in many countries, legislation is leading to reduced pollution and the effects of this can be monitored through historical studies.

Organic pollution

Drammensfjord, Norway, is separated from Oslo Fjord by a sill at 10 m water depth. Below 40–60 m the bottom water is anoxic and slightly brackish. There is a surface brackish layer that extends to sill depth. Two long sediment cores provide a foraminiferal record back to at least AD 900 (Alve, 1991). Three foraminiferal zones span the period up to the middle of the nineteenth century and are interpreted as showing climate change prior to pollution. Originally the fauna was calcareous but increasing river discharge during the 'Little Ice Age' from ~1375 led to more intense stratification of the water and less frequent renewal of the bottom water causing lower oxygen conditions and a change to an agglutinated fauna. From the fourteeth century to the middle of the nine-teenth century *Reophax pilulifer* and *Spiroplectammina biformis* assemblages dominated the northern part and *Adercotryma glomeratum* and *Cribrostomoides kosterensis* assemblages the southern part. In the 1870s wood pulp and paper industries were established and their waste, together with that from humans, led to a rise in the TOC content of the sediment. Faunal changes include a dominance of *Spiroplectammina biformis* in the north and alternating dominance of this species and *Stainforthia fusiformis* in the south. Both of these species tolerate severe dysoxia and even anoxia. There was also a marked increase in test abundance attributed to an increase in nutrients. Thus, the natural envir-onmental change that initiated low oxygen conditions was exacerbated when pollution was added to the system. Similarly, in Frierfjord, southern Norway, well-oxygenated conditions were present throughout the water column in the pre-industrial period (*Cassidulina laevigata* with rare *Stainforthia fusiformis*). With the onset of industrial pollution, dysoxic and anoxic conditions became established. The fauna changed to agglutinated species and in the anoxic areas at depths > 70 m there was an absence of fauna. There must have been a brief period of more frequent water renewal in the last decade of the 1880s when

Stainforthia fusiformis colonised the area. There is not likely to be an influence of sill depth on water exchange as the sill lies at 20 m water depth.

Combined organic and heavy-metal pollution

Bilbao estuary, northern Spain, has received urban and industrial waste for more than 150 years. Over recent decades there has been a marked reduction in the flux of heavy-metal contaminants and of urban waste but the anoxic sediments will act as a long-term source of contaminants. At present, there are few living foraminifera in the area (Cearreta *et al.*, 2000a; 2002c). In the pre-industrial period, there is an indigenous brackish estuarine assemblage of *Ammonia* group and *Haynesina germanica* with subsidiary *Elphidium oceanensis* and an exotic component of *Cibicides lobatulus* and *Rosalina irregularis* transported in from the inner shelf. The same assemblage occurs in the lower part of the industrial-zone sediments which have a high concentration of metals. However, above this (AD 1950–2000), where the metal content becomes very high (e.g., Zn >1000 ppm), foraminifera die out. Thus, severe pollution has annihilated the microfauna (abiotic zone).

10.7 Deformed tests

Deformed tests arise when the growth plan of the test is disrupted to give an abnormal shape (i.e., different from that expected in comparison with other examples of the same species). It ranges from slight deformity (smaller chambers) caused by different rates of growth from one season to another (availability of food, e.g., Murray, 1963) to severe deformity with chambers in the wrong place, distorted spirals and additional apertures. Attempts have been made to classify abnormalities based on morphology without regard to the causes (see review by Geslin *et al.*, 2000). Examples have been known for a long time from modern and fossil sediments and there are good illustrations in the Challenger report (e.g., Plate 5, Figure 15; Plate 17, Figures 1–6; Brady, 1884). In some modern assemblages, deformed tests are conspicuously abundant and the causes of this are of great interest. Explanations put forward to account for deformity include various types of natural environmental stress (salinity ranging from freshwater to hypersaline; acidification leading to test dissolution; low oxygen; trace elements; and hydrodynamic damage followed by regeneration, which may be obvious) and pollution of various types (heavy metals, chemicals, hydrocarbons, thermal, sewage; Geslin *et al.*, 2000). In the majority of cases the explanation of the cause of deformity is based on circumstantial evidence from field data. Up to ~1970 mainly natural causes were invoked but since then the favoured cause has commonly been some aspect of pollution.

Nevertheless, in the lagoons and estuaries of Brazil, the highest incidence of deformity occurs in waters that are either hypersaline or undergo natural seasonal acidification leading to test decalcification followed by regeneration. The lowest abundances of deformed tests occur where pollution is present (Geslin et al., 2002).

The only potentially direct proof of cause and effect is a well-conducted experiment. Some authors have found that forms kept in culture under apparently constant conditions nevertheless grew abnormal chambers (e.g., Peneroplis planatus, Faber and Lee, 1991a) but these do not normally exceed 1% of the live fauna (Stouff et al., 1999b). In laboratory cultures of Ammonia tepida, sometimes instead of a single second chamber two second chambers form together on the proloculus. This can lead to deformity of the test or to twinned test formation (Stouff et al., 1999a). Ammonia grown under hypersaline conditions (50) has a higher incidence of abnormality (Stouff et al., 1999b). Laboratory experiments on Ammonia 'beccarii' suggest that deformity is induced by increased Cu concentrations. However, in experimental colonisation of copper-enriched sediments, Stainforthia fusiformis and Bolivinellina pseudopunctata showed a negative effect at Cu > 900 ppm (reduced reproduction) but there were no more deformed tests developed than in the uncontaminated adjacent sediment (Alve and Olsgard, 1999). In reality, the situation may be complex, depending on the nature of the heavy-metal pollution rather than its quantity.

At the present state of knowledge the precise mechanism by which tests grow is unknown. The shape and arrangement of chambers is probably controlled by the cytoskeleton. The reticulopodia play a role in shell growth and chamber formation. In addition to genetic control of test form, there may be an environmental control via the reticulopodia as their autonomous re-organisation is influenced by environmental parameters. The effects of pollutants on reticulopodial behaviour and physiology have not yet been explored but could be very informative (Travis et al., 2002). Experiments involving mechanical or chemical damage to the cytoskeleton to see if they might lead to deformity were suggested by Bresler and Yanko-Hombach (2000). As yet, such experiments have not been carried out. In forms with a hyaline test deformity seems to be related to imperfect biomineralisation. This may be due to the development of cavities in the wall or to disorganisation of the pattern of crystallites (Geslin et al., 1998). Alternatively, these may not be the cause but merely further manifestations of deformity. The causes of these biomineralisation problems could be linked to the disruption of the cytoskeleton due to natural environmental stress or to pollution or to a combination of both. It certainly seems likely that there are several causes rather than just one. Deformity due to mechanical damage or to dissolution of the test followed by regeneration is probably the most easily recognised. The abundance of deformed tests may be a guide to environmental stress but it is not a guide to the causes of the stress (Geslin *et al.*, 2002). That will not be possible until specific abnormalities can be related to specific causes.

10.8   Summary and conclusions

The wide distribution of benthic foraminifera both in space and geological time and their rapid response to ecological change make them the best microfossil group for studying benthic processes. All applications of benthic foraminifera to interpreting the geological record (stratigraphy, palaeoecology, palaeoceanography) and to monitoring modern environments involve a consideration of their ecology. In stratigraphy, they play a key role in the interpretation of systems tracts and in providing a relative timescale. In palaeoecology, they can be used to recognise broad environments. In palaeoceanography, various proxy relationships have been proposed and are being tested for parameters such as dissolved oxygen, organic production, salinity and temperature. Assemblages can also be used to infer past currents and transport paths.

There is natural variability in the environment on timescales ranging from diurnal to decades or longer. Superimposed on this there may be progressive change which is typically cyclic. Short-term natural variability in the environment is time averaged in the dead assemblages of foraminifera preserved in the sediment (Section 9.3.5). Long-term climatic changes are faithfully recorded by changes in benthic foraminiferal assemblages preserved in the Holocene record as shown by examples from Scandinavian fjords and Chesapeake Bay, USA. Benthic foraminifera also have the potential to be valuable monitors of contemporary environmental change and impact disturbance caused by pollution. However, this potential has yet to be fully realised as few field studies have been sufficiently well-planned and executed to provide reliable results. Historical studies of such changes have been more successful in establishing change from pre-pollution conditions although there is still a major problem in separating the changes brought about by natural climatic variation from those caused by pollution. Comparison of changes in unpolluted and polluted areas may help to overcome this. For instance, in the case of dysoxia/anoxia in fjords, organic pollution often seems to add to a trend which is already happening naturally.

Future improvements in the understanding of species ecology together with technical advances and methods of data analysis will no doubt lead to improved applications of benthic foraminifera to understanding Earth history.

Glossary

Abiotic factors Not related to organisms; includes physical and chemical measures such as temperature and salinity.

Abyssal That part of the ocean having a water depth of approximately 4 to 6 km; mainly abyssal plains.

Aerobic The term relates to life. No oxygen restriction, oxic seawater (Allison *et al.*, 1995), therefore a normal benthic fauna.

Allochthonous Occurring away from the place of origin; exotic.

Anaerobic Relates to life in an anoxic environment (Allison *et al.*, 1995).

Anoxic Relates to water or environment. Absence of oxygen (Allison *et al.*, 1995).

Assemblage The populations of all the species present in a sample.

Association The presence of the same main species in several assemblages.

ATP Adenosine triphosphate; a compound that provides energy for many biochemical cellular processes by undergoing enzymatic hydrolysis to ADP, adenosine diphosphate; ATP is present only in live individuals.

Attached Fixed (sessile) or clinging to a substrate.

Autochthonous Occurring in the place of origin; *in situ*.

Autotroph An organism that uses sunlight and dissolved inorganic nutrients to fix its own organic carbon; i.e., it does not feed on organic matter.

Bathyal That part of the ocean having water depths from the shelf break at ~200 m to ~4 km.

Benthic Mode of life, living on or under the seabed.

Benthos Organisms living on or within the seabed.

Biofilm Sedimentary particles may be covered by a mono- to multilayered veneer of microbes including bacteria and, in the photic zone, diatoms, euglenid flagellates and cyanobacteria. Biofilms provide an important source of food for benthic foraminifera.

Bioherm A build-up of material under the influence of organic processes, e.g., reef or stromatolite.

Biomass The weight (or volume) of living organic matter in a unit area or volume of sediment. $1 \, \mu g \, 10 \, cm^{-2} \, C_{org} = 2 \, \mu g \, 10 \, cm^{-2}$ ash-free dry weight $= 10 \, g \, m^{-2}$ wet weight $= 10 \, mm^3 \, 10 \, cm^{-3} = 0.0033$ ng ATP mm^{-3}.

Biota All the species of plants and animals living in the same area.

Biotic factors Relating to life. Biotic environmental parameters include competition and predation.

Brackish Having a salinity lower than that of normal seawater; here taken as <32; $=$ hyposaline.

Broken-stick model MacArthur (1957) proposed a model for fitting species-abundance data based on niche partitioning (or breaking). Not many datasets fit this model and it has been criticised as being merely a special case of the binomial distribution (Hayek and Buzas, 1997).

Carrying capacity k The maximum number of individuals that a given habitat can support.

CCD Carbonate (or calcite) compensation depth; the depth at which all carbonate is dissolved, i.e., the supply is matched by the rate of dissolution.

Clinging Mode of life. Attached to the substrate by pseudopodia but able to move about.

Continental shelf The terrace extending from the shoreline to the shelf break.

C_{org} Organic carbon.

Critical threshold The value of an environmental parameter beyond which the organism is severely stressed. For most parameters there are lower and upper thresholds.

Dead assemblage The part of the foraminiferal fauna lacking protoplasmic test contents and which therefore does not take up biological stain.

Density Another term for standing crop.

Detritus Recently dead or partially decomposed fragmentary organic matter.

Dissolution The loss of calcium carbonate from calcareous tests through the corrosive action of carbonate-undersaturated water.

Disturbance Damage, displacement or mortality caused by physical, chemical or biotic activity.

Diversity See species diversity.

Dominance The state of being the most abundant species in an assemblage; commonly expressed as % of the most abundant species.

Dysaerobic The term relates to life. Impoverished benthic fauna stressed by dysoxia or suboxia (Allison et al., 1995).

Dysoxic The term relates to water or environment. Low oxygen: 0.2–1.0 ml O_2 $l^{-1} = 9$–45 μM (Allison et al., 1995).

Ecophenotype A morphotype associated with a specific environment.

Ecosystem Comprises the interacting biotic (assemblages of organisms) and abiotic components of the environment.

Endemic Species restricted to a geographic region.

Endosymbiosis Two organisms living together for mutual benefit with one living inside the cell or tissues of the other; e.g., symbiotic algae in a foraminiferal cell.

Environment The physical area (shore, reef, etc.) within which there are habitats.

Epifaunal Living on the sediment surface or another substrate; may be attached or free to move.

Epiphytic Epifaunal organisms living on plant substrates; may be attached or free to move.

Equilibrium species Not showing an increase in abundance in response to short-term environmental changes. Compare with opportunistic.

Estuary The drowned mouth of a river; the meeting of river and sea. If there is a subaerial barrier separating the estuary from the open sea then it is considered a lagoon. If the sediment supply is so large that there is progradation onto the shelf, then it becomes a delta.

Euhaline Seawater of normal salinity: here taken as 32–37.

Euphotic zone Surface waters where there is sufficient solar energy (light) for net primary productivity of phytoplankton to occur by photosynthesis; the base of the zone is taken as the depth to which 1% of the solar energy penetrates.

Euryhaline Able to tolerate a broad range of salinity.

Eurythermal Able to tolerate a broad range of temperature.

Eutrophic Environments in which the nutrient flux is sufficiently high that it supports phytoplankton densities that limit light penetration; production 301–500 g C_{org} m^{-2} y^{-1} (Nixon, 1995).

Eutrophication An increase in nutrient flux, often due to sewage pollution, runoff from fertilised agricultural land, or effluent from fish/shellfish farms; not to be confused with eutrophic, which represents a particular state of nutrient flux; e.g., a eutrophic environment will not show eutrophication unless the rate of nutrient flux is increasing; a change from oligotrophic to mesotrophic can arise from eutrophication (see Nixon, 1995).

Evenness A measure of the distribution of individuals among species in an assemblage. When all species are equally distributed $H(S) = \ln(S)$, where H is the information function and S is the number of species. The ratio e^H/S is also a measure of evenness.

Exotic See allochthonous.

Fisher alpha index A measure of species diversity based on a log-series distribution.

Free living Able to move about, not attached.

Geotropism The response of an organism to gravity.

Habitat The place where an organism lives.

Hadal The part of the ocean with water depths > 6 km; oceanic trenches.

Halocline A rapid increase in salinity with increase in water depth.

Halophyte A land plant able to tolerate salt water.

Heterotroph A predatory, herbivorous or detritus-feeding organism that consumes organic matter.

Hypersaline Having a salinity higher than that of normal seawater; here taken as >37.

Hypertidal A term sometimes used for tides > 6 m.

Hypertrophic Environments in which the surface waters have such a high nutrient flux that phytoplankton is dense; so much organic matter is produced that the benthic communities beneath are limited by hypoxia or anoxia; production $> 500\,g\;C_{org}\;m^{-2}\;y^{-1}$ (Nixon, 1995).

Hyposaline See brackish.

Hypoxia The term relates to water or the environment. A low-oxygen environment with $< 0.1\,ml\;O_2\;l^{-1} = 5\,\mu M$.

Infaunal Living within the sediment.

Information function Species diversity expressed as H, a measure of uncertainty (see Hayek and Buzas, 1997) of the number of species S, where p_i is the species proportion:

$$H(S) = -\sum_{i=1}^{S} p_i \ln p_i.$$

In situ Autochthonous.

k See carrying capacity.

Labile Organic carbon in a form that can be consumed by organisms.

Lagoon A shallow body of water separated from the sea by a subaerial barrier breached by one or more channels. Lagoons may be brackish, normal marine or hypersaline.

Larger foraminifera Having a test > 2 mm in size. The term is usually restricted to forms living in warm shallow tropical waters.

Limit of tolerance The value of an environmental parameter beyond which an organism dies. For most parameters there is an upper and a lower limit of tolerance.

Living assemblage That part of the foraminiferal fauna which is alive at the time of sampling. See stained (living) assemblage.

Macrobenthos Benthic organisms > 500 μm in size.

Macrotidal Tidal range > 4 m. See hypertidal.

Marginal marine A general term for coastal, lagoon, estuary, and marsh environments.

Meiobenthos Infaunal benthic organisms 63–500 μm in size.

Mesotidal Tidal range 2–4 m.

Mesotrophic Moderate abundance of nutrients; intermediate between oligotrophic and eutrophic; production 100–300 g C_{org} m^{-2} y^{-1} (Nixon, 1995).

Microbenthos Tiny (< 100 μm) infaunal benthic organisms.

Microhabitat The habitat defined on a scale of millimetres.

Microtidal Tidal range < 2 m.

Mixotrophic The ability to feed and photosynthesise, often involving endosymbiosis.

Morphogroup/morphotype A group of forms having similar test morphologies rather than taxonomic similarity.

Net primary production Total amount of energy fixed as organic carbon during photosynthesis, less the amount used in respiration, per unit of time.

Niche The habitat with ecological parameters within the range for the successful existence of a species. The potential or *fundamental niche* is the theoretical maximum ecospace and the *realised niche* is that part of ecospace actually occupied by a given species.

Normal marine Salinity ~35. Here taken as 32–37.

Oligotrophic Describes environments in which the nutrient content is low (in the euphotic zone) or food supply is low (below the euphotic zone). Chlorophyll concentrations < 0.1 mg m^{-3} in surface waters; production < 100 g C_{org} m^{-2} y^{-1} (Nixon, 1995); flux of C_{org} < 1 g m^{-2} y^{-1}.

Opportunistic species One rapidly increasing its abundance in response to a favourable change in environmental conditions. Compare with equilibrium species.

Oxic Relates to water or environment. Plentiful oxygen: > 1.0 ml O_2 l^{-1} => 45 μM (Allison *et al.*, 1995).

Oxygen (dissolved in water) To convert ml l^{-1} to μM, divide by 0.0224. To convert μM to ml l^{-1}, multiply by 0.0224.

Oxygen minimum zone, OMZ The zone of contact with the sea floor of waters showing some degree of oxygen depletion. An OMZ invariably develops beneath areas of upwelling. Depending on the intensity of organic flux to the sea floor, the OMZ may be oxic, dysoxic or anoxic.

Pandemic Species having a widespread distribution.

Phototropism The response of an organism to light.

Population A group of individuals of a single species.

Population dynamics Changes, and their causes, in the number of individuals in a population through time.

Postmortem After death.

Primary production Total amount of energy fixed as organic carbon during photosynthesis; expressed as g C_{org} m^2 y^{-1}; in oceans, fixed by plant plankton; in intertidal and very shallow subtidal areas, fixed mainly by benthic micro- and macroalgae or seagrasses.

Proxy Using faunal abundance or geochemistry as an indicator of the past value of an oceanographic variable such as dissolved oxygen.

Pycnocline A rapid increase in density with increase in water depth.

Redox potential The balance between oxidation and reduction processes. Anoxic sediments have negative values.

Salinity The salt content of water. Previously expressed as parts per thousand (‰) based on the salt content or as practical salinity units (psu) based on electrical conductivity. Now values are given as a number without any suffix. There is a continuum from freshwater to very saline water. For convenience, terms are given for sections of this continuum: see brackish, normal marine, hypersaline.

Sessile organism Firmly cemented or permanently attached to a substrate.

Species diversity The number of species present; also termed species richness. Since that number depends on the size of the sample, various univariate measures have been proposed to take this into account. See Fisher alpha index and information function.

Standing crop The number of foraminifera living on or in a unit area or unit volume of sea floor at any one time.

Stained (living) assemblage That part of the foraminiferal fauna with protoplasmic test contents that stain with an appropriate reagent such as rose Bengal. Considered to represent the living assemblage.

Stenohaline Restricted to a narrow range of salinity.

Stenothermal Restricted to a narrow range of temperature.

Stress Living close to the lower or upper threshold of tolerance for one or more environmental parameters.

Stressed environment An environment in which one or more parameters is close to the critical thresholds for the species living there, e.g., a low-oxygen environment.

Suboxic The term relates to water or environment. Low oxygen: 0.0–0.2 ml O_2 $l^{-1} = 0$–9 μM (Allison *et al.*, 1995).

Taphonomic Changes soon after death and, for epifaunal organisms, during the process of burial.

Thermocline The zone of rapid temperature change between the warm mixed surface waters and the underlying cooler waters.

Time-averaged assemblage The accumulation of successive dead assemblages over a period of time to give a mixed, average assemblage.

Time series A suite of samples collected over successive time periods (e.g., monthly) to provide information on cyclic or unidirectional changes.

TOC Total organic carbon (expressed as wt.% of dry sediment).

Total assemblage Stained and unstained tests from the same sample; i.e., an arbitrary mixture (depending on the thickness of the sample) of living and dead assemblages.

Trophic Of, or related to, food.

Upwelling The upward movement into the euphotic zone of deep, cold, nutrient-rich water usually due to Ekman transport. Upwelling promotes vigorous primary production by the phytoplankton.

Variable A measurement taken during sampling, e.g., species abundance, temperature, salinity, etc.

Appendix

The appendix presents ecological data in two sections: information for some of the dominant species; and for selected genera. The abbreviation *s.s.* refers to *sensu stricto* (for species that are commonly misidentified). The source of the information is Chapters 4–7 and is based on the main living occurrences. There may be rare living occurrences beyond this range. Dead occurrences are not considered. References are listed only where there is specific information relevant to a species. Even for dominant species, in many cases there is little reliable ecological information. The numbered section headings link to the appropriate chapter.

Species

4   Marginal marine environments

4.1   Agglutinated

Ammoastuta inepta: brackish marshes and mangals, Texas, USA, to French Guiana.

Ammobaculites balkwilli: restricted to Europe; occasionally on marsh (Cearreta, 1988), mainly low intertidal to subtidal; salinity 15–29 around the Skagerrak-Kattegat (Alve and Murray, 1999).

Ammobaculites crassus: infaunal down to at least 9 cm, subtidal in microtidal estuaries.

Ammobaculites dilatatus: a minor to subsidiary species on brackish marshes; dominant in mangals in French Guiana.

Ammobaculites exiguus: infaunal down to 10 cm on brackish marshes and lagoons, salinity 0–25, temperature 0–27 °C, Massachusetts, USA, to Mexico (Ellison and Murray, 1987).

Ammobaculites exilis: infaunal in lagoons, salinity 34, temperature 17–28 °C, North Carolina to Texas, USA (Buzas and Severin, 1982; Ellison and Murray, 1987).

Ammodiscus gullmarensis: salinity 20–32, temperature 4–14 °C, on organic-rich muddy sediment, dominant in deeper water close to the permanently anoxic layer, Norway (Alve, 1995a).

Ammoscalaria runiana: in Europe, common only in fine to medium sand with <20% mud, and low TOC (0.2–0.7%), salinity 16–18, intertidal-subtidal in estuaries, lagoons and fjords (Alve and Murray, 1999).

Ammotium cassis: infaunal detritivore lying horizontally in the top 5 cm of sediment and possibly indicating the halocline (Olsson, 1976; see also Ellison and Murray, 1987); always subtidal; only recently introduced into the Baltic (Lutze, 1965); transitional water layer, in the lower part or just below the *Zostera* belt (Alve and Murray, 1999); reproduces at temperatures less than 8 °C, the most important control on its distribution (Wefer, 1976b).

Ammotium salsum: infaunal down to 10 cm, detritivore (including bacteria, Matera and Lee, 1972), marshes on Pacific North America but more common on intertidal mudflats in the Gulf of Mexico (Phleger, 1967) and lagoons; can withstand salinities of 60 (Debenay and Pagès, 1987).

Arenoparrella mexicana: epifaunal, and infaunal down to 60 cm, abundant on Gulf of Mexico marshes in *Salicornia–Spartina* zones and mangals, present in European marshes; almost absent from the Pacific seaboard of North America (Phleger, 1967).

Balticammina pseudomacrescens: epifaunal, sometimes clinging to algal filaments; characteristic of low-salinity brackish high marsh closest to land; common in Europe, Atlantic North America and the Gulf of Mexico. Japan (Whittaker, J. E., 2005, personal communication).

Haplophragmoides spp.: three species are recognised but the distinctions between them are uncertain (*bonplandi, manilaensis, wilberti*); *wilberti* is epifaunal and infaunal down to 30 cm, detritivore?; marshes.

Jadammina macrescens: epifaunal, sometimes on decaying leaves, and infaunal down to 60 cm, herbivore or detritivore, widespread on high to mid marsh throughout mid to high latitudes.

Miliammina fusca s.s.: epifaunal in marsh and shallow lagoons and estuaries; infaunal down to 50 cm in marshes, detritivore, strongly euryhaline; widely distributed especially in the intertidal zone.

Paratrochammina (Lepidoparatrochammina) haynesi: attached on gravel substrates and on mollusc shells, in the protection of crevices or ribs; current-swept areas in channels (Murray and Alve, 1993); the primary control on its distribution is the presence of a suitable substrate.

Portatrochammina bipolaris: dominant in an eastern Greenland fjord in the deeper polar water mass (Madsen and Knudsen, 1994).

Pseudothurammina limnetes: infaunal to 40 cm, on brackish marshes, Canada to Mississippi delta.

Reophax moniliformis: detritivore on plant debris and also husbands chloroplasts (Knight and Mantoura, 1985), intertidal to subtidal, euryhaline.

Siphotrochammina lobata: epifaunal, sometimes clinging to algal filaments, restricted to subtropical–tropical marsh/mangal.

Textularia earlandi: dominant at 500 m in a Greenland fjord (Jennings and Helgadottir, 1994); of minor importance in estuaries and lagoons in northwestern Europe.

Tiphotrocha comprimata: epifaunal, free or clinging to algae, and infaunal down to 42.5 cm, herbivore or detritivore, marshes; dominant along the Atlantic seaboard of North America into the Gulf of Mexico, subsidiary in northwestern Europe.

Trochammina inflata: epifaunal and infaunal down to 60 cm, herbivore or detritivore (including bacteria, Matera and Lee, 1972), almost universally present on high to mid marshes throughout the world.

4.2   Porcelaneous

Amphisorus hemprichii: epiphytic on seagrasses and less commonly on algae; has endosymbiotic dinoflagellates; stenohaline (salinity 35–42; Levy, 1991).

Androsina lucasi: epiphytic on green filamentous algae and on rhizomes of *Thalassia*; probably has endosymbionts; euryhaline in Florida Bay (Levy, 1991).

Cornuspira involvens: epifaunal, widespread in marginal marine and shelf environments in low abundance; rarely dominant in Caribbean carbonate lagoons.

Quinqueloculina dimidiata: epifaunal, herbivorous on microalgae (Hohenegger *et al.*, 1989); occurs in the mouths of brackish lagoons in southern England (salinity 31–35) and in the Adriatic (Hohenegger *et al.*, 1989).

Quinqueloculina seminulum: the most commonly identified species of the genus; epifaunal on firm substrates and on sediment; infaunal down to 10 cm (Matera and Lee, 1972); dominant in normal marine lagoons and marshes in the Mediterranean, in fjords in southern Scandinavia, and on the inner shelf.

4.3   Hyaline

Ammonia group: there may be 25–30 species of *Ammonia* (Hayward *et al.*, 2004) so the comments listed here are general; infaunal, herbivore, widespread in marginal marine environments worldwide; common in sediments with highly variable mud and TOC contents, salinity 10–31, marsh to subtidal; tolerates salinities up to 50 in salt works (Zaninetti, 1982, 1984); able to tolerate low oxygen ($<0.1\,\mathrm{ml\,l^{-1}}$) for several days (Moodley and Hess, 1992) and may be a facultative anaerobe (Pawlowski *et al.*, 1995).

Aubignyna perlucida: subsidiary in Venice lagoon, salinity 28–32, temperature 6–27 °C (Donicci *et al.*, 1997); this is a warm-water species.

Buccella frigida: locally subsidiary to dominant in Svalbard fjords at depths >100 m and Russian estuaries, southern limit in northwestern Europe are the English Channel estuaries; brackish lagoons from Canada to Long Island Sound, USA.

Discorinopsis aguayoi: epifaunal, semi-sessile (Hohenegger *et al.*, 1989); warm-water marsh, Greece and South America; on seagrass, intertidal, the Adriatic.

Elphidiella hannai: cold shallow waters of the northeast Pacific and eastern Greenland (Madsen and Knudsen, 1994).

Elphidium albiumbilicatum: infaunal, intertidal to shallow subtidal; southernmost occurrence in Loch Etive, Scotland; probably salinity controlled, 16–26 (Alve and Murray, 1999).

Elphidium clavatum: infaunal–epifaunal, sometimes epiphytic (Lee *et al.*, 1969); in high-latitude fjords and estuaries in surface seasonal water layer, sometimes close to a glacier (Hald and Korsun, 1997), dominant in lagoons along the Atlantic seaboard of North America from Canada to Long Island Sound, USA; Baltic; sometimes regarded as an indicator of cold water but its occurrence as far south as Long Island does not support this view.

Elphidium excavatum: infaunal; common in sediments with highly variable sand, mud and TOC contents, salinity 15–31 (Alve and Murray, 1999), intertidal to subtidal in estuaries, lagoons, continental shelf and slope (down to several hundred metres).

Elphidium galvestonensis: epifaunal, epiphytic; warm-water lagoons and marshes, the Gulf of Mexico; northern limit Long Island Sound, USA.

Elphidium gunteri: dominant on the Gulf of Mexico marshes and lagoons; northern limit Long Island Sound, USA.

Elphidium incertum s.s.: epifaunal, epiphytic; infaunal down to 10 cm (Matera and Lee, 1972); brackish marshes, UK; slightly brackish lagoons from Cape Cod to Long Island Sound, USA.

Elphidium oceanensis: infaunal, detritivore on plant debris (Knight and Mantoura, 1985), widespread in marginal marine environments along the European seaboard, northern limit in Oslo fjord, absent from the Baltic.

Elphidium subarcticum: epifaunal, attached to sand grains (Kitazato, 1994); brackish lagoons from Cape Cod to Virginia, USA; Japan.

Elphidium translucens: epifaunal, epiphytic; dominant to subsidiary in brackish Long Island Sound, USA, and Venice lagoon, Italy; marsh, Greece.

Elphidium williamsoni: infaunal, herbivore, widespread in marginal marine environments along the European seaboard, northern limit Oslo fjord, absent from the Baltic; dominant in a brackish lagoon, Canada, and present in Virginia, USA; low marsh, intertidal to subtidal to 5 m (Alve and Murray, 1999); husbands chloroplasts (Knight and Mantoura, 1985); common in sediments with highly variable mud and TOC contents; euryhaline.

Eoeponidella pulchella: dominant in the Skagerrak–Kattegat at depths of <4 m, salinity 15–27 (Alve and Murray, 1999).

Haynesina germanica: infaunal (Richter, 1964b), widespread in marginal marine environments along the European seaboard including the Mediterranean, northern limit Oslo Fjord, Norway, absent from the Baltic; present in lagoons from Buzzards Bay to Florida, USA; rare in the Gulf of Mexico; husbands chloroplasts (Knight and Mantoura, 1985), herbivore on diatoms (Ward *et al.*, 2003) and cyanobacteria (Hohenegger *et al.*, 1989); common in sediments with highly variable mud and TOC contents (Alve and Murray, 1999); salinity 0–35 (Murray, 1968b) but up to 50 in salt works (Zaninetti, 1984); molecular genetics shows it to be a single species (Langer, 2000).

Haynesina orbiculare: dominant in cool-water lagoon (Canada) and fjords (Svalbard) and present as far south as Long Island Sound, USA.

Nonion depressulus: infaunal–epifaunal, sometimes on seagrass (Hohenegger *et al.*, 1989); herbivore on cyanobacteria and diatoms; an inner-shelf species that penetrates into the intertidal seaward parts of some estuaries (e.g., southern England, northern Spain); northern limit of distribution Oslo Fjord, Norway; low marsh, intertidal to subtidal to 5 m in Skagerrak–Kattegat (Alve and Murray, 1999); intertidal, Adriatic.

Nonionellina labradorica: prefers cold (<1 °C) water of normal salinity >34.5 in high fjords (Greenland, Svalbard, Scandinavia).

Palmerinella gardenislandensis and palmerae: brackish lagoon and delta, the Gulf of Mexico.

Stainforthia fusiformis: infaunal, facultative anaerobe, able to withstand short periods of anoxia; and early coloniser of formerly anoxic sediments; an opportunistic shelf species that thrives under stressed conditions where the salinity exceeds 30 (Alve, 1994; 2003); fjords, lagoons, shelf seas (below fronts, Scott *et al.*, 2003).

5   Shelf environments

5.1   Agglutinated

Clavula cylindrica: shallow infaunal, tolerates dysoxia (Fontanier *et al.*, 2002).

Cribrostomoides jeffreysii: infaunal to epifaunal, attached mobile.

Eggerella advena: mainly arctic–boreal Atlantic North America (records from Europe may be incorrectly identified); extends into lagoons.

Eggerelloides medius: infaunal, positive correlation with %TOC and to a lesser extent with % <63 μm sediment; a strong negative correlation with grain size >1000 μm.

Eggerelloides scaber: infaunal, detritivore, subtidal, typical shelf species but extends into slightly brackish waters in estuaries/lagoons; in Brazil, epiphytic on seagrass (Debenay, 2000).

Lepidodeuterammina ochracea: epifaunal, attached mobile.

Saccammina atlantica: a shelf species that penetrates into the less brackish parts of lagoons on the Atlantic seaboard of USA.

Textularia earlandi: shallow infaunal, tolerates dysoxia (Bernhard *et al.*, 1997).

Textularia sagittula: epifaunal, attached mobile.

Textularia truncata (may be the same as *T. bocki*): epifaunal, normally aperture down on the substrate, attached mobile; able to change its orientation with respect to water currents (Haynes, 1981).

5.2   Porcelaneous

Cornuloculina balkwilli: epifaunal, attached mobile.

Cyclorbiculina compressa: epiphytic on seagrasses and algae, also on coral debris; has endosymbionts; confined to shallow water.

Massilina secans: epifaunal, marine, browses around the holdfasts of algae (Atkinson, 1969).

Miliolinella subrotunda: epifaunal, on plants or sediment; in deep water may construct a detrital tube to lift itself several centimetres above the sediment surface into higher current velocities to enhance its feeding (Altenbach *et al.*, 1993); widespread in slightly brackish to hypersaline lagoons and inner shelf; occasionally found in the bathyal zone.

5.3   Hyaline

Acervulina inhaerens: epifaunal, usually attached and immobile.

Ammonia falsobeccarii: deep infaunal in anoxic sediment, also in surface sediment; colonises macrofaunal burrows to feed on the bacteria in the burrow walls (Fontanier *et al.*, 2002).

Bolivina pacifica: shallow infaunal, oxic but can tolerate dysoxia (Douglas, 1981).

Bolivina seminuda: shallow infaunal, oxic environments (Douglas, 1981).

Bolivina spissa: shallow infaunal, tolerant of dysoxia 22 to <76 μM (0.5 to <1.7 ml l^{-1}, Douglas, 1981).

Brizalina argentea: shallow infaunal, tolerant of dysoxia <18 μM (<0.4ml l^{-1}, Douglas, 1981; Bernhard *et al.*, 1997).

Brizalina striatula: dominant in Indian River lagoon, USA, and Jamaica, and widely present in the Gulf of Mexico and Caribbean; mangal, Florida, USA; probably a shelf species which extends into the more marine parts of marginal environments.

Bulimina aculeata: deep infaunal in anoxic sediment, also in surface sediment; colonises macrofaunal burrows to feed on the bacteria in the burrow walls (Fontanier *et al.*, 2002).

Bulimina marginata: infaunal to at least 4 cm; shows a strong positive correlation with %TOC and a negative correlation with % coarse sand (Murray, 2003b); an opportunist able to respond to high food availability (Jorissen *et al.*, 1992); deep infaunal in anoxic sediment, also in surface sediment; colonises macrofaunal burrows to feed on the bacteria in the burrow walls (Fontanier *et al.*, 2002).

Bulimina tenuata: shallow infaunal, tolerates dysoxia 1–15 μM (Bernhard *et al.*, 1997).

Cassidulina laevigata: infaunal, strong positive correlation with % coarse sand and negative correlation with fine sediment and %TOC off Scotland; typical of boreal shelves bathed in warm Atlantic Water (Mackensen and Hald, 1988); superior competitor in food- and oxygen-rich sediment and resistant to disturbance by macrofaunal predators (de Stigter *et al.*, 1998); organic flux, range 1.5–60, optimum 5–30 g m^{-2} y^{-1} (Altenbach *et al.*, 1999) but sediment grain size may be more important than organic flux (Murray, 2003b).

Cassidulina reniforme: prefers cold (<1 °C) water of normal salinity >34.5 (Korsun and Hald, 1998).

Cassidulina teretis: typical of cold waters (− 11 °C) in high latitudes (Mackensen and Hald, 1988).

Chilostomella ovoidea: infaunal, tolerates dysoxia (Bernhard *et al.*, 1997).

Cibicides lobatulus: epifaunal, usually attached and immobile; especially in high energy.

Elphidium subarcticum: sometimes attached to sediment grains (Poag *et al.*, 1980).

Epistominella vitrea: infaunal, an opportunist able to respond to high food availability (Jorissen *et al.*, 1992).

Gavelinopsis praegeri: epifaunal, attached mobile.

Glabratella ornatissima: epifaunal, clinging to hard substrates during periods of high wave energy and on sediment (especially coarse to fine sand) by late summer, test hemispherical to spherical, robust, and smooth on the spiral side making it resistant to abrasion damage (Erskian and Lipps, 1987).

Loxostomum pseudobeyrichi: shallow infaunal, tolerates dysoxia (Bernhard *et al.*, 1997); test becomes narrower as oxygen decreases.

Nonionella stella: shallow infaunal, tolerates dysoxia and short periods of anoxia (Bernhard *et al.*, 1997).

Nonionoides scaphum: infaunal in suboxic to anoxic sediments (Fontanier *et al.*, 2002).

Planorbulina mediterranensis: epifaunal, attached mobile.

Rosalina anomala: epifaunal, attached mobile.

Rosalina globularis: epifaunal, attached mobile.

Spirillina vivipara: epifaunal, attached mobile.

Suggrunda eckisi: shallow infaunal, 22–67 μM (0.5–1.5 ml l^{-1}, Douglas, 1981).

Uvigerina curticosta: infaunal, occurs in sediments with high TOC (1–6%) especially where oxygen is <1.0 ml l^{-1} (Douglas, 1981).

Valvulineria bradyana: shallow infaunal, tolerates dysoxia (Fontanier *et al.*, 2002).

6   Carbonate environments (larger foraminifera)

Based mainly on Hohenegger (1994, 2000), Hohenegger *et al.* (1999, 2000), Langer and Hottinger (2000) and Renema (2002).

6.1   Porcelaneous

Alveolinella quoyi: epifaunal on sediment and hard substrates, especially coral rubble, sheltering in crevices in high-energy settings such as channels; diatom symbionts; 5–100 m; central Indo-Pacific.

Amphisorus hemprichii: epifaunal/epiphytic on hard substrates including seagrass; dinoflagellate symbionts; test large (up to 1.5 cm); may require high temperatures throughout the year (Renema, 2003); high-light habitats on the upper slope and fore-reef crest, but avoiding high energy; also down to 30–40 m in channels; Indo-Pacific; may not be present in the Caribbean.

Archaias angulatus: epifaunal, epiphytic on seagrass; chlorophyte symbionts; 0–20 m; salinity 35–42; confined to the Atlantic Ocean, mainly Caribbean.

Borelis pulchra and *schlumbergeri*: epifaunal on hard substrates and epiphytic on seagrass; diatom symbionts; back-reef coral rubble and coral sand;

0–40 m; circum-tropical (but not recorded in the Mediterranean or on the Great Barrier Reef, Australia).

Cyclorbiculina compressa: epifaunal; chlorophyte symbionts; reef margins 10–30 m and seagrass (<10 m); confined to the Atlantic Ocean (Lutze and Wefer, 1980).

Marginopora vertebralis: epiphytic; dinoflagellate symbionts; upper euphotic zone, even intertidal (Smith, 1968); mainly Pacific, but extends west to the Seychelles and Maldives in the Indian Ocean; absent from the east coast of Africa and the Red Sea.

Parasorites orbitolinoides: mainly on sand in low-energy settings such as in the protection of reef spurs and channels; test thin and fragile; chlorophyte symbionts; 0–70 m.

Peneroplis planatus: epiphytic on seagrass and small algae; rhodophyte symbionts; mainly 0–10 m.

Sorites marginalis: epiphytic; chlorophyte symbionts?; central Indo-Pacific.

Sorites orbiculus: epiphytic on seagrass and algae; dinoflagellate symbionts; reef moat and lagoon, avoids high-energy areas, 0–30 m; able to tolerate high light; circum-tropical including the Mediterranean.

6.2   *Hyaline (all have diatom symbionts)*

Amphistegina bicirculata: epifaunal mainly on hard substrates; one of the deepest-dwelling *Amphistegina* species, 60–120 m; Indo-Pacific.

Amphistegina gibbosa: epifaunal; Atlantic Ocean.

Amphistegina lessonii: epifaunal on sediment and hard substrates; occurs over a broad depth range of 0–90 m; the umbiliconvex form is adapted to mobile sand; Indo-Pacific and southern Mediterranean.

Amphistegina lobifera: epifaunal on substrates in high-energy settings and on sand in sheltered settings between coral rubble covered in algae; most tolerant of the *Amphistegina* species to high light levels; very shallow water, 0–30 m; tolerates sediment motion; Indo-Pacific.

Amphistegina papillosa: mainly on sandy substrates; one of the deepest of the *Amphistegina* species.

Amphistegina radiata: epifaunal, mainly on hard substrates but sometimes on sediment; below fair-weather wave base, 20–90 m; Indo-Pacific.

Baculogypsina sphaerulata: epifaunal, reef flat pools clinging to algae, <10 m; western Pacific.

Baculogypsinoides spinosus: epifaunal on hard substrates such as coral rubble; a deep form, 15–80 m; core of the Indo-Pacific.

Calcarina defrancii: epifaunal; avoids wave influence at the reef crest, 10–30 m; western Pacific.

Calcarina gaudichaudii: epifaunal, reef flat pools clinging to algae, <10 m; western Pacific.

Calcarina hispida: epifaunal and epiphytic on algae; the most abundant calcarinid, 0–70 m but mainly below fair-weather wave base; western Pacific.

Calcarina spengleri: epifaunal on exposed slopes of reefs and hard substrates where it
attaches itself using protoplasmic plugs at the end of its spines; shallow water;
possibly more abundant where nutrient levels are higher; western Pacific.

Cycloclypeus carpenteri: epifaunal, below storm wave base, 50–90 m; central Indo-Pacific.

Heterocyclina tuberculata: epifaunal on hard and soft substrates down to the base of
the euphotic zone; diatom symbionts; western Indian Ocean (Mauritius and
Mayotte, Somalia) and the Red Sea.

Heterostegina depressa: epifaunal mainly on hard substrates; 2–70% of surface light
intensity; 0–100 m, mainly below fair-weather wave base but also in rock pools
on the reef crest where it seeks shaded settings; circum-tropical and
cosmopolitan.

Heterostegina operculinoides: epifaunal on hard substrates; lower part of the euphotic
zone; 0.45–26% of surface light intensity; Indo-Pacific.

Neorotalia calcar: epifaunal; shallow water (<10 m) on the reef edge or reef flat; more
abundant on the leeward than exposed slope; hides among algae to avoid high
light levels; Indo-Pacific; western-Pacific forms heavily ornamented; Atlantic
occurrences await confirmation.

Operculina ammonoides: epifaunal on sand; low energy and medium light (1.5–68% of
surface intensity); able to tolerate eutrophic conditions; rarely <1 m, mainly
20–70 m; Indian Ocean and western Pacific.

Operculina complanata: epifaunal on sand; 0.2–12% of surface light intensity; 50–90 m.

Operculinella cummingii: epifaunal, coarse to medium sand; 1.2–25% of surface light
intensity; deep; Indo-Pacific.

Operculinella venosus: epifaunal on sand; 2.5–80% of surface light intensity; ranges
slightly deeper than *Operculina* but below fair-weather wave base, 20–85 m;
western Pacific and Indian Ocean.

7   Deep sea

7.1   Agglutinated

Adercotryma glomeratum: infaunal but also epifaunal in phytodetritus (Gooday,
1993; Heinz et al., 2001)

Astrorhiza arenaria: epifaunal; sometimes close to sediment mounds produced by
burrowing crustaceans; probably a suspension feeder (Tendal and Thomsen,
1988; Linke and Lutze, 1993).

Bathysiphon capillare: infaunal; living more-or-less horizontally in the upper 5 cm of
sediment; deposit feeder (Gooday et al., 1992a, 2002); tolerates dysoxia
(Schönfeld, 1997, 2001).

Bathysiphon filiformis and *major*: upright tubes extending out of the sediment; feed on
detrital material from the sediment surface in areas of high organic input
(Gooday et al., 1992b).

Cribrostomoides subglobosus: epifaunal to shallow infaunal; gathers detrital food with
pseudopodia, which extend from the aperture several millimetres across the
sediment surface (Linke and Lutze, 1993).

Cyclammina cancellata: tolerates dysoxia (Schönfeld, 1997).

Cystammina pauciloculata: tolerates dysoxia (Schönfeld, 1997).

Hormosinella guttifer: epifaunal; may respond to phytodetritus inputs (Fontanier *et al.*, 2003).

Reophax bilocularis: epifaunal; occurs in high-productivity areas (Schmiedl *et al.*, 1997) but also tolerates very low organic flux (Altenbach *et al.*, 1999).

Reophax scorpiurus: epifaunal; although it tolerates a wide range in flux of organic matter, it favours areas of high flux (Altenbach *et al.*, 1999).

Rhabdammina abyssorum: epifaunal, lying on the sediment surface where it accumulates detrital material with its pseudopodia; also on rough sediment surfaces, which generate turbulent conditions that may enhance food supply (Linke and Lutze, 1993).

Saccorhiza ramosa: the part of the test above the sediment surface is hook- or T-shaped thus providing support for the pseudopodial net, which is used in suspension feeding (Altenbach *et al.*, 1988).

7.2   Hyaline

Brizalina albatrossi: a facultative anaerobe (Sen Gupta *et al.*, 1997); a high-flux species (Gooday, 2003) and associated with methane seeps.

Brizalina ordinaria: a facultative anaerobe (Sen Gupta *et al.*, 1997).

Bulimina inflata: shallow infaunal (Jorissen *et al.*, 1998).

Bulimina marginata: deep infaunal (Jorissen *et al.*, 1998).

Cancris auriculus: epifaunal; prefers high organic flux (Altenbach *et al.*, 1999).

Cassidulina carinata/laevigata: epifaunal; moderate to high organic flux (Altenbach *et al.*, 1999).

Cassidulina neocarinata: a facultative anaerobe (Sen Gupta *et al.*, 1997).

Chilostomella oolina: deep infaunal (Jorissen *et al.*, 1998).

Chilostomella ovoidea: deep infaunal; tolerates dysoxia (Schönfeld, 2001).

Cibicides refulgens: epifaunal; attached to firm substrates in high-energy areas (such as currents).

Cibicidoides kullenbergi: shallow infaunal (Jorissen *et al.*, 1998); an intermediate-flux species (Gooday, 2003).

Cibicidoides robertsonianus: shallow infaunal (Jorissen *et al.*, 1998).

Epistominella exigua: epifaunal; particularly associated with seasonal deposits of phytodetritus (Gooday, 1993) but has an extremely large tolerance to varying organic flux (Altenbach *et al.*, 1999); opportunistic.

Epistominella rugosa: epifaunal (Jorissen *et al.*, 1998): a facultative anaerobe (Sen Gupta *et al.*, 1997).

Fontbotia wuellerstorfi: epifaunal; attached to hard substrates in elevated positions (up to 14 cm above the sediment on crinoids); suspension feeder (Lutze and Thiel, 1989); low flux (Linke and Lutze, 1993; Gooday, 2003); currents.

Fursenkoina mexicana: deep infaunal (Jorissen *et al.*, 1998).

Gavelinoposis translucens: shallow infaunal (Jorissen *et al.*, 1998); a facultative anaerobe (Sen Gupta *et al.*, 1997).

Globobulimina affinis: infaunal; tolerant of degraded organic matter and feeding
off associated bacteria (Schönfeld, 2001; Fontanier *et al.*, 2002); tolerates
dysoxia (Schönfeld, 1997); abundant under oxic conditions at the sediment
surface (Timm, 1992); high organic flux – does not live where the flux is
$<3.5\,\mathrm{g\ C\ m^{-2}\ y^{-1}}$ (Schönfeld, 1997).

Globobulimina auriculata: infaunal.

Globobulimina pyrula: deep infaunal (Jorissen *et al.*, 1998).

Hoeglundina elegans: epifaunal (Jorissen *et al.*, 1998); shallow infaunal (Fontanier *et al.*,
2002); a high-flux species (Gooday, 2003).

Melonis barleeanum: commonly infaunal but also recorded from the sediment surface
(Linke and Lutze, 1993); tolerant of degraded organic matter and probably
feeding off associated bacteria (Fontanier *et al.*, 2002); tolerates dysoxia and
suboxic pore water but has maximum abundance under oxic conditions near
the sediment surface (Schönfeld, 1997); occurs in absence of oxygen and
decreasing nitrate (Jorissen *et al.*, 1998).

Melonis zaandamae: probably the same as *Melonis barleeanum* (the two may not be
distinct species).

Nonionella iridea: infaunal; occurs in a wide variety of settings from fjords to the
deep sea.

Nuttallides pusillus: epifaunal.

Nuttallides umboniferus: shallow infaunal (Fontanier *et al.*, 2002); tolerates corrosive
bottom waters such as AABW; low flux (Gooday, 2003).

Oridorsalis umbonatus: epifaunal (Jorissen *et al.*, 1998).

Rupertia stabilis: epifaunal; attached to hard substrates in areas of high bottom
currents; suspension feeder (Lutze and Altenbach, 1988; Linke and Lutze, 1993).

Trifarina angulosa: epifaunal; commonly associated with shelf-edge–upper-slope areas
of sand to gravel sediment under the influence of bottom currents (Mackensen
et al., 1985, 1993a; Harloff and Mackensen, 1997).

Trifarina bradyi: a facultative anaerobe (Sen Gupta *et al.*, 1997).

Uvigerina elongatastriata: tolerant of dysoxic conditions (Schönfeld, 1997).

Uvigerina mediterranea: shallow infaunal; most abundant where there is a rich supply
of labile organic matter and slightly less tolerant of dysoxic conditions than
Uvigerina peregrina (Altenbach *et al.*, 1999; Fontanier *et al.*, 2002); superior
competitor in food- and oxygen-rich sediments and resistant to disturbance by
macrofaunal predators (de Stigter *et al.*, 1998).

Uvigerina peregrina: shallow infaunal; most abundant where there is a rich supply of
labile organic matter and a high concentration of bacteria (Altenbach and
Sarnthein, 1989; Altenbach *et al.*, 1999; Fontanier *et al.*, 2002).

Genera

Based on the main living occurrences and excluding rare occurrences in other
environments. The data are given in the following order: mode of life; substrate; mode of
feeding; salinity; temperature; depth; environment. Definitions: hard substrates = rocks, shells,

macrofauna, plants, etc.; S = salinity, brackish 0–32, marine 32–37, hypersaline >37; temperature of the bottom water, cold, temperate, warm (even in tropical areas the water deeper than 100 m is temperate or cold); environment, shelf 0–180 m, bathyal 180–4000 m, abyssal >4000 m.

Acervulina: epifaunal, attached; bryozoa, shells, etc.; passive suspension feeder; marine; temperate to warm; 0–60 m; inner shelf.

Adercotryma: epifaunal or infaunal, free; fine sand; detritivore; marine; cold; shelf-abyssal.

Alabaminella: epifaunal, free; often associated with phytodetritus but also lives on sediment; normal marine; cold; deep sea.

Alveolinella: epifaunal, clinging; algal-covered carbonate gravels; herbivore, symbionts; marine; 18–26 °C; 5–100 m; inner shelf, lagoon.

Alveolophragmium: epifaunal, free; sand; detritivore; marine; 10 °C; 20–700 m; shelf-upper bathyal.

Ammoastuta: infaunal?, free; sediment; detritivore?; brackish; temperate-tropical; marshes.

Ammobaculites: infaunal, free; muddy sediment; detritivore; brackish-marine; temperate-tropical; brackish marshes and lagoons, inner shelf-upper bathyal.

Ammoglobigerina: infaunal?, free; sediment; detritivore?; marine; cold; shelf-upper bathyal.

Ammonia: infaunal, free; muddy sand; herbivore?; brackish, marine, hypersaline; warm temperate-tropical; 0–50 m; brackish and hypersaline lagoons, inner shelf.

Ammoscalaria: infaunal?, free; sediment; detritivore; brackish-marine; cold-temperate; inner shelf.

Ammotium: infaunal, free; muddy sediment; detritivore; brackish; 0–30 °C; 0–10 m; tidal marshes, brackish lagoons and estuaries, enclosed brackish shelf seas.

Amphisorus: epifaunal, clinging; phytal, carbonate sediment; herbivore, symbionts; marine; 19–26 °C; 0–50 m; lagoon, nearshore.

Amphistegina: epifaunal, free; phytal, coarse carbonate; herbivore, symbionts; marine; winter minimum 15 °C, generally >20 °C; 0–130 m; coral reefs, lagoons.

Archaias: epifaunal, clinging; phytal; herbivore, symbionts; marine; >22 °C; 0–20 m; inner shelf.

Arenoparrella: epifaunal and infaunal, free or clinging; muddy sediment; herbivore; brackish-hypersaline; 0–30 °C; mainly intertidal on marshes but some on shelf-upper bathyal.

Articulina: epifaunal, free; phytal or sediment; marine-hypersaline; >20 °C; 0–3500 m; inner shelf-bathyal.

Asterigerina: epifaunal, free; sediment; herbivore?; marine; 0–100 m; subtropical-tropical; inner shelf.

Asterigerinata: epifaunal, free; sediment; herbivore?; marine; temperate; 0–100 m; inner shelf.

Asterorotalia: epifaunal, free; sediment; herbivore?; marine; subtropical-tropical; inner shelf.

Astrononion: epifaunal–infaunal, free–clinging; muddy sediment; detritivore?; marine; cold; shelf–bathyal.

Astrorhiza: epifaunal, sessile; sediment; passive suspension feeder/carnivore; marine; cold–temperate; shelf.

Aubignyna: infaunal?, free; sediment; slightly brackish; temperate to warm; lagoons, inner shelf

Baculogypsina: epifaunal, clinging; phytal; herbivore, symbionts; marine; >25 °C; 0–10 m; coral reefs, lagoons.

Balticammina: epifaunal, free; detritivore?; very euryhaline; uppermost tidal marsh.

Bathysiphon: infaunal, free?; sediment; marine; cold; upper bathyal.

Bigenerina: epifaunal?, free; sediment; omnivore; marine; temperate–warm; shelf.

Bolivina: infaunal–epifaunal, free; muddy sediment; some species tolerate dysoxia; detritivore?; marine; cold–warm; inner shelf–bathyal.

Bolivinella: infaunal?, free; muddy sediment; marine; temperate–cold; shelf–upper bathyal.

Bolivinellina: infaunal, free; muddy sediment; marine–slightly brackish; temperate; inner shelf and marginal marine.

Borelis: epifaunal, free; algal-coated substrates, seagrass and coarse sediment; herbivore, symbionts; marine; 18–26 °C; 0–40 m; lagoon–reef.

Brizalina: infaunal, free; muddy sediment; some species tolerate dysoxia; detritivore; marine; cold–temperate; marginal marine–bathyal.

Buccella: infaunal?, free; muddy sediment; detritivore?; marine; cold–temperate; 0–100 m; lagoons–inner shelf.

Bulimina: infaunal, free; mud–fine sand; some species tolerate dysoxia; detritivore?; marine; cold–temperate; inner shelf–bathyal.

Buliminella: infaunal, free; muddy sediment; detritivore?; marine; temperate; mainly shelf but also lagoons and upper bathyal.

Calcarina: epifaunal, free; sediment; herbivore, symbionts; marine; 18–26 °C; 0–70 m; lagoons, reefs, inner shelf.

Cancris: epifaunal, free; sediment; detritivore?; marine; temperate–subtropical; 50–150 m; shelf.

Cassidella: infaunal, free; mud; detritivore; marine; temperate; shelf.

Cassidulina: infaunal or epifaunal, free; mud, sand; detritivore; marine; cold–temperate; shelf–bathyal.

Chilostomella: deep infaunal, free; mud; tolerates dysoxia; detritivore; marine; outer shelf–bathyal.

Cibicides: epifaunal, attached; hard substrates in high energy; passive suspension feeder?; marine; cold–warm; 0–2000 m; shelf–bathyal.

Cibicidoides: infaunal epifaunal, clinging; hard substrates; passive suspension feeder?; cold; shelf–bathyal.

Cribrostomoides: epifaunal, free or clinging; sand, pebbles; detritivore; S 30–35; <15 °C; shelf–bathyal.

Cuneata: infaunal, free; mud, sand; detritivore; S 30–35; cold; shelf.

Cyclammina: epifaunal?, free; mud, sand; detritivore; marine; >100 m; outer shelf-abyssal.

Cycloclypeus: epifaunal, free; muddy, carbonate sediment; warm; 60–100 m; shelf, below storm-wave base.

Deuterammina: epifaunal, attached mobile; sediment; normal marine; cold-temperate; shelf.

Discorbis: epifaunal, clinging or attached; hard substrates, coarse sand; herbivore; marine; temperate-warm; 0–50 m; inner shelf.

Discorinopsis: infaunal, free; muddy sediment; detritivore; hypersaline; warm; marsh.

Eggerella: infaunal, free; silt, fine sand; detritivore; S 20–37; cold-temperate; shelf-outer lagoon.

Eggerelloides: infaunal, free; sand; detritivore; marine; temperate; shelf.

Ehrenbergina: epifaunal?, free; mud; detritivore?; marine; cold; outer shelf-bathyal.

Eilohedra: infaunal, free; mud; detritivore?; marine; cold; bathyal.

Elphidiella: infaunal?, free; sand; detritivore?; S 30–35; cold; shelf.

Elphidium: (keeled species) epifaunal, free; sand, vegetation; herbivore; marine; temperate-warm; S 30–70; 0–50 m; inner shelf; (non-keeled species) infaunal, free; mud, sand; herbivore; S 0–70; brackish-hypersaline marshes and lagoons, inner shelf (upper bathyal, *Elphidium excavatum* only).

Epistominella: epifaunal to shallow infaunal, free; mud, phytodetritus; detritivore; marine; temperate-cold; shelf-bathyal.

Eponides: epifaunal, free or clinging; sediment or hard substrates; detritivore?; marine; cold-temperate; shelf-abyssal.

Fontbotia: epifaunal, clinging; hard substrates; passive suspension feeder in currents; marine; cold; bathyal.

Fursenkoina: infaunal, free; mud; some species tolerate dysoxia; detritivore; S 30–35; shelf-upper bathyal.

Gaudryina: epifaunal, attached; hard substrates; passive suspension feeder?; marine; temperate; 50–500 m; shelf-upper bathyal.

Gavelinopsis: epifaunal, clinging; hard substrates; passive suspension feeder?; marine; cold-temperate; shelf-bathyal.

Glabratella: epifaunal, clinging or attached; hard substrates; herbivore?; marine-hypersaline; temperate-warm; 0–50 m; hypersaline marshes and lagoons, marine inner shelf.

Globobulimina: deep infaunal, free; facultative anaerobe; mud; tolerates dysoxia; detritivore; cold-temperate; shelf-bathyal.

Globocassidulina: infaunal, free; mud; detritivore?; marine; cold-temperate, shelf-bathyal.

Goesella: epifaunal, free; sediment; detritivore; marine; temperate-warm; shelf-upper bathyal.

Gyroidina: epifaunal, free; mud; detritivore?; marine; cold; shelf-bathyal.

Hanzawaia: epifaunal, clinging; hard substrates; passive suspension feeder?; marine; temperate-warm; inner shelf.

Haplophragmoides: epifaunal to shallow infaunal, free; mud–sand; detritivore?; marine; cold–temperate; marshes–bathyal.

Haynesina: infaunal, free; mud, silt; herbivore, husbands chloroplasts; brackish; cold–warm; marsh, lagoon, inner shelf.

Hemisphaerammina: epifaunal, attached; hard substrates; passive herbivore; marine; cold–temperate; inner shelf.

Heterolepa: epifaunal, clinging?; hard substrates; passive suspension feeder?; marine; cold–temperate; shelf–bathyal.

Heterostegina: epifaunal, free; phytal, muddy carbonate sediment and hard substrates; herbivore, symbionts; marine; 18–26 °C; 0–130 m; shelf, lagoon, seeks shade.

Hoeglundina: infaunal, free; mud; detritivore?; marine; cold; outer shelf–bathyal.

Homotrema: epifaunal, attached; hard substrates in high energy; passive suspension feeder; marine; temperate–warm; inner shelf, reefs.

Hormosina: infaunal, free; mud; detritivore?; marine; cold; bathyal-abyssal.

Hyalinea: epifaunal, free; mud, silt; detritivore?; marine; cold–temperate; 10–1000 m; shelf–upper bathyal.

Islandiella: infaunal, free; mud, silt; detritivore?; marine; <10 °C; >20 m; shelf–upper bathyal.

Jadammina: epifaunal, free; mud, silt; herbivore? or detritivore?; S 0–50; 0–30 °C; high intertidal marsh.

Karreriella: epifaunal, free; mud, silt; detritivore?; marine; <10 °C; outer shelf–bathyal.

Lagenammina: infaunal?, free; sediment; detritivore?; S 30–35; cold-temperate; shelf–bathyal.

Lenticulina: epifaunal, free; mud; detritivore?; marine; cold; outer shelf–bathyal.

Lepidodeuterammina: epifaunal–infaunal, clinging; hard substrates, coarse sand in high energy; herbivore; marine; cold; shelf.

Marginopora: epiphytic, attached; phytal, carbonate sand and gravel; herbivore, symbionts; marine; 18–26 °C; 0–45 m; inner shelf, lagoon, coral reefs.

Martinottiella: epifaunal?, free; mud; detritivore?; marine, cold; >120 m; outer shelf–bathyal.

Massilina: epiphytic, clinging; plants; herbivore; marine; temperate–warm; inner shelf.

Melonis: infaunal, free; mud, silt; some species tolerate dysoxia; detritivore; marine; cold; shelf–bathyal.

Miliammina: infaunal–epifaunal, free; mud, silt, decaying vegetation; detritivore; brackish–hypersaline; 0–30 °C; marshes–upper bathyal.

Miliolinella: epifaunal, clinging; plants and hard substrates; herbivore; S 32–50; 10–30 °C; 0–100 m; hypersaline lagoons, normal marine lagoons and marshes, inner shelf; some deep-sea records.

Neoconorbina: epifaunal, clinging; hard substrates; herbivore?; marine; temperate; inner shelf.

Nonion: infaunal, free; mud, silt; herbivore; S 30–35; cold–warm; 0–180 m; shelf.

Nonionella: infaunal, free; mud; some species tolerate dysoxia; detritivore?; marine; temperate-warm; 10-1000 m; shelf-upper bathyal.

Nonionellina: infaunal, free; mud-sand; detritivore; marine; cold; shelf.

Nonionoides: infaunal, free; muddy sediment; some species tolerate dysoxia; detritivore?; marine; temperate-warm; shelf.

Nuttallides: epifaunal-shallow infaunal, free or clinging; detritivore?; marine; <4 °C; bathyal-abyssal.

Operculina: epifaunal, free; carbonate sediment; herbivore, symbionts; marine-slightly hypersaline; warm; 0-130 m; lagoon-shelf, low energy and medium light.

Operculinella: epifaunal, free; carbonate sediment; herbivore, symbionts; marine; warm; 0-130 m; lagoon to shelf.

Oridorsalis: epifaunal, free; mud; detritivore; marine; <4 °C; bathyal.

Pararotalia: epifaunal, free; sand; herbivore; marine; warm; inner shelf.

Patellina: epifaunal, clinging; hard substrates; herbivore?; marine; cold-warm; 0-100 m; inner shelf.

Peneroplis: epifaunal, clinging; plants and hard substrates; S 35-53; 18-27 °C; 0-70 m; lagoons and innermost shelf.

Planorbulina: epifaunal, attached immobile; hard substrates in high energy; passive suspension feeder?; marine; temperate-warm; 0-50 m; inner shelf.

Planulina: epifaunal, clinging; hard substrates; passive suspension feeder?; marine; cold-warm; shelf-bathyal.

Poroeponides: epifaunal, clinging; hard substrates; herbivore/passive suspension feeder?; marine; temperate-warm; inner shelf.

Portatrochammina: epifaunal, free; sediment; detritivore; marine; cold-temperate; shelf-bathyal.

Pullenia: infaunal, free; mud; detritivore; marine; cold; outer shelf-bathyal.

Pyrgo: *either* epifaunal, free or clinging; plants or sediment; herbivore; marine; temperate-warm; inner shelf; *or* epifaunal, free; sediment; detritivore?; marine; cold; shelf-bathyal.

Quinqueloculina: epifaunal, free or clinging; plants or sediment; herbivore; marine-hypersaline; S 32-65; cold-warm; hypersaline lagoons, marine marsh and shelf, rarely bathyal.

Recurvoides: epifaunal, free; sediment; detritivore; marine; cold; shelf-bathyal.

Reophax: infaunal, free; mud, sand; detritivore; marine; cold-temperate; lagoons, shelf-bathyal.

Rosalina: epifaunal, clinging or attached; hard substrates; herbivore?, omnivore; marine; temperate-warm; 0-100 m; lagoons, inner-mid shelf.

Saccammina: infaunal, free; sand; detritivore; marine; cold-temperate; 0-100 m; inner-mid shelf.

Siphotrochammina: epifaunal, sometimes epiphytic on mangrove roots; brackish; warm, tropical; marsh and mangal.

Sorites: epiphytic, clinging; plants, especially seagrass; herbivore, symbionts; S 37-45; 18-26 °C; 0-70 m; lagoons, nearshore.

Spirilina: epifaunal, clinging; hard substrates; marine; cold-temperate; 0–100 m; inner-mid shelf.

Spirolina: epiphytic, clinging; plants; herbivore; S 37–50; 18–26 °C; lagoons, nearshore.

Spiroloculina: epifaunal, free or clinging; sediment or plants; herbivore; marine-hypersaline; temperate-warm; 0–40 m; lagoons, inner shelf.

Textularia: epifaunal, free or clinging; hard substrates, sediment; detritivore?; marine; cold–warm; 0–500 m; lagoons, shelf-bathyal.

Tiphotrocha: epifaunal, free; sediment; herbivore or detritivore; brackish; temperate; intertidal; marsh.

Trifarina: infaunal, free; mud, sand; detritivore?; marine; cold–temperate; 0–400 m; shelf-upper bathyal.

Triloculina: epifaunal, free or clinging; mud, sand, plants; herbivore, detritivore; marine-hypersaline; S 32–55?; temperate-warm; mainly hypersaline lagoons or marine inner shelf, some bathyal species (cold).

'*Trochammina*': epifaunal or infaunal, free; sediment; herbivore or detritivore; S 0–60; 0–30 °C; 0–6000 m; intertidal-abyssal.

Trochulina: epifaunal or infaunal, free; sediment; detritivore?; marine; temperate-warm; lagoons, shelf.

Uvigerina: mainly infaunal, some epifaunal, free; muddy sediments; detritivore?; marine; cold; 100–4500 m; shelf-abyssal.

References

See also the linked website for references on taxonomy, dead and total assemblages

Adam, P. 1990. *Saltmarsh Ecology*. Cambridge: Cambridge University Press.

Ahmed, M. A. O. S. 1991. Recent benthic foraminifera from Tarut Bay, Arabian Gulf coast of Saudi Arabia. *Journal of Micropalaeontology*, **10**: 33–8.

Ahrens, M. J., Graf, G. and Altnbach, A. V. 1997. Spatial and temporal distribution of benthic foraminifera in the Northeast Water Polynya, Greenland. *Journal of Marine Systems*, **10**: 445–65.

Akers, W. H. 1971. Estuarine foraminiferal associations in the Beaufort area, North Carolina. *Tulane Studies in Geology and Paleontology*, **8**: 147–215.

Akimoto, K., Hattori, M., Uematsu, K. and Kato, C. 2001. The deepest living foraminifera, Challenger Deep, Marianna Trench. *Marine Micropaleontology*, **42**: 95–7.

Akpati, B. N. 1975. Foraminiferal distribution and environmental variables in eastern Long Island Sound, New York. *Journal of Foraminiferal Research*, **5**: 127–44.

Alexander, S. P. and Banner, F. T. 1984. The functional relationship between skeleton and cytoplasm in *Haynesina germanica* (Ehrenberg). *Journal of Foraminiferal Research*, **14**: 159–70.

Alexander, S. P. and Delaca, T. E. 1987. Feeding adaptations of the foraminifera *Cibicides refulgens* living epizoically and parasitically on the Antarctic scallop *Admussium colbecki*. *Biological Bulletin*, **173**: 136–59.

Allen, K., Roberts, S. and Murray, J. W. 1998. Fractal grain distribution in agglutinated foraminifera. *Paleobiology*, **24**: 349–58.

Allen, R. and Roda, R. S. 1977. Benthonic foraminifera from LaHave estuary. *Maritime Sediments*, **13**: 67–72.

Aller, J. Y. 1989. Quantifying sediment disturbance by bottom currents and its effect on benthic communities in a deep sea western boundary zone. *Deep-Sea Research*, **36**: 901–34.

Allison, P. A. and Briggs, D. E. G. 1991. The taphonomy of soft-bodied animals. In Donovan, S. K., ed. *The Processes of Fossilization*. London: Bellhaven Press, pp. 120–40.

Allison, P. A., Wignall, P. B. and Brett, C. E. 1995. Paleo-oxygenation: effects and recognition. In *Marine Palaeoenvironmental Analysis from Fossils*. Geological Society Special Publication, **83**: 97–112.

Almers, H. and Cedhagen, T. 1996. Ecology and hydrodynamic adaptations of the large foraminifera *Discobotellina biperforata* (Hemisphaeramminidae). *Marine Ecology Progress Series*, **140**: 179–86.

Almogi-Labin, A., Perelis-Grossovicz, L. and Raab, M. 1992. Living *Ammonia* from a hypersaline inland pool, Dead Sea area, Israel. *Journal of Foraminiferal Research*, **22**: 257–66.

Almogi-Labin, A., Simon-Tov, R., Rosenfeld, A. and Debard, E. 1995. Occurrence and distribution of the foraminifer *Ammonia beccarii tepida* (Cushman) in water bodies, Recent and Quaternary, of the Dead Sea Rift, Israel. *Marine Micropaleontology*, **26**: 153–9.

Alongi, D. M. 1998. *Coastal Ecosystem Processes*. Boca Raton, Florida: CRC Press.

Altenbach, A. V. 1987. The measurement of organic carbon in foraminifera. *Journal of Foraminiferal Research*, **17**: 106–9.

Altenbach, A. V. 1992. Short term processes and patterns in the foraminiferal response to organic flux rates. *Marine Micropaleontology*, **19**: 119–29.

Altenbach, A. V., Heeger, T., Linke, P., Spindler, M. and Thies, A. 1993. *Miliolinella subrotunda* (Montagu), a miliolid foraminifer building large detritic tubes for a temporary epibenthic lifestyle. *Marine Micropaleontology*, **20**: 293–301.

Altenbach, A. V., Lutze, G. F., Schiebel, R. and Schönfeld, J. 2003. Impact of interrelated and interdependent ecological controls on benthic foraminifera: an example from the Gulf of Guinea. *Palaeogeography, Palaeoclimatology, Palaeoecology*, **197**: 213–38.

Altenbach, A. V., Pflaumann, U., Schiebel, R. *et al.* 1999. Scaling percentages and distributional patterns of benthic foraminifera with flux rates or organic carbon. *Journal of Foraminiferal Research*, **29**: 173–85.

Altenbach, A. V. and Sarnthein, M. 1989. Productivity record in benthic foraminifera. In Berger, W. H., Smetacek, V. S. and Wefer, G., eds. *Production of the Oceans: Present and Past*. New York: John Wiley, pp. 255–69.

Altenbach, A. V. and Struck, U. 2001. On the coherence of organic carbon flux and benthic foraminiferal biomass. *Journal of Foraminiferal Research*, **31**: 79–85.

Altenbach, A. V., Unsöld, G. and Walger, E. 1988. The hydrodynamic environment of *Saccorhiza ramosa* (Brady). *Meyniana*, **40**: 119–32.

Alve, E. 1991. Foraminifera, climate change, and pollution: a study of late Holocene sediments in Drammensfjord, southern Norway. *The Holocene*, **1**: 243–61.

Alve, E. 1994. Opportunistic features of the foraminifer *Stainforthia fusiformis* (Williamson): evidence from Frierfjord, Norway. *Journal of Micropalaeontology*, **13**: 24.

Alve, E. 1995a. Benthic foraminiferal distribution and recolonization of formerly anoxic environments in Drammensfjord, southern Norway. *Marine Micropaleontology*, **25**: 169–86.

Alve, E. 1995b. Benthic foraminiferal responses to estuarine pollution: a review. *Journal of Foraminiferal Research*, **25**: 190–203.

Alve, E. 1996. Benthic foraminiferal evidence of environmental change in the Skagerrak over the past six decades. *Norges geologiske undersøkelse bulletin*, **430**: 85–93.

Alve, E. 1999. Colonization of new habitats by benthic foraminifera: a review. *Earth-Science Reviews*, **46**: 167–85.

Alve, E. 2000. Environmental stratigraphy. A case study reconstructing bottom water oxygen conditions in Frierfjord, Norway, over the past five centuries. In Martin, R. E. 2000. *Environmental Micropaleontology*. New York: Kluwer Academic, pp. 323–50.

Alve, E. 2003. A common opportunistic foraminiferal species as an indicator of rapidly changing conditions in a range of environments. *Estuarine, Coastal and Shelf Science*, **57**: 501–14.

Alve, E. and Bernhard, J. M. 1995. Vertical migratory response of benthic foraminifera to controlled oxygen concentrations in an experimental mesocosm. *Marine Ecology Progress Series*, **116**: 137–51.

Alve, E. and Goldstein, S. T. 2002. Resting stage in benthic foraminiferal propagules: a key feature for dispersal? Evidence from two shallow-water species. *Journal of Micropalaeontology*, **21**: 95–6.

Alve, E. and Goldstein, S. T. 2003. Propagule transport as a key method of dispersal in benthic foraminifera. *Limnology and Oceanography*, **48**: 2163–70.

Alve, E. and Murray, J. W. 1994. Ecology and taphonomy of benthic foraminifera in a temperate mesotidal inlet. *Journal of Foraminiferal Research*, **24**: 18–27.

Alve, E. and Murray, J. W. 1995a. Benthic foraminiferal distribution and abundance changes in Skagerrak surface sediments: 1937 (Höglund) and 1992/1993 data compared. *Marine Micropaleontology*, **25**: 269–88.

Alve, E. and Murray, J. W. 1995b. Experiments to determine the origin and palaeoenvironmental significance of agglutinated foraminiferal assemblages. In Kaminski, M. A., Geroch, S. and Gasiński, M. A., eds. *Proceedings of the Fourth International Workshop on Agglutinated Foraminifera, Kraków, Poland*. Grzybowski Foundation Special Publication, 3: 1–11.

Alve, E. and Murray, J. W. 1997. High benthic fertility and taphonomy of foraminifera: a case study of the Skagerrak, North Sea. *Marine Micropaleontology*, **31**: 157–75.

Alve, E. and Murray, J. W. 1999. Marginal marine environments of the Skagerrak and Kattegat: a baseline study of living (stained) benthic foraminiferal ecology. *Palaeogeography, Palaeoclimatology, Palaeoecology*, **146**: 171–93.

Alve, E. and Murray, J. W. 2001. Temporal variability in vertical distributions of live (stained) intertidal foraminifera, southern England. *Journal of Foraminiferal Research*, **31**: 12–24.

Alve, E. and Nagy, J. 1986. Estuarine foraminiferal distributions in Sandebukta, a branch of Oslo Fjord. *Journal of Foraminiferal Research*, **16**: 261–84.

Alve, E. and Nagy, J. 1990. Main features of foraminiferal distribution reflecting estuarine hydrography in Oslo Fjord. *Marine Micropaleontology*, **16**: 181–206.

Alve, E. and Olsgard, F. 1999. Benthic foraminiferal colonization in experiments with copper-contaminated sediments. *Journal of Foraminiferal Research*, **29**: 186–95.

Anderson, O. R. and Lee, J. J. 1991. Cytology and fine structure. In Lee, J. J. and Anderson, O. R., eds. *Biology of Foraminifera*. London: Academic, pp. 7–40.

Anderson, O. R., Lee, J. J. and Faber, W. W. 1991. Collection, maintenance and culture methods for the study of living foraminifera. In Lee, J. J. and Anderson, O. R., eds. *Biology of Foraminifera*. London: Academic, pp. 335–57.

Andreieff, P., Bouysse, P., Chateauneuf, J. J., Homer, A. l' and Scolari, G. 1971. La couverture sédimentaire meuble du plateau continental externe de la Bretagne méridionale. *Cahiers océanographique*, **4**: 343–81.

Angell, R. W. 1990. Observations on reproduction and juvenile test building in the foraminifer *Trochammina inflata*. *Journal of Foraminiferal Research*, **20**: 246–7.

Anschutz, P., Jorissen, F. J., Chaillou, G., Abu-Zied, R. and Fontanier, C. 2002. Recent turbidite deposition in the eastern Atlantic: early diagenesis and biotic recovery. *Journal of Marine Research*, **60**: 835–54.

Antony, A. 1980. Foraminifera of the Vembanad estuary. *Bulletin of the Department of Marine Science, University of Cochin*, **11**: 25–63.

Arnold, Z. M. 1955. An unusual feature of miliolid reproduction. *Contributions from the Cushman Foundation for In Hedley, R.H. and Adams, C. G. (eds.), Foraminiferal Research*, **6**: 64–96.

Arnold, Z. M. 1964. Biological observations on the foraminifer *Spiroloculina hyalinea* Schultze. *University of California Publications in Zoology*, **72**: 1–93.

Arnold, Z. M. 1974. Field and laboratory techniques for the study of living foraminifera. In Hedley, R. H. and Adams, C. G. (eds.), *Foraminifera*, 1: 153–206.

Asioli, A. 1995. Living (stained) benthic foraminifera distribution in the western Ross Sea (Antarctica). *Palaeopelagos*, **5**: 201–14.

Atkinson, K. 1969. The association of living foraminifera with algae from the littoral zone, south Cardigan Bay, Wales. *Journal of Natural History*, **3**: 517–42.

Austin, W. E. N. and Sejrup, H. P. 1994. Recent shallow water benthic foraminifera from western Norway: ecology and palaeoecological significance. *Cushman Foundation Special Publication*, **32**: 103–25.

Ayala-Castañares, A. 1963. Sistemática y distribución de los foraminíferos recientes de la Laguna de Terminos, Campeche, Mexico. *Boletin Instituto de Geologia, Universidad Nacional, Mexico*, **67**: 1–130.

Ayala-Castañares, A. and Segura, L. R. 1968. Ecologia y distribución de los foraminíferos recientes de la Lagune Madre, Tamaulipas, Mexico. *Boletin Instituto de Geologia, Universidad Nacional, Mexico*, **87**: 1–90.

Ayala-López, A. and Molina-Cruz, A. 1994. Micropalaeontology of the hydrothermal region in the Guaymas Basin, Mexico. *Journal of Micropalaeontology*, **13**: 133–46.

Bahafzallah, A. B. K. 1979. Recent benthic foraminfera from Jiddah Bay, Red Sea (Saudi Arabia). *Neues Jahrbuch für Paläontologie Monatsheft*, **7**: 385–98.

Bandy, O. L. 1956. Ecology of foraminifera in northeastern Gulf of Mexico. *US Geological Survey, Professional Paper*, **274-G**: 179–204.

Banner, F. T. 1971. A new genus of Planorbulinidae. An endoparasite of another foraminifer. *Revista española de micropaleontología*, **3**: 113–28.

Banner, F. T. 1978. Form and function in coiled benthic foraminifera. *The British Micropalaeontologist*, **8**: 11–12.

Banner, F. T. and Culver, S. J. 1978. Quaternary *Haynesina* n. gen. and Paleogene *Protelphidium* Haynes: their morphology, affinities and distribution. *Journal of Foraminiferal Research*, **8**: 177–207.

Banner, F. T., Sheehan, R., and Williams, E. 1973. The organic skeletons of rotaline foraminifera: a review. *Journal of Foraminiferal Research*, **3**: 30–42.

Banner, F. T. and Williams, E. 1973. Test structure, organic skeleton and extrathalamous cytoplasm of *Ammonia* Brünnich. *Journal of Foraminiferal Research*, **3**: 49–69.

Barbero, R. S. and Toffoletto, D. 1996. Dissolution of calcareous foraminifera in microfossil slides. *Micropaleontology*, **42**: 412–14.

Barbosa, C. F., Scott, D. B., Seoane, J. C. S. and Turcq, B. J. 2005. Foraminiferal zonations as base lines for Quaternary sea-level fluctuations in south-southeastern Brazilian mangroves and marshes. *Journal of Foraminiferal Research*, **35**: 22–43.

Barmawidjaja, D. M., Jorissen, F. J., Puskaric, S. and Zwaan, G. J. van der 1992. Microhabitat selection by benthic foraminifera in the northern Adriatic Sea. *Journal of Foraminiferal Research*, **22**: 297–317.

Barnett, P. R. O., Watson, J. and Connelly, D. 1984. A multiple corer for taking virtually undisturbed samples from shelf, bathyal and abyssal sediments. *Oceanologia Acta*, **7**: 399–408.

Basov, I. A. 1974. Biomass of benthic foraminifers in the region of the South Sandwich Trench and Falkland Islands. *Oceanology*, **14**: 277–9.

Basson, P. W., Burchard, J. E., Hardy, J. T and Price, A. R. G. 1977. *Biotopes of the Western Arabian Gulf*. Dahran, Arabian America Oil Company.

Basson, P. W. and Murray, J. W. 1995. Temporal variations in four species of intertidal foraminifera, Bahrain, Arabian Gulf. *Micropaleontology*, **41**: 69–76.

Bates, J. M. and Spencer, R. S. 1979, Modification of foraminiferal trends by the Chesapeake-Elizabeth sewage outfall, Virginia Beach, Virginia. *Journal of Foraminiferal Research*, **9**: 125–40.

Bauch, H. A., Erlenkeuser, H., Bauch, D., Mueller-Lupp, T. and Taldenkova, E. 2004. Stable oxygen and carbon isotopes in modern benthic foraminifera from the Laptev Sea shelf: implications for reconstructing proglacial and profluvial environments in the arctic. *Marine Micropaleontology*, **51**: 285–300.

Belasky, P. 1996. Biogeography of Indo-Pacific larger foraminifera and scleractinian corals: a probabilistic approach to estimating taxonomic diversity, faunal similarity, and sampling bias. *Palaeogeography, Palaeoclimatology, Palaeoecology*, **122**: 119–41.

Belyaeva, N. V. and Khusid, T. A. 1990. Distribution of calcareous foraminifers in Arctic Ocean sediments. In Bleil, U. and Thiede, J., eds. *Geological History of Polar Oceans*. Dordrecht: Kluwer, pp. 311–16.

Bemis, B. E., Spero, H., Bijma, J. and Lea, D. W. 1988. Reevaluation of the oxygen isotopic composition of planktonic foraminifera: experimental results and revised paleotemperature equations. *Paleoceanography*, **13**: 150–60.

Benzie, J. A. H. 1991. Genetic relatedness of foraminiferan (*Marginopora vertebralis*) populations from reefs in the western Coral Sea and the Great Barrier Reef. *Coral Reefs*, **10**: 29–36.

Berger, W. H. and Wefer, G. 1990. Export productivity: seasonality and intermittency, and paleooceanographic implications. *Palaeogeography, Palaeoclimatology, Palaeoecology*, **9**: 245–54.

Bernhard, J. M. 1987. Foraminiferal biotopes in Explorers Cove, McMurdo Sound, Antarctica. *Journal of Foraminiferal Research*, **17**: 286–97.

Bernhard, J. M. 1988. Postmortem vital staining in benthic foraminifera: duration and importance in population and distributional studies. *Journal of Foraminiferal Research*, **18**: 143–6.

Bernhard, J. M. 1989. The distribution of benthic foraminifera with respect to oxygen concentration and organic carbon levels in shallow-water Antarctic sediments. *Limnology and Oceanography*, **34**: 1131–41.

Bernhard, J. M. 1992. Benthic foraminiferal distribution and biomass related to pore-water oxygen content: central California continental slope and rise. *Deep-Sea Research*, **39**: 585–602.

Bernhard, J. M. 1993. Experimental and field evidence of Antarctic foraminiferal tolerance to anoxia and hydrogen sulfide. *Marine Micropaleontology*, **20**: 203–13.

Bernhard, J. M. 1996. Microaerophilic and facultative anaerobic benthic foraminifera: a review of experimental and ultrastructural evidence. *Revue de paléontologie*, **15**: 261–75.

Bernhard, J. M. 2000. Distinguishing live from dead foraminifera: methods review and proper applications. *Micropaleontology*, **46** (Supplement 1): 38–46.

Bernhard, J. M. and Alve, E. 1996. Survival, ATP pool, and ultrastructural characterization of benthic foraminifera from Drammensfjord (Norway): a response to anoxia. *Marine Micropaleontology*, **28**: 5–17.

Bernhard, J. M., Blanks, J. K., Hintz, C. J. and Chandler, G. T. 2004. Use of the fluorescent marker calcein to label foraminiferal tests. *Journal of Foraminiferal Research*, **34**: 96–101.

Bernhard, J. M. and Bowser, S. S. 1992. Bacterial biofilms as a trophic resource for certain benthic foraminifera. *Marine Ecology Progress Series*, **83**: 263–72.

Bernhard, J. M. and Bowser, S. S. 1996. Novel epifluorescence microscopy method to determine life position of foraminifera in sediments. *Journal of Micropalaeontology*, **15**: 68.

Bernhard, J. M. and Bowser, S. S. 1999. Benthic foraminifera of dysoxic sediments: chloroplast sequestration and functional morphology. *Earth-Science Reviews*, **46**: 149–65.

Bernhard, J. M., Buck, K. R. and Barry, J. P. 2001. Monterey cold-seep biota: assemblages, abundance, and ultrastructure of living foraminifera. *Deep-Sea Research I*, **48**: 2233–49.

Bernhard, J. M., Buck, K. R., Farmer, M. A. and Bowser, S. S. 2000. The Santa Barbara Basin is a symbiosis oasis. *Nature*, **403**: 77–80.

Bernhard, J. M., Newkirk, S. G., and Bowser, S. S. 1995. Towards a non-terminal viability assay for foraminiferan protists. *Journal of Eukariote Microbiology*, **42**: 357–67.

Bernhard, J. M. and Reimers, C. E. 1991. Benthic foraminiferal population fluctuations related to anoxia: Santa Barbara Basin. *Biogeochemistry*, **15**: 127–49.

Bernhard, J. M. and Sen Gupta, B. K. 1999. Foraminifera in oxygen-depleted environments. In Sen Gupta, B. K., ed. *Modern Foraminifera*. Dordrecht: Kluwer, pp. 201–16.

Bernhard, J. M., Sen Gupta, B. K. and Borne, P. F. 1997. Benthic foraminiferal proxy to estimate dysaerobic bottom-water oxygen concentrations: Santa Barbara Basin, US Pacific continental margin. *Journal of Foraminiferal Research*, **27**: 301–10.

Bernhard, J. M., Visscher, P. T. and Bowser, S. S. 2003. Submillimeter life positions of bacteria, protsts, and metazoans in laminated sediments of the Santa Barbara Basin. *Limnology and Oceanography*, **48**: 813–28.

Berry, A. J. 1994. Foraminiferal prey in the annual life-cycle of the predatory opisthobranch gastropod *Retusa obtusa* (Montagu). *Estuarine, Coastal and Shelf Science*, **38**: 603–12.

Bertram, M. A. and Cowen, J. P. 1999. Temporal variations in the deep-water colonization rates of small benthic foraminifera: the results of an experiment on Cross Seamount. *Deep-Sea Research I*, **46**: 1021–49.

Bignot, G. and Lamboy, M. 1980. Les foraminiféres épibiontes á test calcaire hyaline des encroûtements polymétalliques de la marge continentale au nord-ouest de la peninsula ibérique. *Revue de micropaléontologie*, **23**: 3–15.

Bilyard, G. R. 1974. The feeding habits and ecology of *Dentalium entale stimpsoni* Henderson (Mollusca: Scaphopoda). *The Veliger*, **17**: 126–38.

Blais-Stevens, A. and Patterson, R. T. 1998. Environmental indicator potential of foraminifera from Saanich Inlet, Vancouver Island, British Columbia, Canada. *Journal of Foraminiferal Research*, **28**: 201–19.

Blanc-Vernet, L. 1969. Contribution à l'étude des foraminifères de Méditerranée. *Receuil des travaux de la station marine d'Endoume*, **64** (48): 1–281.

Blanc-Vernet, L. 1984. Les foraminifères de l'herbier de *Posidonica oceanica* en Méditerranée: analyse des assemblages, aspects régionaux, application aux microfaunes fossils. In Boudouresque, C. F., Jeudy de Grissac, A. and Olivier, J., eds. *International Workshop on Posidonia Oceanica Beds*. Marseille: GIS Posidonie Publications, pp. 3–14.

Bock, W. D. 1970. *Thalassia testudinum*, a habitat and means of dispersal for shallow water benthonic foraminifera. *Transactions of the Gulf Coast Association of Geological Societies*, **19**: 337–40.

Bock, W. D. and Moore, D. R. 1968. A commensal relationship between a foraminifer and a bivalve mollusk. *Gulf Research Report*, **2**: 273–9.

Boetius, A., Jørgensen, R., Amann, R. *et al.* 2002. Microbial systems in sedimentary environments of continental margins. In Wefer, G., Billettt, D., Hebbeln, D. *et al.*, eds. *Ocean Margin Systems*. Berlin: Springer-Verlag, pp. 479–95.

Boltovskoy, E. 1964. Seasonal occurrences of some living foraminifera in Puerto Deseado (Patagonia, Argentina). *Journal du conseil international pour l'exporation de la mer*, **29**: 136–45.

Boltovskoy, E. 1965. *Los foraminíferos recientes*. Buenos Aires: Editorial Universitaria de Buenos Aires.

Boltovskoy, E. 1984. Foraminifera of mangrove swamps. *Physis (Buenos Aires)*, **A42**: 1–9.

Boltovskoy, E. 1990. Variability of foraminifers, some evolutionary trends, and the validity of taxonomic categories. *Acta palaeontologica polonica*, **35**: 3–14.

Boltovskoy, E. 1991. On the destruction of foraminiferal tests (laboratory experiments). *Revue de micropaléontologie*, **34**: 19–25.

Boltovskoy, E., Giussani, G. and Watanabe, S. 1983. Variaciones estacionales y standing crop de los foraminiferos bentonicos de Ushuaia, Tierra del Fuego. *Physis (Buenos Aires)*, **A41**: 113–27.

Boltovskoy, E. and Hincapié de Martínez, S. 1983. foraminíferos del manglar de Tesca, Cartegena, Colombia. *Revista española de micropaleontología*, **15**: 205–20.

Boltovskoy, E. and Lena, H. 1969a. Seasonal occurrences, standing crop and production in benthic foraminifera of Puerto Deseado. *Contributions from the Cushman Foundation for Foraminiferal Research*, **20**: 87–95.

Boltovskoy, E. and Lena, H. 1969b. Les epibiontes de 'Macrocystis' flotante como indicadores hidrologicos. *Neotropicana*, **15**: 135–7.

Boltovskoy, E., Scott, D. B. and Medioli, F. S. 1991. Morphological variations of benthic foraminiferal tests in response to changes in ecological parameters: a review. *Journal of Paleontology*, **65**: 175–85.

Boltovskoy, E. and Totah, V. 1985. Diversity, similarity and dominance in benthic foraminiferal fauna along one transect of the Argentine shelf. *Revue de micropaléontologie*, **28**: 23–31.

Boltovskoy, E. and Wright, R. 1976. *Recent foraminifera*. The Hague: Junk.

Boltovskoy, E. and Zapata, A. 1980. Foraminíferos bentonicos como alimento de otros organismos. *Revista española de micropaleontología*, **12**: 191–8.

Bornhold, B. D. and Giresse, P. 1985. Glauconitic sediments on the continental shelf off Vancouver Island, British Columbia, Canada. *Journal of Sedimentary Petrology*, **55**: 653–4.

Bornmalm, L., Corliss, B. H. and Tedesco, K. 1997. Laboratory observations of rates and patterns of movement of continental margin benthic foraminifera. *Marine Micropaleontology*, **29**: 175–84.

Boudreau, R. E., Patterson, T. R., Dalby, A. P. and McKillop, W. B. 2001. Non-marine occurrence of the foraminifer *Cribroelphidium gunteri* in northern Lake Winnipegosis, Manitoba, Canada. *Journal of Foraminiferal Research*, **31**: 108–19.

Bowser, S. S., Alexander, S. P., Stockton, W. L. and Delaca, T. E. 1992. Extracellular matrix augments mechanical properties of pseudopodia in the carnivorous foraminiferan *Astrammina rara*: role in prey capture. *Journal of Protozoology*, **39**: 724–32.

Bowser, S. S., Bernhard, J. M., Habura, A. and Gooday, A. J. 2002. Structure, taxonomy and ecology of *Astrammina triangularis* (Earland), an allogromiid-like agglutinated foraminifer from Explorers Cove, Antarctica. *Journal of Foraminiferal Research*, **32**: 364–74.

Bowser, S. S. and Delaca, T. E. 1985. Rapid intracellular motility and dynamic membrane events in an Antarctic foraminifer. *Cell Biology International Reports*, **9**: 901–10.

Bowser, S. S., Gooday, A. J., Alexander, S. P. and Bernhard, J. M. 1995. Larger agglutinated foraminifera of McMurdo Sound, Antarctica: are *Astrammina rara* and *Notodendrodes antarcticos* allogromiids incognito? *Marine Micropaleontology*, **26**: 75–88.

Bowser, S. S., McGee-Russell, S. M. and Rieder, C. L. 1985. Digestion of prey in foraminifera is not anomalous: a correlation of light microscopic, cytochemical and HVEM techniques to study phagotrophy in two allogromiids. *Journal of Ultrastructure and Molecular Structure Research*, **94**: 149–60.

Bowser, S. S. and Travis, J. L. 2000. Methods for structural studies of reticulopodia, the vital foraminiferal "soft part". *Micropaleontology*, **46** (Supplement 1): 47–56.

Bradshaw, J. S. 1957. Laboratory studies on the rate of growth of the foraminifera, 'Streblus beccarii' (Linné). *Journal of Paleontology*, **31**: 1138–47.

Bradshaw, J. S. 1961. Laboratory experiments on the ecology of foraminifera. *Contributions from the Cushman Foundation for Foraminiferal Research*, **12**: 7–102.

Bradshaw, J. S. 1968. Environmental parameters and marsh foraminifera. *Limnology and Oceanography*, **13**: 26–38.

Brady, H. B. 1884. Report on the foraminifera dredged by H.M.S. Challenger during the years 1873–1876. *Report on the Scientific Results of the Voyage of H.M.S. Challenger During the Years 1873–1876, Zoology*, **9**: 1–800.

Brasier, M. D. 1975a. Morphology and habitat of living benthic foraminiferids from Caribbean carbonate environment. *Revista española de micropaleontología*, **7**: 567–9.

Brasier, M. D. 1975b. The ecology of sediment-dwelling and phytal foraminifera from the lagoons of Barbuda, West Indies. *Journal of Foraminiferal Research*, **5**: 42–62.

Brasier, M. D. 1975c. The ecology and distribution of recent foraminifera from the reefs and shoals around Barbuda, West Indies. *Journal of Foraminiferal Research*, **5**: 193–210.

Brasier, M. D. 1986. Why do plants and animals biomineralize? *Paleobiology*, **12**: 241–50.

Bresler, V. and Yanko, V. 1995. Chemical ecology: a new approach to the study of living benthic epiphytic foraminifera. *Journal of Foraminiferal Research*, **25**: 267–79.

Bresler, V. M. and Yanko-Hombach, V. V. 2000. Chemical ecology of foraminifera. Parameters of health, environmental pathology, and assessment of environmental quality. In. Martin, R. E., ed. *Environmental Micropaleontology. The Application of Microfossils to Environmental Geology*. New York: Kluwer, pp. 217–54.

Brodsky, A. 1928. Foraminifera (Polythalamia) in Grundwasser der Wüste Kara-Kum (Turkmenistan). *Acta universitatis asiæ mediæ series VIII-a zoologia*, **5**: 3–16.

Bromley, R. G. 1996. *Trace Fossils. Biology, Taphonomy and Applications*, 2nd edn. London: Chapman and Hall.

Brönnimann, P., Lutze, G. F. and Whittaker, J. E. 1989a. *Balticammina pseudomacrescens*, a new brackish water trochamminid from the western Baltic Sea, with remarks on the wall structure. *Meyniana*, **41**: 167–77.

Brönnimann, P., van Dover, C. L. and Whittaker, J. E. 1989b. *Abyssotherma pacifica*, n. gen., n. sp., a recent remaneicid (Foraminiferida, Remaneicacea) from the East Pacific Rise. *Micropaleontology*, **35**: 142–9.

Brooks, A. L. 1967. Standing crop, vertical distribution, and morphometrics of *Ammonia beccarii* (Linné). *Limnology and Oceanography*, **12**: 667–84.

Buchanan, J. B. and Hedley, R. H. 1960. A contribution to the biology of *Astrorhiza limicola* (Foraminifera). *Journal of the Marine Biological Association of the UK*, **39**: 549–60.

Buzas, M. A. 1965. The distribution and abundance of foraminifera in Long Island Sound. *Smithsonian Miscellaneous Collections*, **149**: 1–89.

Buzas, M. A. 1968a. Foraminifera from the Hadly Harbor complex, Massachusetts. *Smithsonian Miscellaneous Collections*, **152**: 1–26.

Buzas, M. A. 1968b. On the spatial distribution of foraminifera. *Contributions from the Cushman Foundation for Foraminiferal Research*, **19**: 1–11.

Buzas, M. A. 1969. Foraminiferal species densities and environmental variables in an estuary. *Limnology and Oceanography*, **14**: 411–22.

Buzas, M. A. 1970. Spatial homogeneity: statistical analysis of unispecies and multispecies populations of foraminifera. *Ecology*, **51**: 874–9.

Buzas, M. A. 1974. Vertical distribution of *Ammobaculites* in the Rhode River, Maryland. *Journal of Foraminiferal Research*, **4**: 144–7.

Buzas, M. A. 1977. Vertical distribution of foraminifera in the Indian River, Florida. *Journal of Foraminiferal Research*, **7**: 234–7.

Buzas, M. A. 1978. Foraminifera as prey for benthic deposit feeders: results of predator exclusion experiments. *Journal of Marine Science*, **36**: 617–25.

Buzas, M. A. 1982. Regulation of foraminiferal densities by predation in the Indian River, Florida. *Journal of Foraminiferal Research*, **12**: 66–71.

Buzas, M. A. 1989. The effect of quartz versus calcareous sand on the densities of living foraminifera. *Micropaleontology*, **35**: 135–41.

Buzas, M. A. 1990. Another look at confidence limits for species proportions. *Journal of Paleontology*, **64**: 842–3.

Buzas, M. A. 1993. Colonization rate of foraminifera in the Indian River, Florida. *Journal of Foraminiferal Research*, **23**: 156–61.

Buzas, M. A. and Carle, K. 1979. Predators of foraminifera in the Indian River, Florida. *Journal of Foraminiferal Research*, **9**: 336–40.

Buzas, M. A., Collins, L. S, Richardson, S. L. and Severin, K. P. 1989. Experiments on predation, substrate preference, and colonization of benthic foraminifera at the shelfbreak off the Ft. Pierce Inlet, Florida. *Journal of Foraminiferal Research*, **19**: 146–52.

Buzas, M. A. and Culver, S. J. 1980. Foraminifera: distribution of provinces in the western North Atlantic. *Science*, **209**: 687–9.

Buzas, M. A. and Culver, S. J. 1986. Geographic origin of benthic foraminiferal species. *Science*, **232**: 775–6.

Buzas, M. A. and Culver, S. J. 1989. Biogeographic and evolutionary patterns of continental margin benthic foraminifera. *Paleobiology*, **15**: 11–9.

Buzas, M. A. and Culver, S. J. 1991. Species diversity and dispersal in benthic foraminifera. *Bioscience*, **41**: 483–9.

Buzas, M. A. and Culver, S. J. 1994. Species pool and dynamics of marine paleocommunities. *Science*, **264**: 1439–41.

Buzas, M. A. and Culver, S. J. 1999. Understanding regional species diversity through the log series distribution of occurrences. *Biodiversity Research*, **8**: 187–95.

Buzas, M. A. and Culver, S. J. 2001. On the relationship between species distribution–abundance–occurrence and species duration. *Historical Biology*, **15**: 151–9.

Buzas, M. A., Culver, S. J. and Jorissen, F. J. 1993. A statistical evaluation of the microhabitats of living (stained) infaunal benthic foraminifera. *Marine Micropaleontology*, **20**: 311–20.

Buzas, M. A. and Hayek, L. E. C. 1996. Biodiversity resolution: an integrated approach. *Biodiversity Letters*, **3**: 40–3.

Buzas, M. A. and Hayek, L. E. C. 1998. SHE analysis for biofacies identification. *Journal of Foraminiferal Research*, **28**: 233–9.

Buzas, M. A. and Hayek, L. E. C. 2000. A case for long-term monitoring of the Indian River lagoon, Florida: foraminiferal densities, 1977–1996. *Bulletin of Marine Science*, **67**: 805–14.

Buzas, M. A., Hayek, L. C., Reed, S. A. and Jett, J. A. 2002. Foraminiferal densities over five years in the Indian River lagoon, Florida: a model of pulsating patches. *Journal of Foraminiferal Research*, **32**: 68–93.

Buzas, M. A., Koch, C. F., Culver, S. J. and Sohl, N. F. 1982. On the distribution of species occurrence. *Paleobiology*, **8**: 143–50.

Buzas, M. A. and Severin, K. P. 1982. Distribution and systematics of foraminifera in the Indian River, Florida. *Smithsonian Contributions to the Marine Sciences*, **16**: 1–73.

Buzas, M. A. and Severin, K. P. 1993. Foraminiferal densities and pore water chemistry in the Indian River, Florida. *Smithsonian Contributions to the Marine Sciences*, **36**: 1–38.

Buzas, M. A., Smith, R. K. and Beem, K. A. 1977. Ecology and systematics of foraminifera in two *Thalassia* habitats, Jamaica, West Indies. *Smithsonian Contributions to Paleobiology*, **31**: 1–139.

Castignetti, P. 1996. A time-series study of foraminiferal assemblages of the Plym estuary, south-west England. *Journal of the Marine Biological Association of the UK*, **76**: 569–78.

Castignetti, P. and Manley, C. J. 1998. Benthic foraminiferal depth distribution within the sediments of a modern ria. *Terra Nova*, **10**: 37–41.

Cato, I., Olsson, I. and Rosenberg, R. 1980. Recovery and decontamination of estuaries. In Olauson, E. and Cato, I., eds. *Chemistry and Biogeochemistry of Estuaries*. Chichester: Wiley, pp. 403–40.

Cearreta, A. 1988. Population dynamics of benthic foraminifera in the Santoña estuary, Spain. *Revue de paléobiologie*, volume spécial, **2**: 721–4.

Cearreta, A. 1989. Foraminiferal assemblages in the ria of San Vicente de la Barquera (Cantabria, Spain). *Revista española de micropaleontología*, **21**: 67–80.

Cearreta, A., Irabien, M. J., Leorri, E. *et al.* 2000a. Recent anthropogenic impacts on the Bilbao estuary, northern Spain: geochemical and macrofaunal evidence. *Estuarine, Coastal and Shelf Science*, **50**: 571–92.

Cearreta, A., Irabien, M. J., Ulibarri, I. *et al.* 2002b. Recent salt marsh development and natural regeneration of reclaimed areas in the Plentzia estuary, N. Spain. *Estuarine, Coastal and Shelf Science*, **54**: 863–86.

Cearreta, A., Irabien, M. J., Leorri, E. *et al.* 2002c. Environmental transformation of the Bilbao estuary, N Spain: microfaunal and geochemical proxies in the recent sedimentary record. *Marine Pollution Bulletin*, **44**: 487–503.

Cearreta, A. and Murray, J. W. 1996. Holocene paleoenvironmental and sea-level changes in the Santoña estuary, Spain. *Journal of Foraminiferal Research*, **26**: 289–99.

Cebulski, D. E. 1961. *Distribution of Foraminifera in the Barrier Reef and Lagoon of British Honduras*. College Station, TX: Texas A & M Research Foundation.

Cebulski, D. E. 1969. Foraminiferal populations and faunas in barrier-reef tract and lagoon, British Honduras. *American Association of Petroleum Geologists, Memoir*, **11**: 311–28.

Cedhagen, T. 1988. Position in the sediment and feeding of *Astrorhiza limicola* Sandahl, 1857 (Foraminiferida). *Sarsia*, **73**: 43–7.

Cedhagen, T. 1989. A method for disaggregating clay concretions and eliminating formalin smell in the processing of sediment samples. *Sarsia*, **74**: 221–2.

Cedhagen, T. 1991. Retention of chloroplasts and bathymetric distribution in the sublittoral foraminiferan *Nonionella labradorica*. *Ophelia*, **33**: 17–30.

Cedhagen, T. 1994. Taxonomy and biology of *Hyrrokkin sarcophaga* gen. et sp. n., a parasitic foraminiferan (Rosalinidae). *Sarsia*, **79**: 65–82.

Cedhagen, T. 1996. Foraminifer as food for cephalaspideans (Gastropoda: Opisthobranchia), with notes on secondary tests around calcareous foraminiferans. *Phuket Marine Biological Center Special Publication*, **16**: 279–90.

Cedhagen T. and Frimanson, H. 2002. Temperature dependence of pseudopodial organelle transport in seven species of foraminifera and its functional consequences. *Journal of Foraminiferal Research*, **32**: 434–9.

Cedhagen, T., Goldstein, S. T. and Gooday, A. J. 2002. Biology and diversity of allogromiid foraminifera. *Journal of Foraminiferal Research*, **32**: 331–452.

Cedhagen, T. and Tendal, O. S. 1989. Evidence of test detachment in *Astrorhiza limicola*, and two consequential synomyms: *Amoeba gigantea* and *Megamoebomyxa argillobia* (Foraminiferida). *Sarsia*, **74**: 195–200.

Chai, J. and Lee, J.J. 2000. Recognition, establishment and maintenance of diatom endosymbiosis in foraminifera. *Micropaleontology*, **46** (Supplement 1): 182–95.

Chandler, T. G. 1989. Foraminifera may structure meiobenthic communities. *Oecologia*, **81**: 354–60.

Chang, S. K. and Lee, K. S. 1984. A study on the recent foraminifera of the intertidal flats of Asan Bay, Korea. *Journal of the Geological Society of Korea*, **20**: 171–88.

Chang, Y. M. and Kaesler, R. L. 1974. Morphological variation of the foraminifer *Ammonia beccarii* (Linné) from the Atlantic coast of the United States. *Kansas University Paleontological Contributions Paper*, **69**: 1–23.

Chester, C. M. 1993. Comparative feeding biology of *Acteocina canaliculata* (Say, 1826) and *Haminoea solitaria* (Say, 1822) (Opisthobranchia: Cephalaspidea). *American Malacological Bulletin*, **10**: 93–101.

Christiansen, O. 1964. *Spiculosiphon radiata*, a new foraminifera from northern Norway. *Astarte*, **25**: 1–8.

Christiansen, O. 1971. Notes on the Biology of Foraminifera. *Vie et milieu*, Supplement **22**: 465–78.

Clarke, K. R. and Warwick, R. M. 1994. *Change in Marine Communities: an Approach to Statistical Analysis and Interpretation*. Plymouth: Natural Environment Research Council.

Clements, W. H. and Newman, M. C. 2002. *Community Ecotoxicology*. Chichester: Wiley.

Closs, D. and Madeira, M. L. 1966. Foraminifera from the Parangua Bay, State of Paraná, Brasil. *Boletim da Universidade Federal do Paraná, Zoologia*, **II**: 139–61.

Closs, D. and de Medeiros, V. M. F. 1965. New observations on the ecological subdivision of the Patos lagoon in southern Brasil. *Boletim da Instituto de Ciencas Naturais*, **24**: 1–33.

Collen, J. 1996. Recolonization of reef flat by larger foraminifera, Funafuti, Tuvalu. *Journal of Micropalaeontology*, **15**: 130.

Collen, J. D. and Newell, P. 1999. *Fissurina* as an ectoparasite. *Journal of Micropalaeontology*, **18**: 10.

Collins, E. S., Scott, D. B. and Gayes, P. T. 1999. Hurricane records on the South Carolina coast: can they be detected in the sediment record? *Quaternary International*, **56**: 15–26.

Collins, E. S., Scott, D. B., Gayes, P. T and Medioli, F. 1995. Foraminifera in Winyah Bay and North Inlet marshes, South Carolina: relationship to local pollution sources. *Journal of Foraminiferal Research*, **25**: 212–23.

Collins, L. S. 1991. Regional versus physiographic effects on morphologic variability within *Bulimina aculeata* and *B. marginata*. *Marine Micropaleontology*, **17**: 155–70.

Collison, P. 1980. Vertical distribution of foraminifera off the coast of Northumberland, England. *Journal of Foraminiferal Research*, **10**: 75–8.

Colom, G. 1974. Foraminíferos Ibéricos. *Investigación pesquera*, **38**: 1–245.

Cooper, W. C. 1961. Intertidal foraminifera of the California and Oregon coast. *Contributions from the Cushman Foundation for Foraminiferal Research*, **12**: 47–63.

Corliss, B. H. 1985. Microhabitats of benthic foraminifera within deep-sea sediments. *Nature*, **314**: 435–8.

Corliss, B. H. 1991. Morphology and habitat preferences of benthic foraminifera from the northwest Atlantic Ocean. *Marine Micropaleontology*, **17**: 195–236.

Corliss, B. H. and Chen, C. 1988. Morphotype patterns of Norwegian Sea deep-sea benthic foraminifera and ecological implications. *Geology*, **16**: 716–19.

Corliss, B. H. and Emerson, S. 1990. Distribution of rose Bengal stained deep-sea benthic foraminifera from the Nova Scotian continental margin and Gulf of Maine. *Deep-Sea Research*, **37**: 381–400.

Corliss, B. H. and Fois, E. 1990. Morphotype analysis of deep-sea benthic foraminifera from the northwest Gulf of Mexico. *Palaios*, **5**: 589–605.

Corliss, B. H. and Honjo, S. 1981. Dissolution of deep-sea benthonic foraminifera. *Micropaleontology*, **27**: 356–78.

Cornelius, N. and Gooday, A. J. 2004. 'Live' (stained) deep-sea benthic foraminiferans in the western Weddell Sea: trends in abundance, diversity and taxonomic composition along a depth transect. *Deep-Sea Research II*, **51**: 1571–602.

Corner, G. D., Steinsund, P. I. and Aspeli, R. 1996. Distribution of recent benthic foraminifera in a subarctic fjord-delta: Tana, Norway. *Marine Geology*, **134**: 113–25.

Coulbourn, W. T. and Lutze, G. F. 1988. Benthic foraminifera and their relation to the environment offshore of northwest Africa: a multivariate statistical analysis. *Revue de paléobiologie*, volume spécial, **2**: 755–64.

Craib, J. S. 1965. A sampler for taking short undisturbed marine cores. *Journal du conseil permanent international pour l' exploration de la mer*, **30**: 34–9.

Cronin, T. M., Dwyer, G. S., Kamiya, T., Schwede, S. and Willard, D. A. 2003. Medieval warm period, little ice age and 20th century temperature variability from Chesapeake Bay. *Global and Planetary Change*, **36**: 17–29.

Cronin, T., Willard, D., Karlsen, A. *et al.* 2000. Climatic variability in the eastern United States over the past millennium from Chesapeake Bay sediments. *Geology*, **28**: 3–6.

Culver, S. J. and Banner, F. T. 1978. Foraminiferal assemblages as Flandrian palaeoenvironmental indicators. *Palaeogeography, Palaeoclimatology, Palaeoecology*, **24**: 53–72.

Culver, S. J. and Buzas, M. A. 1981a. Recent benthic foraminiferal provinces on the Atlantic continental margin of North America. *Journal of Foraminiferal Research*, **11**: 217–40.

Culver, S. J. and Buzas, M. A. 1981b. Foraminifera: distribution of provinces in the Gulf of Mexico. *Nature*, **290**: 328–9.

Culver, S. J. and Buzas, M. A. 1982. Recent benthic foraminiferal provinces between Newfoundland and Yucatan. *Bulletin of the Geological Society of America*, **93**: 269–77.

Culver, S. J. and Buzas, M. A. 1983. Recent benthic foraminiferal provinces in the Gulf of Mexico. *Journal of Foraminiferal Research*, **13**: 21–31.

Culver, S. J. and Buzas, M. A. 1995. The effects of anthropogenic habitat disturbance, habitat destruction, and global warming on shallow marine benthic foraminifera. *Journal of Foraminiferal Research*, **25**: 204–11.

Culver, S. J. and Buzas, M. A. 1999. Biogeography of neritic foraminifera. In Sen Gupta, B. K., ed. *Modern Foraminifera*. Dordrecht: Kluwer, pp. 93–102.

Culver, S. J. and Buzas, M. A. 2000. Global latitudinal species diversity gradient in deep-sea benthic foraminifera. *Deep-Sea Research I*, **47**: 259–75.

Culver, S. J., Woo, H. J., Oertel, G. F. and Buzas, M. A. 1996. Foraminifera of coastal depositional environments, Virginia, USA: distribution and taphonomy. *Palaios*, **11**: 459–86.

Cushman, J. A. 1922. Shallow water foraminifera of the Tortugas region. *Papers of the Tortugas Laboratory*, **17**: 1–85.

Cushman, J. A. 1948. *Foraminifera. Their Classification and Economic Use*, 4th edn. Cambridge: Harvard.

Dale, A. L. and Dale, B. 2002. Application of ecologically based statistical treatments to micropalaeontology. In: Haslett, S. K., ed. *Quaternary Environmental Micropalaeontology*. London: Arnold, pp. 259–86.

Daniels, C. H. 1970. Quantitative ökologische Beobachtungenen an Foraminiferen im Limski Kanal bei Rovinj/Jugoslawien (nördliche Adria). *GöttingerArbeiten zur Geologie und Paläontologie*, **8**: 1–109.

Daniels, C. H. 1971. Jahreszeitliche ökologische Beobachtungen an Foraminiferen im Limski kanal bei Rovinj/Jugoslawien (nördliche Adria). *Geologische Rundschau*, **60**: 192–204.

Debenay, J. P. 1990. Recent foraminiferal assemblages and their distribution relative to environmental stress in the paralic environments of West Africa (Cape Timiris to Ebrie Lagoon). *Journal of Foraminiferal Research*, **20**: 267–82.

Debenay, J. P. 2000. Foraminifers of tropical paralic environments. *Micropaleontology*, **46** (Supplement 1): 153–60.

Debenay, J. P., Duleba, W., Bonetti, C. *et al.* 2001a. *Pararotalia cananeiaensis* n. sp.: indicator of marine influence and water circulation in Brazilian coastal and paralic environments. *Journal of Foraminiferal Research*, **31**: 152–63.

Debenay, J. P., Geslin, E., Eichler, B. B. *et al.* 2001b. Foraminiferal assemblages in a hypersaline lagoon, Araruama (R. J.) Brazil. *Journal of Foraminiferal Research*, **31**: 133–51.

Debenay, J. P. and Guillou, J. J. 2002. Ecological transitions indicated by foraminiferal assemblages in paralic environments. *Estuaries*, **25**: 1107–20.

Debenay, J. P., Guiral, D. and Parra, M. 2002. Ecological factors acting on the microfauna in mangrove swamps. The case of foraminiferal assemblages in French Guiana. *Estuarine, Coastal and Shelf Science*, **55**: 509–33.

Debenay, J. P., Guiral, D. and Parra, M. 2004. Behaviour and taphonomic loss in foraminiferal assemblages of mangrove swamps of French Guiana. *Marine Geology*, **208**: 295–314.

Debenay, J. P. and Pagès, J. 1987. Foraminifères et thécamoebiens de l'estuaire hyperhalin du fleuve Casamance (Sénégal). *Revue d'hydrobiologie tropicale*, **20**: 233–56.

DeLaca, T. E. 1982. Use of dissolved amino acids by the foraminifer *Notodendrodes antarctikos*. *American Zoologist*, **22**: 683–90.

DeLaca, T. E., Bernhard, J. M., Reilly, A. A., and Bowser, S. S. 2002. *Notodendrodes hyalinosphaira* (sp. nov.): structure and autecology of an allogromiid-like agglutinated foraminifer. *Journal of Foraminiferal Research*, **32**: 177–87.

DeLaca, T. E., Karl, D. M. and Lipps, J. H. 1981. Direct use of dissolved organic carbon by agglutinated benthic foraminifera. *Nature*, **289**: 287–9.

DeLaca, T. E. and Lipps, J. H. 1972. The mechanism and adaptive significance of attachment and substrate pitting in the foraminiferan *Rosalina globularis* d'Orbigny. *Journal of Foraminiferal Research*, **2**: 68–72.

Denne, R. A. and Sen Gupta, B. K. 1989. Effects of taphonomy and habitat on the record of benthic foraminifera in modern sediments. *Palaios*, **4**: 414–23.

de Rijk, S. D. 1995. Salinity control on the distribution of salt marsh foraminifera (Great Marshes, Massachusetts). *Journal of Foraminiferal Research*, **25**: 156–66.

de Rijk, S., Jorissen, F. J., Rohling, E. J. and Troelstra, S. R. 2000. Organic flux control on bathymetric zonation of Mediterranean foraminifera. *Marine Micropaleontology*, **40**: 151–60.

de Rijk, S. D., Troelstra, S. R. and Rohling, E. J. 1999. Benthic foraminiferal distribution in the Mediterranean Sea. *Journal of Foraminiferal Research*, **29**: 93–103.

de Rijk, S. D. and Troelstra, S. R. 1999. The application of a foraminiferal actuo-facies model to salt marsh cores. *Palaeogeography, Palaeoclimatology, Palaeoecology*, **149**: 59–66.

de Stigter, H. C., Jorissen, F. J. and Zwaan, G. J. van der 1998. Bathymetric distribution and microhabitat partitioning of live (rose Bengal stained) benthic foraminifera along a shelf to bathyal transect in the southern Adriatic Sea. *Journal of Foraminiferal Research*, **28**: 40–65.

de Stigter, H. C., Zwaan, G. J. van der and Langone, L. 1999. Differential rates of benthic foraminiferal test production in surface and subsurface sediments in the southern Adriatic Sea. *Palaeogeography, Palaeoclimatology, Palaeoecology*, **149**: 67–88.

Dettmering, C., Röttger, R, Hohenegger, J. and Schmaljohann, R. 1998. The trimorphic life cycle in foraminifera: observations from culture allow new evaluations. *European Journal of Protistology*, **34**: 3638

Dieckmann, G., Hemleben, C. and Spindler, M. 1987. Biogenic and mineral inclusions in a green iceberg from the Weddell Sea, Antarctica. *Polar Biology*, **7**: 31–3.

Diz, P., Frances, G., Costas, S., Souto, C. and Allejo, I. 2004. Distribution of benthic foraminifera in coarse sediments, Ría de Vigo, NW Iberian margin. *Journal of Foraminiferal Research*, **34**: 258–75.

Dobson, M. and Haynes, J. R. 1973. Association of foraminifera with hydroids on the deep shelf. *Micropaleontology*, **19**: 78–90.

Donnici, S. and Serandrei Barbero, R. 2002. The benthic foraminiferal communities of the northern Adriatic continental shelf. *Marine Micropaleontology*, **44**: 93–123.

Donnici, S., Serandrei Barbero, R. and Taroni, G. 1997. Living benthic foraminifera on the Lagoon of Venice (Italy): population dynamics and its significance. *Micropaleontology*, **43**: 440–54.

Douglas, R. G. 1981. Paleoecology of continental margin basins: a modern case history from the borderland of Southern California. Short Cource Notes, Society of Economic Paleontologists and Mineralogists, California State University, Fullerton, CA: Pacific Section Society for Sedimentary Geology, pp. 121–56.

Douglas, R. G. and Heitman, H. L. 1979. Slope and basin benthic foraminifera of the California borderland. *Society of Economic Paleontologists and Mineralogists Special Publication*, **27**: 231–46.

Douglas, R. G., Liestman, J., Walch, C., Blake, G. and Cotton, M. L. 1980. The transition from live to sediment assemblages in benthic foraminifera from the southern California borderland. *Quaternary depositional environments of the Pacific*. Society of Economic Paleontologists and Mineralogists, Pacific Section Symposium, **4**: 257–80.

Douglas, R. G. and Woodruff, F. 1981. Deep sea benthic foraminifera. In Emiliani, C., ed. *The Sea*, vol. 7, pp. 1233–327. New York: Wiley International.

Dover, C. L. van, Berg, C. J. and Turner, R. D. 1988. Recruitment of marine invertebrates to hard substrates at deep-sea hydrothermal vents on the East Pacific Rise and Galapagos spreading center. *Deep-Sea Research*, **35**: 1833–59.

Dudley, W. C. 1976. Cementation and iron concentration in foraminifera on manganese nodules. *Journal of Foraminiferal Research*, **6**: 202–7.

Dugolinsky, B. K., Margolis, S. V. and Dudley, W. C. 1977. Biogenic influence on growth of manganese nodules. *Journal of Sedimentary Petrology*, **47**: 428–45.

Duguay, L. E. 1983. Comparative laboratory and field studies on calcification and carbon fixation in foraminiferal-algal associations. *Journal of Foraminiferal Research*, **13**: 252–61.

Duleba, W. and Debenay, J. P. 2003. Hydrodynamic circulation in the estuaries of Estação ecológica Juréia-Itiatins, Brazil, inferred from the foraminifera and thecamoebiam assemblages. *Journal of Foraminiferal Research*, **33**: 62–93.

Dulk, M. den, Reichart, G. J., Memon, G. M. *et al.* 1998. Benthic foraminiferal response to variations in surface water productivity and oxygenation in the northern Arabian Sea. *Marine Micropaleontology*, **35**: 43–66.

Dulk, M. den, Reichart, G. J., Heyst, S. van and Zwaan, G. J. van der 2000. Benthic foraminifera as proxies of organic matter flux and bottom water oxygenation? A case history from the northern Arabian Sea. *Palaeogeography, Palaeoclimatology, Palaeoecology*, **161**: 337–59.

Earland, A. 1934. The Falklands sector of the Antarctic (excluding South Georgia). *Discovery Reports*, **10**: 1–208.

Ebrahim, M. T. 2000. Impact of anthropogenic environmental change on larger foraminifera. In Martin, R. E., ed. *Environmental Micropaleontology. The Application of Microfossils to Environmental Geology*. New York: Kluwer, pp. 105–19.

Echols, R. J. 1971. Distribution of foraminifera in sediments of the Scotia Sea area, Antarctic waters. *Antarctic Research Series*, **15**: 93–168.

Edelman-Furstenberg, Y., Scherbacher, M. Hemleben, C. and Almogi-Labin, A. 2001. Deep-sea benthic foraminifera from the central Red Sea. *Journal of Foraminiferal Research*, **31**: 48–59.

Edwards, M., Beaugrand, G., Reid, P. C., Rowden, A. A. and Jones, M. B. 2002. Ocean climate anomalies and the ecology of the North Sea. *Marine Ecology Progress Series*, **239**: 1–10.

Edwards, R. I. 2001. Mid- to late Holocene relative sea-level change in Poole Harbour, southern England. *Journal of Quaternary Science*, **16**: 221–35.

Edwards, R. J. and Horton, B. P. 2004. Reconstructing relative sea-level using UK salt marsh foraminifera. *Marine Geology*, **169**: 41–56.

Edwards, R. J., Plassche, O. van der, Gehrels, W. R. and Wright, A. J. 2004. Assessing sea-level from Connecticut, USA, using a foraminiferal transfer function for tide level. *Marine Micropaleontology*, **51**: 239–55.

Eichler, B. B., Debenay, J. P., Bonetti, C. and Duleba, W. 1995. Répartition des foraminifères benthiques dans le zone sud-ouest du système estuarien-lagunaire d'Iguape-Cananéia (Brésil). *Boletim do Instituto Oceaográfico de USP, São Paulo*, **43**: 1–17.

Eldredge, N. 1990. Hierarchy and macroevolution. In: Briggs, D. E. G. and Crowther, P. R., eds. *Palaeobiology: a Synthesis*. Oxford: Blackwell, pp. 124–9.

Ellison, R. L. 1972. *Ammobaculites*, foraminiferal proprietor of Chesapeake Bay estuaries. *Geological Society of America, Memoir* **133**: 247–62.

Ellison, R. L. 1984. Foraminifera and meiofauna on an intertidal mudflat, Cornwall, England: populations; respiration and secondary production; and energy budget. *Hydrobiologia*, **109**: 131–48.

Ellison, R. L. and Murray, J. W. 1987. Geographical variation in the distribution of certain agglutinated foraminifera along the North Atlantic margins. *Journal of Foraminiferal Research*, **17**: 123–31.

Ellison, R. L. and Peck, G. E. 1983. Foraminiferal recolonization on the continental shelf. *Journal of Foraminiferal Research*, **13**: 231–41.

Elmgren, R. 1973. Methods for sampling sublittoral soft bottom meiofauna. *Oikos*, **15** (Supplement): 112–20.

Ernst, S. R., Duijnstee, I. A. P., Jannink, N. T. and Zwaan, G. J. van der 2000. An experimental mesocosm study of microhabitat preferences and mobility in benthic foraminifera: preliminary results. In *Proceedings of the Fifth International Workshop on Agglutinated Foraminifera*. Grzybowski Foundation Special Publication, 7: 101–4.

Erskian, M. G. and Lipps, J. H. 1987. Population dynamics of the foraminiferan *Glabratella ornatissima* (Cushman) in northern California. *Journal of Foraminiferal Research*, **17**: 240–56.

Evans, J. R., Austin, W. E. N., Brew, D. S., Wilkinson, I. P. and Kennedy, H. A. 2002. Holocene shelf sea evolution offshore northeast England. *Marine Geology*, **191**: 147–64.

Faber, W. W. 1991. Distribution and substrate preference of *Peneroplis planatus* and *P. arietinus* from the *Halophila* meadow near Wadi Taba, Eilat, Israel. *Journal of Foraminiferal Research*, **21**: 218–21.

Faber, W. W. and Lee, J. J. 1991a. Feeding and growth of the foraminifer *Peneroplis planatus* (Fichtel and Moll) Montfort. *Symbiosis*, **10**: 63–82.

Faber, W. W. and Lee, J. J. 1991b. Histochemical evidence for digestion in *Heterostegina depressa* and *Operculina ammonoides*. *Endocytobiosis and Cell Research*, **8**: 53–9.

Fatella, F. and Taborda, R. 2002. Confidence limits of species proportions in microfossil assemblages. *Marine Micropaleontology*, **45**: 169–74.

Fenchel, T. and Finlay, B. J. 1995. *Ecology and Evolution in Anoxic Worlds*. Oxford: Oxford University Press.

Fenchel, T. and Reidl, R. J. 1970. The sulfide system: new biotic community underneath the oxidized layer in marine sand bottoms. *Marine Biology*, **7**: 155–68.

Feyling-Hanssen, R. W. 1972. The foraminifer *Elphidium excavatum* (Terquem) and its variant forms. *Micropaleontology*, **18**: 337–54.

Filipsson, H. L. and Nordberg, K. 2004a. Climate variations, an overlooked factor influencing the recent marine environment. An example from Gullmar Fjord, Sweden, illustrated by benthic foraminifera and hydrographic data. *Estuaries*, **27**: 867–81.

Filipsson, H. L. and Nordberg, K. 2004b. A 200-year environmental record of a low-oxygen fjord, Sweden, elucidated by benthic foraminifera, sediment characteristics and hydrographic data. *Journal of Foraminiferal Research*, **34**: 277–93.

Finger, K. L. and Lipps, J. H. 1981. Foraminiferal decimation and repopulation in an active volcanic caldera, Deception Island, Antarctica. *Micropaleontology*, **27**: 111–39.

Fisher, R. A., Corbet, A. S., and Williams, C. B. 1943. The relationship between the number of species and the number of individuals in a random sample of an animal population. *Journal of Animal Ecology*, **12**: 42–58.

Flach, E. C. 2003. Factors controlling soft bottom macrofauna along and across European continental margins. In Wefer, G., Billett, D., Hebbeln, D. *et al.*, eds. *Ocean Margin Systems*. Berlin: Springer-Verlag, pp. 351–63.

Fontanier, C., Jorissen, F. J., Licari, L. *et al.* 2002. Live benthic foraminiferal faunas from the Bay of Biscay: faunal density, composition, and microhabitats. *Deep-Sea Research I*, **49**: 751–85.

Fontanier, C., Jorissen, F. J., Chaillou, G. *et al.* 2003. Seasonal and interannual variability of benthic foraminiferal faunas as 550 m depth in the Bay of Biscay. *Deep-Sea Research I*, **50**: 457–94.

Frankel, L. 1970. A technique for investigating microorganism associations. *Journal of Paleontology*, **44**: 575–7.

Freiwald, A. 1995. Bacteria-induced carbonate degradation: a taphonomic case study of *Cibicides lobatulus* from a high-boreal setting. *Palaios*, **10**: 337–46.

Freiwald, A. 2003. Reef-forming cold-water corals. In Wefer, G., Billett, D., Hebbeln, D. *et al.*, eds. *Ocean Margin Systems*. Berlin: Springer-Verlag. pp. 365–85.

Freiwald, A. and Schönfeld, J. 1996. Substrate pitting and boring pattern of *Hyrrokkin sarcophaga* Cedhagen, 1994 (Foraminifera) in a modern deep-water coral reef mound. *Marine Micropaleontology*, **28**: 199–207.

Fuhrman, J. A. and Noble, R. T. 1995. Viruses and protists cause bacterial mortality in coastal seawater. *Limnology and Oceanography*, **40**: 1236–42.

Fujita, K., Nishi, H and Saito, T. 2000. Population dynamics of *Marginopora kudakajimensis* Gudmundsson (Foraminifera: Soritidae) in the Ryukyu Islands, the subtropical northwest Pacific. *Marine Micropaleontology*, **38**: 267–84.

Gage, J. D. and Tyler, P. A. 1991. *Deep-Sea Biology*. Cambridge: Cambridge University Press.

Gao, S. and Collins, M. 1995. Net sand transport directions in a tidal inlet, using foraminiferal tests as natural tracers. *Estuarine, Coastal and Shelf Science*, **40**: 681–97.

Gamito, S. L., Berge, J. A. and Gray, J. S. 1988. The spatial distribution patterns of the foraminiferan *Pelosina* cf. *arborescens* Pearcey in a mesocosm. *Sarsia*, **73**: 33–8.

Gehrels, W. R. 1994. Determining relative sea-level change from salt-marsh foraminifera and plant zones on the coast of Maine, USA. *Journal of Coastal Research*, **10**: 990–1009.

Gehrels, W. R. and Plassche, O. van der 1999. The use of *Jadammina macrescens* (Brady) and *Balticammina pseudomacrescens* Brönnimann, Lutze and Whittaker (Protozoa: Foraminiferida) as sea-level indicators. *Palaeogeography, Palaeoclimatology, Palaeoecology*, **149**: 89–101.

George, M. and Murray, J. W. 1977. Glauconite in Celtic Sea sediments. *Proceedings of the Ussher Society*, **4**: 94–101.

Gerlach, S. A. 1978. Food-chain relationships in subtidal silty sand marine sediments and the role of meiofauna in stimulating bacterial productivity. *Oecologia*, **33**: 55–69.

Geslin, E., Debenay, J. P., Duleba, W. and Bonetti, C. 2002. Morphological abnormalities of foraminiferal tests in Brazilian environments: comparison between polluted and non-polluted areas. *Marine Micropaleontology*, **45**: 151–68.

Geslin, E., Debenay, J. P. and Lesourd, M. 1998. Abnormal wall textures and test deformation in *Ammonia* (hyaline foraminifer). *Journal of Foraminiferal Research*, **28**: 148–56.

Geslin, E., Heinz, P. and Hemleben, C. 2004. Behaviour of *Bathysiphon* sp. and *Siphonammina bertholdii* n. gen. n. sp. under controlled oxygen conditions in the laboratory: implications for bioturbation. In *Proceedings of the Sixth International Workshop on Agglutinated Foraminifera*. Grzybowski Foundation Special Publication, 8: 105–18.

Geslin, E., Stouff, V., Debenay, J. P. and Lesourd, M. 2000. Environmental variation and foraminiferal test abnormalities. In Martin, R. E., ed. *Environmental Micropaleontology. The Application of Microfossils to Environmental Geology*. New York: Kluwer, pp. 191–215.

Glenn, C., McManus, J. W., Aliño, P. M., Talaue, L. L. and Banzon, V. F. 1981. Distribution of live foraminifera on a portion of Apo Reef, Mindoro, Philippines. *Proceedings of the Fourth International Coral Reef Symposium, Manila*, **2**: 775–80.

Goldstein, S. T. 1997. Gametogenesis and the antiquity of reproductive pattern in the Foraminiferida. *Journal of Foraminiferal Research*, **27**: 319–28.

Goldstein, S. T. 1999. Foraminifera: a biological overview. In Sen Gupta, B. K., ed. *Modern Foraminifera*. Dordrecht: Kluwer, pp. 7–55.

Goldstein, S. T. and Bernhard, J. M. 1997. Biology of Foraminiferida: applications in paleoceanography, paleobiology, and environmental sciences. *Journal of Foraminiferal Research*, **27**: 253–328.

Goldstein, S. T. and Corliss, B. H. 1994. Deposit feeding in selected deep-sea and shallow-water benthic foraminifera. *Deep-Sea Research*, **41**: 229–41.

Goldstein, S. T. and Harben, E. B. 1993. Taphonomic implications of infaunal foraminiferal assemblages in a Georgia salt marsh, Sapelo Island. *Micropaleontology*, **39**: 53–62.

Goldstein, S. T. and Moodley, L. 1993. Gametogenesis and the life cycle of the foraminifer *Ammonia beccarii* (Linné) forma *tepida* (Cushman). *Journal of Foraminiferal Research*, **23**: 213–20.

Goldstein, S. T., Watkins, G. T. and Kuhn, R. M. 1995. Microhabitats of salt marsh foraminifera: St. Catharines Island, Georgia, USA. *Micropaleontology*, **26**: 17–29.

Golubic, S., Campbell, S. E., Drobne, K. et al. 1984. Microbial endoliths: a benthic imprint in the sedimentary record and a paleobathymetric cross-reference with foraminifera. *Journal of Paleontology*, **58**: 351–61.

Golubic, S., Perkins, R. D. and Lucas, K. J. 1975. Boring organisms in carbonate substrates. In Frey, R., ed. *The Study of Trace Fossils*. New York: Springer-Verlag, pp. 229–59.

Gooday, A. J. 1984. Records of deep-sea rhizopod tests inhabited by metazoans in the north-east Atlantic. *Sarsia*, **69**: 45–53.

Gooday, A. J. 1986a. Meiofaunal foraminiferans from the bathyal Porcupine Seabight: size structure, taxonomic composition, species diversity and vertical distribution in the sediment. *Deep-Sea Research*, **33**: 1345–73.

Gooday, A. J. 1986b. Soft-shelled foraminifera in meiofaunal samples from the bathyal northeast Atlantic. *Sarsia*, **71**: 275–87.

Gooday, A. J. 1988. A response by benthic foraminifera to the deposition of phytodetritus in the deep sea. *Nature*, **332**: 70–3.

Gooday, A. J. 1990. *Tinogullmia riemanni* sp. nov. (Allogromiina; Foraminiferida), a new species associated with organic detritus in the deep-sea. *Bulletin of the British Museum Natural History (Zoology)*, **56**: 93–103.

Gooday, A. J. 1993. Deep-sea benthic foraminiferal species which exploit phytodetritus: characteristic features and controls on distribution. *Marine Micropaleontology*, **22**: 187–205.

Gooday, A. J. 1994. The biology of deep-sea foraminifera: a review of some advances and their applications in paleoceanography. *Palaios*, **9**: 14–31.

Gooday, A. J. 1996. Epifaunal and shallow infaunal foraminiferal communities at three abyssal NE Atlantic sites subject to differing phytodetritus input regimes. *Deep-Sea Research I*, **43**: 1395–421.

Gooday, A. J. 1999. Biovidersity of foraminifera and other protists in the deep sea: scales and patterns. *Belgian Journal of Zoology*, **129**: 61–80.

Gooday, A. J. 2002. Biological responses to seasonally varying fluxes of organic matter to the ocean floor: a review. *Journal of Oceanography*, **58**: 305–32.

Gooday, A. J. 2003. Benthic foraminifera (Protista) as tools in deep-water palaeoceanography: environmental influences on faunal characteristics. *Advances in Marine Biology*, **46**: 1–90.

Gooday, A. J., Bernhard, J. M. and Bowser, S. S. 1995. The taxonomy and ecology of *Crithionina delacai* sp. nov., an abundant large agglutinated foraminifer from Explorers Cove, Antarctica. *Journal of Foraminiferal Research*, **25**: 290–8.

Gooday, A. J., Bernhard, J. M., Levin, L. A. and Suhr, S. B. 2000. Foraminifera in the Arabian Sea oxygen minimum zone and other oxygen-deficient settings: taxonomic composition, diversity, and relation to metazoan faunas. *Deep-Sea Research II*, **47**: 25–54.

Gooday, A. J., Bett, B. J. and Pratt, D. N. 1993. Direct observation of episodic growth in an abyssal xenophyophore (Protista). *Deep-Sea Research I*, **40**: 2131–43.

Gooday, A. J., Bett, B. J., Shires, R. and Lambshead, J. D. 1998. Deep-sea benthic foraminiferal species diversity in the NE Atlantic and NW Arabian sea: a synthesis. *Deep-Sea Research II*, **45**: 165–201.

Gooday, A. J., Bowser, S. S. and Bernhard, J. M. 1996. Benthic foraminiferal assemblages in Explorers Cove, Antarctica: a shallow-water site with deep-sea characteristics. *Progress in Oceanography*, **37**: 117–66.

Gooday, A. J., Carstens, M., and Thiel, H. 1995b. Micro- and nannoforaminifera from the abyssal northeast Atlantic sediments: a preliminary report. *Internationale Revue der gesamten Hydrobiologie*, **80**: 361–83.

Gooday, A. J. and Cook., P. L. 1984. An association between komokiacean foraminifers (Protozoa) and paludicelline ctenostomes (Bryozoa) from the abyssal northeast Atlantic. *Journal of Natural History*, **18**: 765–84.

Gooday, A. J. and Haynes, J. R. 1983. Abyssal foraminifera, including two new genera, encrusting the interior of *Bathysiphon rusticus* tubes. *Deep-Sea Research*, **30**: 591–614.

Gooday, A. J., Holzmann, M., Guiard, J., Cornelius, N. and Pawlowski, J. 2004. A new monothalamous foraminiferan from 1000 to 6000 m water depth in the Weddell Sea: morphological and molecular characterisation. *Deep-Sea Research II*, **51**: 1603–16.

Gooday, A. J. and Hughes, J. A. 2002. Foraminifera associated with phytodetritus deposits at a bathyal site in the northern Rockall Trough (NE Atlantic). *Marine Micropaleontology*, **46**: 83–110.

Gooday, A. J., Hughes, J. A. and Levin, L. A. 2001a. The foraminiferan macrofauna from three North Carolina (USA) slope sites with contrasting carbon flux: a comparison with the metazoan macrofauna. *Deep-Sea Research I*, **48**: 1709–39.

Gooday, A. J., Kitazato, H., Hori, S. and Toyofuku, T. 2001b. Monothalamous soft-shelled foraminifera at an abyssal site in the north Pacific: a preliminary report. *Journal of Oceanography*, **57**: 377–84.

Gooday, A. J. and Lambshead, P. J. D. 1989. Influence of seasonally deposited phytodetritus on benthic foraminiferal populations in the bathyal northeast Atlantic: the species response. *Marine Ecology Progress Series*, **58**: 53–67.

Gooday, A. J., Levin, L. A., Linke, P. and Heeger, T. 1992a. The role of benthic foraminifera in deep-sea food webs and carbon cycling. In Rowe, G. T. and Pariente, V., eds. *Deep-Sea Food Chains and the Global Carbon Cycle*. Dordrecht: Kluwer, pp. 63–91.

Gooday, A. J., Levin, L. A., Thomas, C. L. and Hecker, B. 1992b. The distribution and ecology of *Bathysiphon filiformis* Sars and *B. major* de Folin (Protista, Foraminiferida) on the continental slope off North Carolina. *Journal of Foraminiferal Research*, **22**: 129–46.

Gooday, A. J., Pond, D. W. and Bowser, S. S. 2002. Ecology and nutrition of the large agglutinated foraminiferan *Bathysiphon capillare* in the bathyal NE Atlantic: distribution within the sediment profile and lipid biomarker composition. *Marine Ecology Progress Series*, **245**: 69–82.

Gooday, A. J. and Rathburn, A. E. 1999. Temporal variability in living deep-sea benthic foraminifera: a review. *Earth-Science Reviews*, **46**: 187–212.

Gooday, A. J., Shires, R. and Jones, A. R. 1997. Large deep-sea agglutinated foraminifera: two differing kinds of organization and their possible ecological significance. *Journal of Foraminiferal Research*, **27**: 278–91.

Gooday, A. J. and Tendal, O. S. 1988. New xenophyophores (Protista) from the bathyal and abyssal north-east Atlantic Ocean. *Journal of Natural History*, **22**: 413–34.

Gooday, A. J. and Turley, C. M. 1990. Responses by benthic organisms to inputs of organic material to the ocean floor: a review. *Philosophical Transactions of the Royal Society London*, A331: 119–38.

Goudie, A. S and Sperling, C. H. B. 1977. Long distance transport of foraminiferal tests by wind in the Thar Desert, northwest India. *Journal of Sedimentary Petrology*, **47**: 630–3.

Grabert, B. 1971. Zur Eignung von Foraminiferen als Indikatoren für Sandwanderung. *Sonderdruck aus der Deutschen hydrographischen Zeitschrift*, **24**: 1–14.

Graf, G., Gerlach, S. A., Linke, P. *et al.* 1995. Benthic–pelagic coupling in the Norwegian–Greenland Sea and its effect on the geological record. *Geologische Rundschau*, **84**: 49–58.

Graney, R. L., Giesy, J. P. and Clark, J. R., 1995. Field studies. In Rand, G. M., ed. *Fundamentals of Aquatic Toxicology*, 2nd edn. Washington: Taylor and Francis, pp. 257–305.

Grant, K., Hoare, T. B., Ferrall, K. W. and Steinker, D. C. 1973. Some habitats for foraminifera, Coupon Bight, Florida. *The Compass*, **59**: 11–16.

Gray, J. S. 1981. *The Ecology of Marine Sediments. An Introduction to the Structure and Function of Benthic Communities.* Cambridge: Cambridge University Press.

Green, M. A. and Aller, R. C. 1998. Seasonal patterns of carbonate diagenesis in nearshore terrigenous muds: relation to spring phytoplankton bloom and temperature. *Journal of Marine Research*, **56**: 1097–123.

Green, M. A., Aller, R. C. and Aller, J. Y. 1993. Carbonate dissolution and temporal abundances of foraminifera in Long Island Sound sediments. *Limnology and Oceanography*, **38**: 331–45.

Gross, O. 2000. Influence of temperature, oxygen and food availability on the migrational activity of bathyal benthic foraminifera: evidence of microcosm experiments. *Hydrobiologia*, **426**: 123–37.

Gross, O. 2002. Sediment interactions of foraminifera: implications for food degradation and bioturbation processes. *Journal of Foraminiferal Research*, **32**: 414–24.

Grossman, E. 1984a. Carbon isotopic fractionation in live benthic foraminifera – comparison with inorganic precipitate studies. *Geochimica cosmochimica acta*, **48**: 1505–12.

Grossman, E. 1984b. Stable isotope fractionation in live benthic foraminifera from the southern California borderland. *Palaeogeography, Palaeoclimatology, Palaeoecology*, **47**: 301–27.

Grossman, E. 1987. Stable isotopes in modern benthic foraminifera: a study of vital effect. *Journal of Foraminiferal Research*, **17**: 48–61.

Grzymski, J., Schofield, O. M., Falkowski, P. G. and Bernhard, J. M. 2002. The function of plastids in the deep-sea benthic foraminifer, *Nonionella stella*. *Limnology and Oceanography*, **47**: 1569–80.

Guilbault, J. P., Clague, J. J. and LaPointe, M. 1996. Foraminiferal evidence for the amount of coseismic subsidence during a late Holocene earthquake on Vancouver Island, west coast of Canada. *Quaternary Science Reviews*, **15**: 913–37.

Gustafsson, M., Dahllöff, I., Blanck, H. *et al.* 2000. Benthic foraminiferal tolerance to tri-*n*-butyltin (TBT) pollution in an experimental mesocosm. *Marine Pollution Bulletin*, **40**: 1072–5.

Gustafsson, M. and Nordberg, K. 1999. Benthic foraminifera and their response to hydrography, periodic hypoxic conditions and primary production in the Kolöfjord on the Swedish west coast. *Journal of Sea Research*, **41**: 163–78.

Gustafsson, M. and Nordberg, K. 2000. Living (stained) benthic foraminifera and their response to the seasonal hydrographic cycle, periodic hypoxia and to primary production in Havstens Fjord on the Swedish west coast. *Estuarine, Coastal and Shelf Science*, **51**: 743–61.

Gustafsson, M. and Nordberg, K. 2001. Living (stained) benthic foraminiferal response to primary production and hydrography in the deepest part of the Gullmar Fjord, Swedish west coast, with comparisons to Höglund's 1927 material. *Journal of Foraminiferal Research*, **31**: 2–11.

Haake, F. W. 1962. Untersuchungen an der Foraminiferen-Fauna im Wattgebiet zwischen Langeoog und dem Festland. *Meyniana*, **12**: 25–64.

Haake, F. W. 1967. Zum Jahresgang von Populationen einer Foraminiferen-Art in der Westlichen Ostsee. *Meyniana*, **17**: 13–27.

Haake, F. W. 1970. Zur Tiefenverteilung von Miliolinen (Foram.) im Persischen Golf. *Paläontologische Zeitschrift*, **44**: 196–200.

Haake, F. W. 1975. Miliolinen (Foram.) in Oberflächensedimenten des Persischen Golfes. *"Meteor" Forschungs-Ergebnisse, Reihe C*, **21**: 15–51.

Haake, F. W. 1977. Living benthic foraminifera in the Adriatic Sea: influence of water depth and sediment. *Journal of Foraminiferal Research*, **7**: 62–75.

Haake, F. W. 1980a. Sedimentologische und faunistiche Untersuchungen an Watten in Taiwan. III. Mikropaläontologische Untersuchungen. *Senckenbergiana maritima*, **12**: 247–55.

Haake, F. W. 1980b. Benthische Foraminiferen in Oberflächensedimenten und Kernen des Ostatlantiks vor Senegal/Gambia (Westafrika). *"Meteor" Forschungs-Ergebnisse, Reihe C*, **32**: 1–29.

Haake, F. W. 1982. Occurrence of living and dead salt marsh foraminifera in the interior of northern Germany. *Senckenbergiana maritima*, **14**: 217–25.

Haig, D. W. and Burgin, S. 1982. Brackish-water foraminiferids from the Purari river delta, Papua New Guinea. *Revista española de micropaleontología*, **14**: 359–66.

Hald, M. and Korsun, S. 1997. Distribution of modern benthic foraminifera from fjords of Svalbard, European Arctic. *Journal of Foraminiferal Research*, **27**: 101–22.

Hald, M. and Steinsund, P. I. 1996. Benthic foraminifera and carbonate dissolution in the surface sediments of the Barents and Kara seas. *Berichte zur Polarforschung*, **212**: 285–307.

Hald, M., Steinsund, P. I., Dokken, T. *et al.* 1994. Recent and Late Quaternary distributions of *Elphidium excavatum* f. *clavatum* in Arctic seas. *Cushman Foundation Special Publication*, **32**: 141–53.

Halicz, E., Noy, N. and Reiss, Z. 1984. Foraminifera from Shura Arwashie mangrove (Sinai). In: Por, F. D. and Dor, I., eds. *Hydrobiology of the Mangal*. Junk, The Hague, pp. 145–49.

Hallock, P. 1981a. Algal symbiosis: a mathematical analysis. *Marine Biology*, **62**: 249–55.

Hallock, P. 1981b. Light dependence in *Amphistegina*. *Journal of Foraminiferal Research*, **11**: 40–6.

Hallock, P. 1985. Why are larger foraminifera large? *Paleobiology*, **11**: 195–208.

Hallock, P. 1987. Fluctuations in the trophic resource continuum: a factor in global diversity cycles? *Paleoceanography*, **2**: 457–71.

Hallock, P. 1999. Symbiont-bearing foraminifera. In Sen Gupta, B. K., ed. *Modern Foraminifera*. Dordrecht: Kluwer, pp. 123–39.

Hallock, P. 2000a. Larger foraminifera as indicators of coral-reef vitality. In Martin, R. E., ed. *Environmental Micropaleontology. The Application of Microfossils to Environmental Geology*. New York: Kluwer, pp. 121–50.

Hallock, P. 2000b. Symbiont-bearing foraminifera: harbingers of global change? *Micropaleontology*, **46** (Supplement 1): 95–104.

Hallock, P. 2001. Coral reefs, carbonate sediments, and global change. In Stanley, G. D., ed. *The History and Sedimentology of Ancient Reef Systems*. New York: Kluwer, pp. 387–427.

Hallock, P., Cottey, T. L., Forward, L. B. and Halas, J. 1986a. Population biology and sediment production of *Archaias angulatus* (Foraminiferida) in Largo Sound, Florida. *Journal of Foraminiferal Research*, **16**: 1–8.

Hallock, P., Forward, L. B. and Hansen, H. J. 1986b. Influence of environment on the test shape of *Amphistegina*. *Journal of Foraminiferal Research*, **16**: 224–31.

Hallock, P. and Peebles, M. W. 1993. Foraminifera with chlorophyte symbionts: habitats of six species in the Florida Keys. *Marine Micropaleontology*, **20**: 277–92.

Hallock, P., Röttger, R. and Wetmore, K. 1991. Hypotheses on form and function in foraminifera. In Lee, J. J. and Anderson, O. R., eds. *Biology of Foraminifera*. London: Academic, pp. 41–72.

Hallock, P. and Talge, H. K. 1993. Symbiont loss ('bleaching') in the reef-dwelling benthic foraminifer *Amphistegina gibbosa* in the Florida Keys in 1991–92. In Ginsburg, R. N., ed. *Global Aspects of Coral Reefs: Health, Hazards and History*. Miami: University of Miami, pp. 8–14.

Hallock, P. and Talge, H. K. 1994. A predatory foraminifer, *Floresina amphiphaga*, n. sp., from the Florida Keys. *Journal of Foraminiferal Research*, **24**: 210–13.

Hallock, P., Talge, H. K., Cockey, E. M. and Muller, R. G. 1995. A new disease in reef-dwelling foraminifera: implications for coastal sedimentation. *Journal of Foraminiferal Research*, **25**: 280–6.

Hallock, P., Talge, H. K., Smith, K. and Cockey, E. M. 1992. Bleaching in a reef-dwelling foraminifer, *Amphistegina gibbosa*. *Proceedings of the 7th International Coral Reef Symposium, Guam*, **1**: 44–9.

Haman, D. 1969. Seasonal occurrence of *Elphidium excavatum* (Terquem) in Llandanwg lagoon (North Wales, UK). *Contributions from the Cushman Foundation for Foraminiferal Research*, **20**: 139–42.

Hannah, F. and Rogerson, A. 1997. The temporal and spatial distribution of foraminifera in marine benthic sediments of the Clyde Sea area, Scotland. *Estuarine and Coastal Shelf Science*, **44**: 377–83.

Hannah, F., Rogerson, A. and Laybourn-Parry, J. 1994. Respiration rates and biovolumes of common benthic foraminifera (Protozoa). *Journal of the Marine Biological Association of the UK*, **74**: 301–12.

Hansen, A. and Knudsen, K. L. 1995. Recent foraminiferal distribution in Freemansundet and early Holocene stratigraphy on Edgeøya, Svalbard. *Polar Research*, **14**: 215–38.

Harloff, J. and Mackensen, A. 1997. Recent benthic foraminiferal associations and ecology of the Scotia Sea and Argentine Basin. *Marine Micropaleontology*, **31**: 1–29.

Harman, R. A. 1964. Distribution of foraminfera in the Santa Barbara Basin, California. *Micropaleontology*, **10**: 81–96.

Harney, J. N., Hallock, P. and Talge, H. K. 1998. Observations of a trimorphic life cycle in *Amphistegina gibbosa* populations from the Florida Keys. *Journal of Foraminiferal Research*, **28**: 141–7.

Haslett. S. K. 2002. *Quaternary Environmental Micropalaeontology*. London: Arnold.

Haward, J. B. and Haynes, J. R. 1976. *Chlamys opercularis* (Linnaeus) as a mobile substrate for foraminifera. *Journal of Foraminiferal Research*, **6**: 30–8.

Hayek, L. A. C. and Buzas, M. A. 1997. *Surveying Natural Populations*. New York: Columbia.

Haynes, J. R. 1965. Symbiosis, wall structure and habitat in foraminifera. *Contribution to the Cushman Foundation for Foraminiferal Research*, **16**: 40–3.

Haynes, J. R. 1981. *Foraminifera*. London: Macmillan.

Haynes, J. R. 1992. Supposed pronounced ecophenotypy in foraminifera. *Journal of Micropalaeontology*, **11**: 59–63.

Hayward, B. W. 1979. An intertidal *Zostera* pool community at Kawerua, Northland and its foraminiferal microfauna. *Tane*, **25**: 173–86.

Hayward, B. W. 1982. Associations of benthic foraminifera (Protozoa: Sarcodina) of inner shelf sediments around the Cavalli Islands, north-east New Zealand. *New Zealand Journal of Marine and Freshwater Research*, **16**: 27–56.

Hayward, B. W., Grenfell, H. R. and Scott, D. B. 1999. Tidal range of marsh foraminifera for determining former sea-level heights in New Zealand. *New Zealand Journal of Geology and Geophysics*, **42**: 395–413.

Hayward, B. W., Holzmann, M. Grenfell, H. R., Pawlowski, J. and Triggs, C. M. 2004a. Morphological distinction of molecular types in *Ammonia* – towards a taxonomic revision of the world's most commonly misidentified foraminifera. *Marine Micropaleontology*, **50**: 237–71.

Hayward, B. W., Scott, G. H., Grenfell, H. R., Carter, R. and Lipps, J. H. 2004b. Techniques for estimation of tidal elevation and confinement (~salinity) histories of sheltered harbours and estuaries using benthic foraminifera; examples from New Zealand. *The Holocene*, **14**: 218–32.

Heeger, T. 1988. Virus-like particles and cytopathological effects in *Elphidium excavatum clavatum*, a benthic foraminiferan. *Diseases of Aquatic Organisms*, **4**: 233–6.

Heeger, T. 1990. Electronenmikroskopische Untersuchungen der Ernähungsbiologie benthischer Foraminiferen. *Berichte aus dem Sondersforschungsbereich*, **313**: 1–139.

Heinz, P., Kitazato, H., Schmiedl, G. and Hemleben, C. 2001. Response of deep-sea benthic foraminifera from the Mediterranean Sea to simulated phytoplankton pulses under laboratory conditions. *Journal of Foraminiferal Research*, **31**: 210–27.

Hemleben, C., Anderson, O. R., Berthold, W. and Spindler, M. 1986. Calcification and chamber formation in foraminifera: a brief overview. In Leadbeater, B. S. C. and Riding, R., eds. Biomineralization in lower plants and animals. *The Systematics Association*, Special Volume 30: 237–49.

Hemleben, C. and Kitazato, H. 1995. Deep-sea foraminifera under long time observation in the laboratory. *Deep-Sea Research I*, **42**: 827–32.

Herbert, D. G. 1991. Foraminiferivory in a *Puncturella* (Gastropoda: Fissurellidae). *Journal of Molluscan Studies*, **57**: 127–9.

Herguera, J. C. 1992. Deep-sea foraminiferal and biogenic opal: glacial to post-glacial productivity changes in the western Pacific. *Marine Micropaleontology*, **19**: 79–98.

Herguera, J. C. and Berger, W. H. 1991. Paleoproductivity from benthic foraminifera abundance: glacial to postglacial change in the west-equatorial Pacific. *Geology*, **19**: 1173–6.

Heron-Allen, E. 1915. Contributions to the study of the bionomics and reproductive processes of the foraminifera. *Philosophical Transactions of the Royal Society of London, Series B*, **206**: 227–79.

Hess, S. 1998. Verteilungsmuster rezenter benthischer Foraminiferen im Südchinesischen Meer. *Berichte-Reports, Geologisch-Paläontologisches Institut und Museum, Universität Kiel*, **91**: 1–173.

Hess, S. and Kuhnt, W. 1996. Deep-sea benthic foraminiferal recolonization of the 1991 Mt. Pinatubo ash layer in the South China Sea. *Marine Micropaleontology*, **28**: 171–97.

Hess, S., Kuhnt, W., Hill, S., Kaminski, M. A., Holbourn, A. and de Leon, M. 2001. Monitoring the recolonization of the Mt. Pinatuno 1991 ash layer by benthic foraminifera. *Marine Micropaleontology*, **43**: 119–42.

Hickman, C. S. and Lipps, J. H. 1983. Foraminiferivory: selective ingestion of foraminifera and test alterations produced by the neogastropod *Olivella*. *Journal of Foraminiferal Research*, **13**: 108–14.

Hilborn, R. and Mangel, M. 1997. *The Ecological Detective: Confronting Models with Data*. Princeton: University Press.

Hinrichs, K. U. and Boetius, A. 2002. The aerobic oxidation of methane: new insights in microbial ecology and biogeochemistry. In Wefer, G., Billettt, *et al.*, eds. *Ocean Margin Systems*. Berlin: Springer-Verlag, pp. 457–77.

Hippensteel, S. P. and Martin, R. E. 1999. Foraminifera as an indicator of overwash deposits, barrier island sediment supply, and barrier island evolution: Folly Island, South Carolina. *Palaeogeography, Palaeoclimatology, Palaeoecology*, **149**: 115–25.

Hippensteel, S. P., Martin, R. E., Nikitina, D. and Pizzuto, J. 2000. The formation of Holocene marsh foraminiferal assemblages, Middle Atlantic coast, USA: implications for Holocene sea-level change. *Journal of Foraminiferal Research*, **30**: 272–93.

Hippensteel, S. P., Martin, R. E., Nikitina, D. and Pizzuto, J. 2002. Interannual variation of marsh foraminiferal assemblages, (Bombay Hook National Wildlife Refuge, Smyrna, De): do foraminiferal assemblages have a memory? *Journal of Foraminiferal Research*, **32**: 97–109.

Höglund, H. 1947. Foraminifera in the Gullmar Fjord and the Skagerak. *Zoologiska bidrag från Uppsala*, **26**: 1–328.

Hohenegger, J. 1994. Distribution of living larger foraminifera NW of Sesoku-Jima, Okinawa, Japan. *PSZNI, Marine Ecology*, **15**: 291–334.

Hohenegger, J. 1995. Depth estimation by proportions of living larger foraminifera. *Marine Micropaleontology*, **26**: 31–47.

Hohenegger, J. 2000. Coenoclines of larger foraminifera. *Micropaleontology*, **46** (Supplement 1): 127–51.

Hohenegger, J., Pillar, W. and Baal, C. 1989. Reasons for spatial microdistributions of foraminifera in an intertidal pool (northern Adriatic Sea). *PSZNI, Marine Ecology*, **10**: 43–78.

Hohenegger, J., Pillar, W. and Baal, C. 1993. Horizontal and vertical spatial distribution of foraminifers in the shallow subtidal Gulf of Trieste, northern Adriatic Sea. *Journal of Foraminiferal Research*, **23**: 79–101.

Hohenegger, J., Yordanova, E. and Hatta, A. 2000. Remarks on West Pacific Nummulitidae (Foraminifera). *Journal of Foraminiferal Research*, **30**: 3–28.

Hohenegger, J., Yordanova, E., Nakano, Y. and Tatzreiter, F. 1999. Habitats of larger foraminifera on the upper reef slope of Sesko Island, Okinawa, Japan. *Marine Micropaleontology*, **36**: 109–68.

Hollaus, S. S. and Hottinger, L. 1997. Temperature dependance of endosymbiotic relationships? Evidence from the depth range of Mediterranean *Amphistegina lessonii* (Foraminiferida) truncated by the thermocline. *Eclogae geologicae helvetiae*, **90**: 591–7.

Holzmann, M. 2000. Species concept in foraminifera: *Ammonia* as a case study. *Micropaleontology*, **46** (Supplement 1): 21–37.

Holzmann, M., Hohenegger, J., Hallock, P., Pillar, W. E. and Pawlowski, J. 2001. Molecular phylogeny of large miliolid foraminifera (Soritacea Ehrenberg 1839). *Marine Micropaleontology*, **43**: 57–74.

Holzmann, M. and Pawlowski, J. 1997. Molecular, morphological and ecological evidence for species recognition in *Ammonia* (Foraminifera). *Journal of Foraminiferal Research*, **27**: 311–18.

Horton, B. P. 1999. The distribution of contemporary intertidal foraminifera at Cowpen marsh, Tees estuary, UK: implications for studies of Holocene sea-level change. *Palaeogeography, Palaeoclimatology, Palaeoecology*, **149**: 127–49.

Horton, B. P. and Edwards, R. J. 2003. Seasonal distributions of foraminifera and their implications for sea-level studies, Cowpen marsh, UK. *SEPM Special Publication*, **75**: 21–30.

Horton, B. P. and Edwards, R. J. 2005. The application of local and regional transfer functions to the reconstruction of Holocene sea levels, north Norfolk, England. *The Holocene*, **15**: 216–28.

Horton, B. P., Edwards, R. J. and Lloyd, J. M. 1999. A foraminiferal-based transfer function: implications for sea-level studies. *Journal of Foraminiferal Research*, **29**: 117–29.

Horton, B. P., Whittaker, J. E., Thomson, K. H. *et al.* 2005. The development of a moderm foraminiferal data set for sea-level reconstructions, Wakatobi Marine National Park, southeast Sulawesi, Indonesia. *Journal of Foraminiferal Research*, **35**: 1–14.

Hottinger, L. 1980. Répartition comparée des grands foraminifères de la Mer Rouge et de l'Océan Indien. *Annali dell' Università de Ferrara, NS IX*, **6** (supplimento): 1–13.

Hottinger, L. 1984. Foraminifères de grande taille: signification des structures complexes de la coquille. In Proceedings of Benthos '83, 2e symposium international sur les foraminifères benthiques, Pau (France), 11th–15th April, pp. 309–15.

Hottinger, L. 1990. Significance of diversity in shallow benthic foraminifera. *Atti del quarto simposia di ecologia e paleocologia delle comunità bentoniche Sorrento*. Torino: Museo Regionale de Scienze Naturali, pp. 35–50.

Hottinger, L. C. 2000. Functional morphology of benthic foraminiferal shell, envelopes beyond measure. *Micropaleontology*, **46** (Supplement 1): 57–86.

Hovland, M. and Mortensen, P. B. 1999. *Norske korallrev og prosesser I havbunnen*. Bergen: John Grieg Forlag.

Howarth, R. J. and Murray, J. W. 1969. The Foraminiferida of Christchurch harbour, England: a reappraisal using multivariate techniques. *Journal of Paleontology*, **43**: 600–75.

Hueni, C. M., Anepohl, J., Gevirtz, J. and Casey, R. 1978. Distribution of living benthonic foraminifera as indicators of oceanographic processes of the south Texas outer continental shelf. *Transactions of the Gulf Coast Association of Geological Societies*, **28**: 183–200.

Hughes, J. A. and Gooday, A. J. 2004. Associations between living benthic foraminifera and dead tests of *Syringammina fragilissima* (Xenophyophorea) in the Darwin Mounds region (NE Atlantic). *Deep-Sea Research I*, **51**: 1741–58.

Hughes, J. A., Gooday, A. J. and Murray, J. W. 2000. Distribution of live foraminifera at three oceanographically dissimilar sites in the northeast Atlantic: preliminary results. *Hydrobiologia*, **440**: 227–38.

Hunt, A. S. and Corliss, B. H. 1993. Distribution and microhabitats of living (stained) benthic foraminifera from the Canadian Arctic archipelago. *Marine Micropaleontology*, **20**: 321–45.

Husum, K. and Hald, M. 2004. Modern foraminiferal distribution in the subarctic Malangen Fjord and adjoining shelf, northern Norway. *Journal of Foraminiferal Research*, **34**: 34–48.

Ikeya, N. 1970. Population ecology of benthonic foraminifera in Ishikari Bay, Hokkaido, Japan. *Records of Oceanographic Works in Japan*, **10**: 173–91.

Ikeya, N. 1971. Species diversity of recent benthonic foraminifera off the Pacific coast of north Japan. *Reports of Faculty of Science, Shizuoka University*, **6**: 179–201.

Ishman, S. E. and Webb, P. N. 2003. Cryogenic taphonomy of supra-ice shelf foraminiferal assemblages, McMurdo Sound, Antarctica. *Journal of Foraminiferal Research*, **33**: 122–31.

James, N. P. 1997. The cool-water carbonate depositional realm. In *Cool-Water Carbonates*. SEPM Special Publication, 56: 1–20.

Jannink, N. T. 2001. Seasonality, biodiversity and microhabitats in benthic foraminiferal communities. *Geologica ultraiectina*, **203**: 1–190.

Jannink, N. T., Zachariasse, W. J. and Zwaan, G. J. van der 1998. Living (rose Bengal stained) benthic foraminifera from the Pakistan continental margin (northern Arabian Sea). *Deep-Sea Research*, **I, 45**: 1483–513.

Jennings, A. E. and Helgadottir, G. 1994. Foraminiferal assemblages from the fjords and shelf of eastern Greenland. *Journal of Foraminiferal Research*, **24**: 123–44.

Jennings, A. E., Weiner, N. J., Helgadottir, G. and Andrews, T. 2004. Modern foraminiferal faunas of the southwestern to northern Iceland shelf: oceanographic and environmental controls. *Journal of Foraminiferal Research*, **24**: 180–207.

Jensen, P. 1982. A new meiofaunal splitter. *Annales zoologici fennici*, **19**: 233–6.

Jepps, M. W. 1942. Studies on *Polystomella* Lamarck (Foraminifera). *Journal of the Marine Biological Association of the United Kingdom*, **25**: 607–66.

John, A. W. G. 1987. The regular occurrence of *Reophax scottii* Chaster, a benthic foraminifera, in plankton samples from the North Sea. *Journal of Micropalaeontology*, **6**: 61–3.

Jonasson, K. E. and Schröder-Adams, C. J. 1996. Encrusting agglutinated foraminifera on indurated sediment at a hydrothermal venting area on the Juan de Fuca Ridge, northeast Pacific Ocean. *Journal of Foraminiferal Research*, **26**: 137–49.

Jones, G. D. and Ross, C. A. 1979. Seasonal distribution of foraminifera in Samish Bay, Washington. *Journal of Paleontology*, **53**: 245–57.

Jones, R. W. and Charnock, M. A. 1985. "Morphogroups" of agglutinating foraminifera. Their life positions and feeding habits and potential applicability in (paleo)ecological studies. *Revue de paléobiologie*, **4**: 311–20.

Jones, R. W. and Simmons, M. D., eds. 1999. *Biostatigraphy in Production and Development Geology*. Geological Society, London, Special Publication, 152: 5–22.

Jorissen, F. J. 1999. Benthic foraminiferal microhabitats below the sediment–water interface. In Sen Gupta, B. K., ed. *Modern Foraminifera*. Dordrecht: Kluwer, pp. 161–79.

Jorissen, F. J., Barmawidjaja, D. M., Puskaric, S. and Zwaan, G. J. van der. 1992. Vertical distribution of benthic foraminifera in the northern Adriatic Sea: the relation with organic flux. *Marine Micropaleontology*, **19**: 131–46.

Jorissen, H. J., Buzas, M. A., Culver, S. J. and Kuehl, S. A. 1994. Vertical distribution of living benthic foraminifera in submarine canyons off New Jersey. *Journal of Foraminiferal Research*, **24**: 28–36.

Jorissen, F. J., de Stigter, H. C. and Widmark, J. G. V. 1995. A conceptual model explaining benthic foraminiferal microhabitats. *Marine Micropaleontology*, **26**: 3–15.

Jorissen, F. J. and Wittling, L. 1999. Ecological evidence from live–dead comparisons of benthic foraminiferal faunas off Cape Blanc (northwest Africa). *Palaeogeography, Palaeoclimatology, Palaeoecology*, **149**: 151–70.

Jorissen, F. J., Wittling, L., Peypouquet, J. P., Rabouille, C. and Relexans, J. C. 1998. Live benthic foraminiferal faunas off Cape Blanc, NW Africa: community structure and microhabitats. *Deep-Sea Research I*, **45**: 2157–88.

Kaiho, K. 1994. Benthic foraminiferal dissolved-oxygen index and dissolved-oxygen levels in the modern ocean. *Geology*, **22**: 719–22.

Kaiho, K. 1999. Effect of organic carbon flux and dissolved oxygen on the benthic foraminiferal oxygen index (BFOI). *Marine Micropaleontology*, **37**: 67–76.

Kaminski, M. A., Al Hassawi, A. M. and Kuhnt, W. 1997. Seasonality in microhabitats of rose Bengal stained deep-sea benthic foraminifera from the New Jersey continental margin. In *Contributions to the Micropaleontology and Paleoceanography of the Northern North Atlantic.* Grzybowski Foundation Special Publication, **5**: 245–8.

Kaminski, M. A., Boersma, A., Tyszka, J. and Holbourn, A. E. L. 1995. Response of deep-water agglutinated foraminifera to dysoxic conditions in the California borderland basins. In *Proceedings of the Fourth International Workshop on Agglutinated Foraminifera, Kraków, Poland, September 12–19, 1993.* Grzybowski Foundation Special Publication, **3**: 131–40.

Kaminski, M. A., Grassle, J. F. and Whitlatch, R. B. 1988. Life history and recolonization among agglutinated foraminifera in the Panama Basin. *Abhandlungen der Geologischen Bundesanstalt*, **41**: 229–43.

Kaminski, M. A. and Kuhnt, W. 1995. Tubular agglutinated foraminifera as indicators of organic carbon flux. In *Proceedings of the Fourth International Workshop on Agglutinated Foraminifera, Kraków, Poland, September 12–19, 1993.* Grzybowski Foundation Special Publication, **3**: 141–4.

Kaminski, M. A. and Wetzel, A. 2004a. Sediment disturbance caused by a suspension-feeding tubular agglutinated foraminifer. In *Proceedings of the Sixth International Workshop on Agglutinated Foraminifera.* Grzybowski Foundation Special Publication, **8**: 294.

Kaminski, M. A. and Wetzel, A. 2004b. A tubular protozoan predator: a burrow selectively filled with tubular agglutinated protozoans (Xenophyophorea, Foraminifera) in the abyssal South China Sea. In *Proceedings of the Sixth International Workshop on Agglutinated Foraminifera.* Grzybowski Foundation Special Publication, **8**: 287–93.

Karlsen, A. W., Cronin, T. M., Ishman, S. *et al.* 2000. Historical trends in Chesapeake Bay dissolved oxygen based on benthic foraminifera from sediment cores. *Estuaries*, **23**: 488–508.

Kennett, J. P. 1982. *Marine Geology.* Englewood Cliff: Prentice-Hall Inc.

Kingsford, M. and Battershill, C. 1998. *Studying Temperate Marine Environments. A Handbook for Ecologists.* Christchurch: Canterbury University Press.

Kitazato, H. 1981. Observations of behaviour and mode of life of benthic foraminifers in a laboratory. *Geoscience Reports of Shizuoka University*, **6**: 61–71.

Kitazato, H. 1988. Locomotion of some benthic foraminifera in and on sediments. *Journal of Foraminiferal Research*, **18**: 344–9.

Kitazato, H. 1990. Pseudopodia of benthic foraminifera and their relationships to the test morphology. In Takayanagi, Y. and Saito, T., eds. *Studies in Benthic Foraminifera. Proceedings of the Fourth International Symposium on Benthic Foraminifera Sendai, 1990.* Tokyo: Tokai University, pp. 103–8.

Kitazato, H. 1994. Foraminiferal microhabitats in four marine environments around Japan. *Marine Micropaleontology*, **24**: 29–41.

Kitazato, H. 1995. Recolonization by deep-sea benthic foraminifera: possible substrate preferences. *Marine Micropaleontology*, **26**: 65–74.

Kitazato, H. and Ohga, T. 1995. Seasonal changes in deep-sea benthic foraminiferal populations: results of long-term observations at Sagami Bay, Japan. In Sakai, H. and Nozaki, Y., eds. *Biogeochemical Processes and Ocean Flux in the Western Pacific*. Tokyo: Terra Scientific Publishing Company, pp. 331–42.

Kitazato, H., Shirayama, Y., Nakatsuka, T. *et al.* 2000a. Seasonal phytodetritus deposition and responses of bathyal benthic foraminiferal populations in Sagami Bay, Japan: preliminary results from "Project Sagami 1996–1999". *Marine Micropaleontology*, **40**: 135–49.

Kitazato, H., Tsuchiya, M. and Takahara, K. 2000b. Recognition of breeding populations in foraminifera: an example using the genus *Glabratella*. *Paleontological Research*, **4**: 1–15.

Kloos, D. P. 1982. Destruction of tests of the foraminifer *Sorites orbiculus* by endolithic microorganisms in a lagoon on Curaçao (Netherlands Antilles). *Geologie en Mijnbouw*, **61**: 201–5.

Knight, R. and Mantoura, R. F. C. 1985. Chlorophyll a and carotenoid pigments in foraminifera and their symbiotic algae: analysis in high-performance liquid chromatography. *Marine Ecology Progress Series*, **23**: 241–9.

Kontrovitz, M., Snyder, S. W. and Brown, R. J. 1978. A flume study of foraminifera tests. *Palaeogeography, Palaeoclimatology, Palaeoecology*, **23**: 141–50.

Korsun, S. 1999. Benthic foraminifera in the Ob estuary, west Siberia. *Berichte zur Polarforschung*, **300**: 59–70.

Korsun, S. 2002. Allogromiids in foraminferal assemblages on the western Eurasian Arctic shelf. *Journal of Foraminiferal Research*, **32**: 400–13.

Korsun, S. and Hald, M. 1998. Modern benthic foraminifera off Novaya Zemlya tidewater glaciers. *Arctic and Alpine Research*, **30**: 61–77.

Korsun, S. and Hald, M. 2000. Seasonal dynamics of benthic foraminifera in a glacially fed fjord of Svalbard, European arctic. *Journal of Foraminiferal Research*, **30**: 251–71.

Korsun, S., Hald, M., Panteleeva, N. and Tarasov, G. 1998. Biomass of foraminifera in the St. Anna trough, Russian arctic continental margin. *Sarsia*, **83**: 419–31.

Korsun, S. A., Pogodina, I. A., Tarasov, G. A. and Matishov, G. G. 1994. Foraminifera of the Barents Sea: Hydrobiology and Palaeoecology. Apatity, Kola Science Center Publication, 1–140. In Russian with English abstract.

Kotler, E., Martin, R. E. and Liddell, W. D. 1992. Experimental analysis of abrasion and dissolution-resistance of modern reef dwelling foraminifera for the preservation of biogenic carbonate. *Palaios*, **7**: 244–76.

Krüger, R., Röttger, R., Lietz, R. and Hohenegger, J. 1997. Biology and reproductive processes of the larger foraminiferan *Cycloclypeus carpenteri* (Protozoa, Nummulitidae). *Archiv für Protistenkunde*, **147**: 307–21.

Kucera, M. and Malmgren, B. A. 1998. Logratio transformation of compositional data: a resolution of the constant sum constraint. *Marine Micropaleontology*, **34**: 117–20.

Kuhnt, W., Collins, E. S. 1995. Fragile abyssal foraminifera from the northwestern Sargasso Sea: distribution, ecology, and paleoceanographic significance. In *Proceedings of the Fourth International Workshop on Agglutinated Foraminifera*. Grzybowski Foundation Special Publication, 3: 159–72.

Kuhnt, W., Collins, E. and Scott, D. B. 2000. Deep water agglutinated foraminiferal assemblages across the Gulf Stream: distribution patterns and taphonomy. In *Proceedings of the Fifth International Workshop on Agglutinated Foraminifera*. Grzybowski Foundation Special Publication, 7: 261–98.

Kuhnt, W., Hess, S. and Jian, Z. 1999. Quantitative composition of benthic foraminiferal assemblages as a proxy indicator for organic carbon flux rates in the South China Sea. *Marine Geology*, **156**: 123–57.

Lambshead, P. J. D. and Gooday, A. J. 1990. The impact of seasonally deposited phytodetritus on epifaunal and shallow infaunal benthic foraminiferal populations in the bathyal northeast Atlantic: the assemblage response. *Deep-Sea Research*, **32**: 885–97.

Lampitt, R. S. 1996. Snow falls in the open ocean. In Summerhayes, C. P. and Thorpe, S. A., eds. *Oceanography. An Illustrated Guide*. London: Manson Publishing Ltd., pp. 96–112.

Langer, M. 1993. Epiphytic foraminifera. *Marine Micropaleontology*, **20**: 235–65.

Langer, M. 2000. Comparative molecular analysis of small-subunit ribosomal 18s DNA sequences from *Haynesina germanica* (Ehrenberg, 1840), a common intertidal foraminifer from the North Sea. *Neues Jahrbuch für Geologie und Paläontologie*, **11**: 641–50.

Langer, M. R. and Bagi, H. 1994. Tubicolous polychaetes as substrates for epizoic foraminifera. *Journal of Micropalaeontology*, **13**: 132.

Langer, M. R. and Bell, C. J. 1995. Toxic foraminifera: innocent until proven guilty. *Marine Micropaleontology*, **24**: 205–14.

Langer, M. R., Frick, H. and Silk, M. T. 1998. Photophile and sciaphile foraminiferal assemblages from marine plant communities of Lavezzi Islands (Corsica, Mediterranean Sea). *Revue de paléobiologie*, **17**: 525–30.

Langer, M. R. and Gehring, C. A. 1993. Bacteria farming: a possible feeding strategy of some smaller, motile foraminifera. *Journal of Foraminiferal Research*, **23**: 40–6.

Langer, M. R. and Hottinger, L. 2000. Biogeography of selected 'larger' foraminifera. *Micropaleontology*, **46** (Supplement 1): 105–27.

Langer, M. R., Hottinger, L. and Huber, B. 1989. Functional morphology in low-diverse benthic foraminiferal assemblages from tidal flats in the North Sea. *Senckenbergiana maritima*, **20**: 81–99.

Langer, M. R., Lipps, J. H. and Moreno, G. 1995. Predation on foraminifera by the dentaliid scaphopod *Fissidentalium megathyris*. *Deep-Sea Research*, **42**: 849–57.

Langer, M. R., Silk, M. T. and Lipps, J. H. 1997. Global ocean carbonate and carbon dioxide production: the role of reef foraminifera. *Journal of Foraminiferal Research*, **27**: 271–7.

Langezaal, A. M., Ernst, S. R., Haese, R. R., Bergen, P. F. van and Zwaan, G. J. van der 2003. Disturbance of intertidal sediments: the response of bacteria and foraminifera. *Estuarine, Coastal and Shelf Science*, **58**: 249–64.

Lankford, R. R. 1959. Distribution and ecology of foraminifera from east Mississippi delta margin. *Bulletin of the Association of Petroleum Geologists*, **43**: 2068–99.

Lankford, R. R. and Phleger, F. P. 1973. Foraminifera from the nearshore turbulent zone, western North America. *Journal of Foraminiferal Research*, **3**: 101–32.

Lawton, J. H. 1999. Are there general rules in ecology? *Oikos*, **84**: 177–92.

Lear, C. H., Rosenthal, Y. and Slowey, N. 2002. Benthic foraminiferal Mg/Ca paleothermometry: a revised core-top calibration. *Geochimica et Cosmochemica Acta*, **66**: 3375–87.

Le Calvez, J. 1947. *Entosolenia marginata* (Walker and Boys), foraminifère apogamique ectoparasite d'une autre foraminifère: *Discorbis villardeboanus* (d'Orbigny). *Comptes rendus hebdomadaire des séances de l'Academie des Sciences, Paris*, **224**: 1448–50.

Le Campion, J. 1970. Contribution á l'étude des foraminifers du Bassin d'Arcachon et du proche océan. *Bulletin Institute Géologique du Bassin Aquitaine*, **8**: 3–98.

Leckie, R. M. and Olson, H. C. 2003. Foraminifera as proxies for sea-level change in siliciclastic margins. *SEPM Special Publication*, **75**: 5–19.

Lee, J. J. 1974. Towards understanding the niche of foraminifera. In Hedley, R. H. and Adams, C. G., eds. *Foraminifera*, vol. 1: pp. 207–60.

Lee, J. J. 1980. Nutrition and physiology of the foraminifera. In *Biochemistry and Physiology of Protozoa*, 2nd edn. **vol. 3, pp**. 43–66.

Lee, J. J. 1983. Perspectives on algal endosymbionts in larger foraminifera. *International Review of Cytology*, **11**: 49–77 (supplement).

Lee, J. J. and Anderson, O. R. 1991a. *Biology of Foraminifera*. London: Academic.

Lee, J. J. and Anderson, O. R. 1991b. Symbiosis in Foraminifera. In Lee, J. J. and Anderson, O. R., eds. *Biology of Foraminifera*. London: Academic, pp. 157–220.

Lee, J. J., Anderson, O. R., Karim, B. and Beri, J. 1991a. Additional insight into the structure and biology of *Abyssotherma pacifica* (Brönnimann, van Dover and Whittaker) from the East Pacific Rise. *Micropaleontology*, **37**: 303–12.

Lee, J. J. and Bock, W. D. 1976. The importance of feeding in two species of soritid foraminifera with algal symbionts. *Bulletin of Marine Science*, **26**: 530–7.

Lee, J. J., Faber, W. W., Anderson, O. R. and Pawlowski, J.1991b. Life-cycles of foraminifera. In Lee, J. J. and Anderson, O. R., eds. *Biology of Foraminifera*. London: Academic, pp. 285–334.

Lee, J. J., Faber, W. W. and Lee, R. E. 1991c. Granular reticulopodial digestion – a possible preadaptation to benthic foraminiferal symbiosis? *Symbiosis*, **10**: 47–61.

Lee, J. J. and Hallock, P. 2000. Advances in the biology of foraminifera. *Micropaleontology*, **46** (Supplement 1): 1–198.

Lee, J. J., Lanners, E. and Ter Kuile, B. 1988. Retention of chloroplasts by the foraminifer *Elphidium crispum. Symbiosis*, **5**: 45–60.

Lee, J. J. and Lee, R. E. 1990. Chloroplast retention in elphidiids (Foraminifera). In Nardon, P., Gianinazzi-Pearson, V., Grenier, A. M., Margulis, L. and

Smith, D. C., eds. *Endocytobiology IV*. Paris: Institut National de la Recherche Agronomique, pp. 215–20.

Lee, J. J., Leedale, G. F. and Bradbury, P. 2000. *An Illustrated Guide to the Protozoa*, 2nd edn. Lawrence: Society of Protozoologists.

Lee, J. J., McEnery, M. E. and Garrison, J. R. 1980. Experimental studies of larger foraminifera and their symbionts from the Gulf of Elat on the Red Sea. *Journal of Foraminiferal Research*, **10**: 31–47.

Lee, J. J., McEnery, M. E., Pierce, S., Freudenthal, H. D. and Muller, W. A. 1966. Tracer experiments in feeding littoral foraminifera. *Journal of Protozoology*, **13**: 659–70.

Lee, J. J., Morales, J., Bacus, S. *et al.* 1997. Progress in characterizing the endosymbiotic dinoflagellates of soritid foraminifera and related studies of some stages in the life cycle of *Marginopora vertebralis*. *Journal of Foraminiferal Research*, **27**: 254–63.

Lee, J. J., Morales, J., Symons, A. and Hallock, P. 1995. Diatom symbionts in larger foraminifera from Caribbean hosts. *Marine Micropaleontology*, **26**: 99–105.

Lee, J. J., Muller, W. A., Stone, R. J., McEnery, M. E. and Zucker, W. 1969. Standing crop of foraminifera in sublittoral epiphytic communities of a Long Island salt marsh. *Marine Biology*, **4**: 44–61.

Lee, J. J., Tietjen, J. H., Mastropaolo, C. and Rubin, H. 1977. Food quality and the heterogeneous spatial distribution of meiofauna. *Helgoländer wissenschalftliche Meeresuntersuchungen, Kiel*, **30**: 272–82.

Lee, J. J., Tietjen, J. H., Stone, R. J., Muller, W. A., Rullman, J. and McEnery, M. 1970. The cultivation and physiological ecology of salt marsh epiphytic communities. *Helgoländer wissenschafliche Meeresuntersuchangen, Kiel*, **20**: 136–56.

Lefévre, M. 1984. Répartition de la biocénose zooplanctonique autour de l'île de Mooréa (Polynésie français). *Journal de recherche océanographique*, **9**: 20–2.

LeFurgey, A. and St. Jean, J. 1976. Foraminifera in brackish-water ponds designed for waste control and aquaculture studies in North Carolina. *Journal of Foraminiferal Research*, **6**: 274–94.

Lehmann, G. and Röttger, R. 1997. Techniques for the concentration of foraminifera from coastal salt meadows. *Journal of Micropalaeontology*, **16**: 144.

Lena, H. and L'Hoste, S. G. 1975. Foraminiferos de agues salobres (Mar Chiquita, Argentina). *Revista española de micropaleontología*, **7**: 539–48.

Lenihan, H. S. and Micheli, F. 2001. Soft-sediment communities. In Bertness, M. D., Gaines, S. D. and Hay, M. E., eds. *Marine Community Ecology*. Sunderland, MA: Sinauer Associates, pp. 253–87.

Leutenegger, S. 1983. Specific host–symbiont relationship in larger foraminifera. *Micropaleontology*, **29**: 111–25.

Leutenegger, S. 1984. Symbiosis in benthic foraminifera: specificity and host adaptations. *Journal of Foraminiferal Research*, **14**: 16–35.

Leutenegger, S. and Hansen, H. J. 1979. Ultrastructural and radiotracer studies of pore function in foraminifera. *Marine Biology*, **54**: 11–16.

Levin, L. A. 2003. Oxygen minimum zone benthos: adaptation and community response to hypoxia. *Oceanography and Marine Biology Annual Review*, **41**: 1–45.

Levin, L. A., Childers, S. E. and Smith, C. R. 1991. Epibenthic, agglutinating foraminiferans in the Santa Catalina Basin and their response to disturbance. *Deep-Sea Research*, **38**: 465–83.

Levin, L. A., Etter, R. J., Rex, M. A. *et al.* 2001. Environmental influences on regional deep-sea species diversity. *Annual Review of Ecology and Systematics*, **32**: 51–93.

Levin, L. A. and Gooday, A. J. 1992. Possible roles for xenophyophores in deep-sea carbon recycling. In Rowe, G. A. and Pariente, V., eds. *Deep-Sea Food Chains and the Global Carbon Cycle*. Dordrecht: Kluwer, pp. 93–104.

Levy, A. 1991. Peuplements actuels et thanatocénoses á Soritidae et Peneroplidae des Keys de Floride (USA). *Oceanologia acta*, **14**: 515–24.

Li, C., Jones, B. and Blanchon, P. 1997. Lagoon-shelf sediment exchange by storm-evidence from foraminiferal assemblages, east coast of Grand Cayman, British West Indies. *Journal of Sedimentary Research*, **A67**: 17–25.

Li, C., Jones, B. and Kalbfleisch, W. B. C. 1998. Carbonate sediment transport pathways based on foraminifera: a case study from Frank Sound, Grand Cayman, British West Indies. *Sedimentology*, **45**: 109–20.

Li, Q., Davies, P. J., McGowran, B. and Linden, T. van der 2003. Foraminiferal ecostratigraphy of late Oligocene sequences, southeastern Australia: patterns and inferred sea levels at third-order and Milankovitch cycles. *Society of Economic Paleontologists and Mineralogists Special Publication*, **75**: 147–71.

Licari, L. N., Schumacher, S., Wenzhöfer, F., Zabel, M. and Mackensen, A. 2003. Communities and microhabitats of living benthic foraminifera from the tropical east Atlantic: impact of different productivity regimes. *Journal of Foraminiferal Research*, **33**: 10–31.

Lidz, L. 1966. Planktonic foraminifera in the water column of the mainland shelf off Newport Beach, California. *Limnology and Oceanography*, **11**: 257–63.

Linke, P. 1992. Metabolic adaptations of deep-sea benthic foraminifera to seasonally varying food input. *Marine Ecology Progress Series*, **81**: 51–63.

Linke, P., Altenbach, A. V., Graf, G. and Heeger, T. 1995. Response of deep-sea benthic foraminifera to a simulated sedimentation event. *Journal of Foraminiferal Research*, **25**: 75–82.

Linke, P. and Lutze, G. F. 1993. Microhabitat preferences of benthic foraminifera: a static concept or a dynamic adaptation to optimise food acquisition? *Marine Micropaleontology*, **20**: 215–34.

Lipps, J. H. 1982. Biology/paleobiology of foraminifera. In Broadhead, R. W., ed. *Foraminifera. Notes for a Short Course*. New Orleans: Paleontological Society, pp. 1–21.

Lipps, J. H. 1983. Biotic interactions in benthic foraminifera. In Trevesz, M. J. S. and McCall, P. L., eds. *Biotic Interactions in Recent and Fossil Benthic Communities*. New York: Plenum, pp. 331–76.

Lipps, J. H. 1988. Predation on foraminifera by coral reef fish: taphonomic and evolutionary implications. *Palaios*, **3**: 315–26.

Lipps, J. H. and DeLaca, T. E. 1980. Shallow-water foraminiferal ecology, Pacific
    Ocean. In Field, M. E., Bouma, A. H., Colburn, I.P. *et al.*, eds, *Quaternary
    Depositional Environments of the Pacific Coast, Pacific Coast Paleogeography Symposium.*
    Society of Economic Paleontologists and Mineralogists, pp. 325–40.

Lipps, J. H. and Severin, K. P. 1985. *Alveolinella quoyi*, a living fusiform foraminifera, at
    Motupore Island, Papua New Guinea. *Science in New Guinea*, **11**: 126–37.

Lister, J. J. 1895. Contributions to the life history of foraminifera. *Philosophical
    Transactions of the Royal Society of London, Series B*, **186**: 401–53.

Lloyd, J. M. and Evans, J. R. 2002. Contemporary and fossil foraminifera from
    isolation basins in northwest Scotland. *Journal of Quaternary Science*,
    17: 431–43.

Locklin, J. A. and Maddocks, R. F. 1981. Recent foraminifera around petroleum
    production platforms on the southwest Louisiana shelf. *Transactions of the Gulf
    Coast Association of Geological Societies*, **32**: 377–97.

Loeblich, A. R. and Tappan, H. 1964. *Sarcodina, chiefly 'Thecamoebians' and
    Foraminiferida.* In *Treatise on Invertebrate Paleontology, Part C, Protista 2.* Lawrence,
    KA: Geological Society of America and University of Kansas Press.

Loeblich, A. R. and Tappan, H. 1987. *Foraminiferal Genera and Their Classification.* New
    York: Van Nostrand Reinhold, vols. 1 and 2.

Lohrenz, S. E., Wiesenburg, D. A., Arnone, R. A. and Chen, X. 1999. What controls
    primary production in the Gulf of Mexico? In Kumpf, H., Steidinger, K. and
    Sherman, K., eds. *The Gulf of Mexico Large Marine Ecosystem.* Oxford: Blackwell
    Science, pp. 171–95.

Loose, T. L. 1970. Turbulent transport of benthonic foraminifera. *Contributions to the
    Cushman Foundation for Foraminiferal Research*, **21**: 161–7.

Lopez, R. 1979. Algal chloroplasts in the protoplasm of three species of benthic
    foraminifera: taxonomic affinity, viability and persistence. *Marine Biology*, **53**:
    201–11.

Loubere, P. 1989. Bioturbation and sedimentation rate control of benthic microfossil
    taxon abundances in surface sediments: a theoretical approach to the analysis
    of species microhabitats. *Marine Micropaleontology*, **14**: 317–25.

Loubere, P., Gary, A. and Lage, M. 1993. Generation of the benthic foraminiferal
    assemblage: theory and preliminary data. *Marine Micropaleontology*, **20**: 165–81.

Loubere, P., Meyers, P. and Gary, A. 1995. Benthic foraminiferal microhabitat
    selection, carbon isotope values, and association with larger animals: a test
    with *Uvigerina peregrina*. *Journal of Foraminiferal Research*, **25**: 83–95.

Loubere, P. and Qian, H. 1997. Reconstructing paleoecology and
    paleoenvironmental variables using factor analysis and regression: some
    limitations. *Marine Micropaleontology*, **31**: 205–17.

Lueck, K. L. O. and Snyder, S. W. 1997. Lateral variations among populations of
    stained benthic foraminifera in surface sediments of the North Carolina
    continental shelf (USA). *Journal of Foraminiferal Research*, **27**: 20–41.

Lutze, G. F. 1964. Statistical investigations on the variability of *Bolivina argentea* Cushman. *Contributions from the Cushman Foundation for Foraminiferal Research*, **15**: 105–16.

Lutze, G. F. 1965. Zur Foraminiferen-Fauna der Ostsee. *Meyniana*, **15**: 75–142.

Lutze, G. F. 1968. Jahresgang der Foraminiferen-Fauna in der Bottsand-Lagune (westliche Ostsee). *Meyniana*, **18**: 13–30.

Lutze, G. F. 1974a. Foraminiferen der Kieler Bucht (westliche Ostsee). 1."Hausgartengebiet" des Sondersforschungsbereichs 95 der Universität Kiel. *Meyniana*, **26**: 9–22.

Lutze, G. F. 1974b. Benthische Foraminiferen in Oberflächen-Sedimenten des Persischen Golfes. Teil 1: Arten. *"Meteor" Forschungs-Ergebnisse, Reihe C*, **17**: 1–66.

Lutze, G. F. 1980. Depth distribution of benthic foraminifera on the continental margin off NW Africa. *"Meteor" Forschungs-Ergebnisse, Reihe C*, **32**: 31–80.

Lutze, G. F. and Altenbach, A. 1988. *Rupertina stabilis* (Wallich), a highly adapted, suspension-feeding foraminifer. *Meyniana*, **40**: 55–69.

Lutze, G. F. and Altenbach, A. 1991. Technik und Signifikanz der Lebendfarbung benthischer Foraminiferen mit Bengalrot. *Geologisches Jahrbuch*, **A128**: 251–65.

Lutze, G. F. and Coulbourn, W. T. 1984. Recent benthic foraminifera from the continental margin of northwest Africa: community structure and distribution. *Marine Micropaleontology*, **8**: 361–401.

Lutze, G. F., Grabert, B. and Seibold, E. 1971. Lebenbeobachtungen an Groß-Foraminiferen (*Heterostegina*) aus dem Persischen Golf. *"Meteor" Forschungs-Ergebnisse, Reihe C*, **6**: 21–40.

Lutze, G. F., Mackensen, A. and Wefer, G. 1983. Foraminiferen der Kieler Bucht, 2. Salinitätsansprüche von *Eggerella scabra* (Williamson). *Meyniana*, **35**: 55–65.

Lutze, G. F. and Salomon, B. 1987. Foraminiferen-Verbreitung zwischen Norwegen und Grönland: ein Ost-West Profil. *Bericht Sonderforschungsbereich 313, Universität Kiel*, **6**: 69–78.

Lutze, G. F. and Thiel, H. 1989. Epibenthic foraminifera from elevated microhabitats: *Cibicidoides wuellerstorfi* and *Planulina ariminensis*. *Journal of Foraminiferal Research*, **19**: 153–8.

Lutze, G. F. and Wefer, G. 1980. Habitat and asexual reproduction of *Cyclorbiculina compressa* (Orbigny), Soritidae. *Journal of Foraminiferal Research*, **10**: 251–60.

Lynts, G. W. 1966. Variation of foraminiferal standing crop over short lateral distances in Buttonwood Sound, Florida Bay. *Limnology and Oceanography*, **11**: 562–6.

Lynts, G. W. 1971. Distribution and model studies on foraminifera living in Buttonwood Sound, Florida Bay. *Miami Geological Society, Memoir 1*: 73–245.

MacArthur, R. H. 1957. On the relative abundance of bird species. *Proceedings of the National Academy of Science*, **43**: 293–5.

Mackensen, A. 1987. Benthische Foraminiferen auf dem Island-Schottland Rücken: Umwelt-Anzeiger an der Grenze zweier ozeanischer-Raüme. *Paläontologische Zeitschrift*, **61**: 149–79.

Mackensen, A. 2001. Oxygen and carbon stable isotope tracers of Weddell Sea water masses: new data and some paleoceanographic implications. *Deep-Sea Research, I*, **48**: 1401–22.

Mackensen, A. and Douglas, R. G. 1989. Down-core distribution of live and dead deep-water benthic foraminifera in box cores from the Weddell Sea and California continental borderland. *Deep-Sea Research*, **36**: 879–900.

Mackensen, A., Fütterer, D. K., Grobe, H. and Schmiedl, G. 1993a. Benthic foraminiferal assemblages from the South Atlantic Polar Front region between 35 and 57° S: distribution, ecology and fossilization potential. *Marine Micropaleontology*, **22**: 33–69.

Mackensen, A., Grobe, H., Kuhn, G. and Fütterer, D. K. 1990. Benthic foraminiferal assemblages from the eastern Weddell Sea between 68 and 73° S: distribution, ecology and fossilization potential. *Marine Micropaleontology*, **16**: 241–83.

Mackensen, A. and Hald, M. 1988. *Cassidulina teretis* Tappan and *C. laevigata* d'Orbigny: their modern and late Quaternary distribution in northern seas. *Journal of Foraminiferal Research*, **18**: 16–24.

Mackensen, A., Hubberten, H. W., Bickert, T., Fisher, G. and Füttere, D. K. 1993b. The δ^{13}C in benthic foraminiferal tests of *Fontbotia wuellerstorfi* (Schwager) relative to the δ^{13}C of dissolved inorganic carbon in Southern Ocean deep water: implications for glacial ocean circulation models. *Paleoceanography*, **8**: 587–610.

Mackensen, A. and Licari, L. 2003. Carbon isotopes of live benthic foraminifera from the South Atlantic: sensitivity to bottom water carbonate saturation state and organic matter rain rates. In Wefer, G., Mulitza, S. and Ratmeyer, V., eds. *The South Atlantic in the Late Quaternary: reconstruction of material budgets and current systems*. Berlin: Springer-Verlag, pp. 623–44.

Mackensen, A., Schumacher, S., Radke, J. and Schmidt, D. N. 2000. Microhabitat preferences and stable carbon isotopes of endobenthic foraminifera: clue to quantitative reconstruction of oceanic new production? *Marine Micropaleontology*, **40**: 233–58.

Mackensen, A., Sejrup, H. P. and Jansen, E. 1985. The distribution of living benthic foraminifera on the continental slope and rise off southwest Norway. *Marine Micropaleontology*, **9**: 275–306.

MacKenzie, F. T., Kulm, L. D., Cooley, R. L. and Barbhart, J. T. 1965. *Homotrema rubrum* (Lamarck), a sediment transport indicator. *Journal of Sedimentary Petrology*, **35**: 265–72.

Madsen, H. B. and Knudsen, K. L. 1994. Recent foraminifera in shelf sediments of the Scoresby Sund fjord, east Greenland. *Boreas*, **23**: 495–504.

Maiklem, W. R. 1968. Some hydraulic properties of bioclastic carbonate grains. *Sedimentology*, **10**: 101–9.

Maldonado, A. and Murray, J. W. 1975. The Ebro delta, sedimentary environments and development, with comments on the foraminifera. In: Maldonado, A., ed. *Les deltas de la Méditerranée du nord, Excursion 16. IXth International Congress of Sedimentology, Nice, July 1975*. International Association of Sedimentologists, pp. 19–58.

Malmgren, B. A. 1984. Analysis of the environmental influence on the morphology of *Ammonia beccarii* (Linné) in southern European salinas. *Geobios*, **17**: 737–46.

Manley, C. J. and Shaw, S. R. 1997. Geotaxis and phototaxis in *Elphidium crispum* (Protozoa: Foraminiferida). *Journal of the Marine Biological Association of the UK*, **77**: 959–67.

Marshall, P. R. 1976. Some relationships between living and total foraminiferal faunas on Pedro Bank, Jamaica. *Maritime Sediments, Special Publication*, **1**: 61–70.

Marszalek, D. S., Wright, R. C., and Hay, W. W. 1969. Function of the test in foraminifera. *Transactions of the Gulf Coast Association of Geological Societies*, **19**: 341–52.

Martin, P. A., Lea, D. W., Rosenthal, Y. *et al.* 2002a. Quaternary deep sea temperature histories derived from benthic foraminiferal Mg/Ca. *Earth and Planetary Science Letters*, **198**: 193–209.

Martin, R. E. 1986. Habitat and distribution of the foraminifer *Archaias angulatus* (Fichtel and Moll) (Miliolina, Soritidae) northern Florida Keys. *Journal of Foraminiferal Research*, **16**: 201–6.

Martin, R. E. 1988. Benthic foraminiferal zonation in deep-water carbonate platform environments, northern Little Bahama Bank. *Journal of Paleontology*, **62**: 1–8.

Martin, R. E. 1999. *Taphonomy. A Process Approach.* Cambridge: Cambridge University Press.

Martin, R. E. 2000. *Environmental Micropaleontology. The Application of Microfossils to Environmental Geology.* New York: Kluwer.

Martin, R. E., Hippensteel, S. P., Nikitina, D. and Pizzuto, J. E. 2002b. Artificial time-averaging of marsh foraminiferal assemblages linking the temporal scales of ecology and paleoecology. *Paleobiology*, **28**: 263–77.

Martin, R. E. and Liddell, W. D. 1991. The taphonomy of foraminifera in modern carbonate environments: implications for the formation of foraminiferal assemblages. In Donovan S. K., ed. *The Processes of Fossilisation.* London: Belhaven Press, pp. 170–93.

Martin, R. E. and Steinker, D. C. 1973. Evaluation of techniques of recognition of living foraminifera. *Compass*, **50**: 26–30.

Martin, R. E., Wehmiller, J. F., Harris, M. S. and Liddell, W. D. 1996. Comparative taphonomy of bivalves and foraminifera from Holocene tidal flat sediments, Bahia la Choya, Sonora, Mexico (northern Gulf of California): taphonomic grades and temporal resolution. *Paleobiology*, **22**: 80–90.

Martin, R. E. and Wright, R. C. 1988. Information loss in the transition from life to death assemblages of foraminifera in back reef environments, Key Largo, Florida. *Journal of Paleontology*, **62**: 399–410.

Matera, N. J. and Lee, J. J. 1972. Environmental factors affecting the standing crop of foraminifera in sublittoral and psammolittoral communities of a Long Island salt marsh. *Marine Biology*, **14**: 89–103.

Mateu, G. 1968. Contribución al conocimiento de los foraminíferos que sirven de alimento a las *Holoturias. Boletin de la Sociedad de Historia Natural de Baleares*, **14**: 5–17.

Matoba, Y. 1970. Distribution of recent shallow water foraminifera of Matsushima Bay, Miyagi Prefecture, northeast Japan. *Science Reports of the Tohoku University, Sendai, Second Series (Geology)*, **42**: 1–85.

Matoba, Y. 1976a. Foraminifera from off Noshiro, Japan, and postmortem destruction of tests in the Japan Sea. *Progress in Micropaleontology*, 169–89.

Matoba, Y. 1976b. Distribution of foraminifera around the Oga Peninsula. *Reports on the Influence of the Waste from Steel Industries on Environments, Akita Prefecture*, **3**: 182–216. [in Japanese]

Matsushita, S. and Kitazato, H. 1990. Seasonality in the benthic foraminiferal community and the life history of *Trochammina hadai* Uchio in Hamana Lake, Japan. In Hemleben, C., Kaminski, M. A., Kuhnt, W. and Scott, D. B., eds. *Paleoceanography and Taxonomy of Agglutinated Foraminifera*. Dordrecht: Kluwer, pp. 695–715.

Maybury, C. A. 1996. Crevice foraminifer from abyssal south east Pacific manganese nodules. In Moguilevsky, A. and Whatley, R., eds. *Microfossils and Oceanic Environments*. Aberystwyth: Aberystwyth University Press, pp. 281–93.

Maybury, C. A. and Evans, K. R. 1994. Pennsylvanian phylloid alage reinterpreted as shallow-water xenophyophores. *Lethaia*, **27**: 29–33.

Maybury, C. A. and Gwynn, I. A. 1993. Wet processing of recent calcareous foraminifera: methods for preventing dissolution. *Journal of Micropalaeontology*, **12**: 67–9.

McCorkle, D. C., Corliss, B. H. and Farnham, C. A. 1997. Vertical distributions and stable isotope compositions of live (stained) foraminifera from the North Carolina and California continental margins. *Deep-Sea Research I*, **44**: 983–1024.

McCorkle, D. C., Emerson, S. R. and Quay, P. D. 1985. Stable carbon isotopes in marine porewaters. *Earth and Planetary Science Letters*, **74**: 13–26.

McCorkle, D. C. and Keigwin, L. D. 1994. Depth profiles of δ^{13}C in bottom water and core-top *C. wuellerstorfi* on the Ontong-Java Plateau and Emperor Seamounts. *Paleoceanography*, **9**: 197–208.

McCorkle, D. C., Keigwin, L. D., Corliss, B. H. and Emerson, R. R. 1990. The influence of microhabitats on the carbon isotopic composition of deep-sea benthic foraminifera. *Paleoceanography*, **5**: 161–85.

McCormick, J. M., Severin, K. P. and Lipps, J. H. 1994. Summer and winter distribution of foraminifera in Tomales Bay, Northern California. *Cushman Foundation for Foraminiferal Research, Special Publication*, **32**: 69–101.

McGlasson, R. H. 1959. Foraminiferal biofacies around Santa Catalina Island, California. *Micropaleontology*, **5**: 217–40.

McLusky, D. S. and Elliott, M. 2004. *The Estuarine Ecosystem: Ecology, Threats and Management*, 3rd edn. Oxford: Oxford University Press.

Meyer-Reil, L. A. and Köster, M. 1991. Fine-scale distribution of hydrolytic activity associated with foraminiferans and bacteria in deep-sea sediments in the Norwegian–Greenland Sea. *Kieler Meereforschungen*, **8**: 121–6.

Mikhalevich, V. I. 2004. The general aspects of the distribution of Antarctic foraminifera. *Micropaleontology*, **50**: 179–94.

Miller, A. A., Scott, D. B. and Medioli, F. 1982. *Elphidium excavatum* (Terquem): ecophenotypic versus subspecific variation. *Journal of Foraminiferal Research*, **12**: 116–44.

Miller, A. A. L. 1996. Foraminiferal distribution on selected bank areas of the Scotian shelf and Holocene history of inner Chedabucto Bay: foraminiferal evidence; as a contribution to assessment of the marine aggregate potential of the Scotian shelf. *Marine Geological Exploration and Offshore Services, Technical Report*, **9**: 1–57.

Miskov-Nodland, K., Buhl-Mortensen, L. and Høisæter, T. 1999. Has the fauna in the deeper parts of the Skagerrak changed?: a comparison of the present amphipod fauna with observations from 1933/37. *Sarsia*, **84**: 137–55.

Molina-Cruz, A. and Ayala-López, A. 1988. Influence of the hydrothermal vents on the distribution of benthic foraminifera from the Guayamas Basin, Mexico. *Geo-Marine Letters*, **8**: 49–56.

Moodley, L. 1990a. "Squatter" behaviour in soft-shelled foraminifera. *Marine Micropaleontology*, **16**: 149–53.

Moodley, L. 1990b. Southern North Sea seafloor and subsurface distribution of living benthic foraminifera. *Netherlands Journal of Sea Research*, **27**: 51–71.

Moodley, L., Boschker, H. T. S., Middelburg, J. J. *et al.* 2000. Ecological significance of benthic foraminifera: ^{13}C labelling experiments. *Marine Ecology Progress Series*, **202**: 289–95.

Moodley, L., Heip, C. H. R. and Middelburg, J. J. 1998. Benthic activity in sediment of the northwestern Adriatic Sea: sediment oxygen consumption, macro- and microfauna dynamics. *Journal of Sea Research*, **40**: 263–80.

Moodley, L. and Hess, C. 1992. Tolerance of infaunal benthic foraminifera for low and high oxygen conditions. *Biological Bulletin*, **183**: 94–8.

Moodley, L., Schaub, B. E. M., Zwaan, G. J. van der and Herman, P. M. J. 1998a. Tolerance of benthic foraminifera (Protista: Sarcodina) to hydrogen sulphide. *Marine Ecology Progress Series*, **169**: 77–86.

Moodley, L., Zwaan, G. J. van der, Herman, P. M. J., Kempers, L., and Breugel, P. van 1997. Differential response of benthic meiofauna to anoxia with special reference to foraminifera (Protista: Sarcodina). *Marine Ecology Progress Series*, **158**: 151–63.

Moodley, L., Zwaan, G. J. van der, Rutten, G. M. W., Boom, R. C. E. and Kempers, A. J. 1998b. Subsurface activity of benthic foraminifera in relation to porewater oxygen content: laboratory experiments. *Marine Micropaleontology*, **34**: 91–106.

Moore, P. G. 1985. *Cibicides lobatulus* (Protozoa: Foraminifera) epizoic on *Astacilla logicornis* (Crustacea: Isopoda) in the North Sea. *Journal of Natural History*, **19**: 129–33.

Morse, J. W. and Arvidson, R. S. 2002. The dissolution kinetics of major sedimentary carbonate minerals. *Earth Science Reviews*, **58**: 85–119.

Moulinier, M. 1967. Repartition des foraminifères benthiques dans les sediments de la Baie de Seine entre le Cotentin at le meridien de Oistrehem. *Cahiers océanographiques*, **19**: 477–94.

Moulinier, M. 1972. Étude des foraminifères des côtes nord et ouest de Bretagne. *Travaux du Laboratoire de Géologie, École Normale Supérieure, Paris*, **6**: 1–225.

Mudroch, A. and McKnight, S. D. 1994. *Handbook of Techniques for Aquatic Sediments Sampling*, 2nd edn. Boca Raton: Lewis.

Muller, W. A. 1975. Competition for food and other niche-related studies of salt-marsh foraminifera. *Marine Biology*, **31**: 339–51.

Muller, W. A. and Lee, J. J. 1969. Apparent indispensability of bacteria in foraminiferal nutrition. *Journal of Protozoology*, **16**: 471–8.

Mullineaux, L. S. 1987. Organisms living on manganese nodules and crusts: distribution and abundance at three north Pacific sites. *Deep-Sea Research*, **34**: 165–84.

Mullineaux, L. S. 1988a. The role of settlement in structuring a hard-substratum community in the deep sea. *Journal of Experimental Marine Biology and Ecology*, **120**: 247–61.

Mullineaux, L. S. 1988b. Taxonomic notes on large agglutinated foraminifers encrusting manganese nodules, including a description of a new genus, *Chondrodapis* (Komokiacea). *Journal of Foraminiferal Research*, **18**: 46–53.

Mullineaux, L. S. 1989. Vertical distributions of the epifauna on manganese nodules: implications for settlement and feeding. *Limnology and Oceanography*, **34**: 1247–62.

Mullineaux, L. S., Mill, S. W. and Goldman, E. 1988. Recruitment variation during a pilot colonization study of hydrothermal vents (9°50′ N, East Pacific Rise). *Deep-Sea Research II*, **45**: 441–64.

Murosky, M. W. and Snyder, S. W. 1994. Vertical distribution of stained benthic foraminifera in sediments of southern Onslow Bay, North Carolina continental shelf. *Journal of Foraminiferal Research*, **24**: 158–70.

Murray, J. W. 1963. Ecological experiments on Foraminiferida. *Journal of the Marine Biological Association of the UK*, **43**: 621–42.

Murray, J. W. 1965a. On the Foraminiferids of the Plymouth region. *Journal of the Marine Biological Association, UK*, **45**: 481–505.

Murray, J. W. 1965b. The Foraminiferida of the Persian Gulf. Part 1. *Rosalina adhaerens sp. nov. Annals and Magazine of Natural History, Series 13*, **8**: 77–9.

Murray, J. W. 1965c. The Foraminiferida of the Persian Gulf. 2. The Abu Dhabi region. *Palaeogeography, Palaeoclimatology, Palaeoecology*, **1**: 307–32.

Murray, J. W. 1965d. Significance of benthic foraminiferids in plankton samples. *Journal of Paleontology*, **39**: 156–7.

Murray, J. W. 1966a. The Foraminiferida of the Persian Gulf. 3. The Halat al Bahrani region. *Palaeogeography, Palaeoclimatology, Palaeoecology*, **2**: 59–68.

Murray, J. W. 1966b. The Foraminiferida of the Persian Gulf. 4. Khor al Bazam. *Palaeogeography, Palaeoclimatology, Palaeoecology*, **2**: 153–69.

Murray, J. W. 1966c. The Foraminiferida of the Persian Gulf. 5. The shelf off the Trucial Coast. *Palaeogeography, Palaeoclimatology, Palaeoecology*, **2**: 267–78.

Murray, J. W. 1967a. Production in benthic foraminiferids. *Journal of Natural History*, **1**: 71–68.

Murray, J. W. 1967b. Transparent and opaque foraminiferid tests. *Journal of Paleontology*, **41**: 791.

Murray, J. W. 1968a. Living foraminifers of lagoons and estuaries. *Micropaleontology*, **14**: 435–55.

Murray, J. W. 1968b. The living Foraminiferida of Christchurch harbour, England. *Micropaleontology*, **14**: 83–96.

Murray, J. W. 1969. Recent foraminifers of the Atlantic continental margin of the United States. *Micropaleontology*, **15**: 401–9.

Murray, J. W. 1970a. Foraminifers of the western approaches to the English Channel. *Micropaleontology*, **16**: 471–85.

Murray, J. W. 1970b. The Foraminiferida of the Persian Gulf 6. Living forms in the Abu Dhabi region. *Journal of Natural History*, **4**: 55–67.

Murray, J. W. 1970c. The Foraminifera of the hypersaline Abu Dhabi lagoon, Persian Gulf. *Lethaia*, **3**: 51–68.

Murray, J. W. 1973. *Distribution and Ecology of Living Benthic Foraminiferids*. London: Heinemann.

Murray, J. W. 1976a. Comparative studies of living and dead benthic foraminiferal distributions. In Hedley, R. H. and Adams, C. G., eds. *Foraminifera*: **2**, 45–109.

Murray, J. W. 1979. Recent benthic foraminiferids of the Celtic Sea. *Journal of Foraminiferal Research*, **9**: 193–209.

Murray, J. W. 1980. The Foraminifera of the Exe estuary. *Devonshire Association Special Volume*, **2**: 89–115.

Murray, J. W. 1982. Benthic foraminifera: the validity of living, dead or total assemblages for the interpretation of palaeoecology. *Journal of Micropalaeontology*, **1**: 137–40.

Murray, J. W. 1983. Population dynamics of benthic foraminifera: results from the Exe estuary, England. *Journal of Foraminiferal Research*, **13**: 1–12.

Murray, J. W. 1984. Paleogene and Neogene benthic foraminifers from Rockall Plateau. *Initial Reports of the Deep Sea Drilling Project*, **81**: 503–34.

Murray, J. W. 1985. Recent foraminifera from the North Sea (Forties and Ekofisk areas) and the continental shelf west of Scotland. *Journal of Micropalaeontology*, **4**: 117–25.

Murray, J. W. 1986. Living and dead Holocene foraminifera of Lyme Bay, southern England. *Journal of Foraminiferal Research*, **16**: 347–52.

Murray, J. W. 1987a. Benthic foraminiferal assemblages: criteria for the distinction of temperate and subtropical environments. In Hart, M. B., ed. *Micropalaeontology of Carbonate Environments*. Chichester: Ellis Horwood, British Micropalaeontological Society, pp. 9–20.

Murray, J. W. 1987b. Biogenic indicators of suspended sediment transport in marginal marine environments: quantitative examples from SW Britain. *Journal of the Geological Society, London*, **144**: 127–33.

Murray, J. W. 1991. *Ecology and Palaeoecology of Benthic Foraminifera*. Harlow: Longman.

Murray, J. W. 1992. Distribution and population dynamics of benthic foraminifera from the southern North Sea. *Journal of Foraminiferal Research*, **22**: 114–28.

Murray, J. W. 2000a. The enigma of the continued use of total assemblages in ecological studies of benthic foraminifera. *Journal of Foraminiferal Research*, **30**: 244–45; 2002, *ibid.*, 32: 200; 2002 (erratum).

Murray, J. W. 2000b. When does environmental variability become environmental change? The proxy record of benthic foraminifera. In Martin, R. E., ed. *Environmental Micropaleontology. The Application of Microfossils to Environmental Geology*. New York: Kluwer, pp. 7–37.

Murray, J. W. 2001. The niche of benthic foraminifera, critical thresholds and proxies. *Marine Micropaleontology*, **41**: 1–7.

Murray, J. W. 2003a. Patterns in the cumulative increase in species from foraminiferal time-series. *Marine Micropaleontology*, **48**: 1–21.

Murray, J. W. 2003b. Foraminiferal assemblage formation in depositional sinks on the continental shelf margin west of Scotland. *Journal of Foraminiferal Research*, **33**: 101–21.

Murray, J. W. 2003c. An illustrated guide to the benthic foraminifera of the Hebridean shelf, west of Scotland, with notes on their mode of life. *Palaeontologia electronica*, **5** (issue 2) article 1 (1.4 Mb). http//www-odp.tamu.edu/paleo/2002_2/guide/issue2_02.htm.

Murray, J. W. 2004. The Holocene palaeoceanographic history of Muck Deep, Hebridean shelf, Scotland: has there been a change of wave climate in the past 12 000 years? *Journal of Micropalaeontology*, **23**: 153–61.

Murray, J. W. and Alve, E. 1993. The habitat of the foraminifer *Paratrochammina (Lepidoparatrochammina) haynesi*. *Journal of Micropalaeontology*, **12**: 34.

Murray, J. W. and Alve, E. 1999a. Natural dissolution of shallow water benthic foraminifera: taphonomic effects on the palaeoecological record. *Palaeogeography, Palaeoclimatology, Palaeoecology*, **146**: 195–209.

Murray, J. W. and Alve, E. 1999b. Taphonomic experiments on marginal marine foraminiferal assemblages: how much ecological information is preserved? *Palaeogeography, Palaeoclimatology, Palaeoecology*, **149**: 183–97.

Murray, J. W. and Alve, E. 2000a. Major aspects of foraminiferal variability (standing crop and biomass) on a monthly scale in an intertidal zone. *Journal of Foraminiferal Research*, **30**: 177–91.

Murray, J. W. and Alve, E. 2000b. Do calcareous dominated shelf foraminiferal assemblages leave worthwhile ecological information after their dissolution? *Proceedings of the Fifth International Workshop on Agglutinated Foraminifera, Plymouth, England, September 1997*. Grzybowski Foundation Special Publication, 7: 311–31.

Murray, J. W., Alve, E. and Cundy, A. 2003. The origin of modern agglutinated foraminiferal assemblages: evidence from a stratified fjord. *Estuarine, Coastal and Shelf Science*, **58**: 677–97.

Murray, J. W. and Bowser, S. S. 2000. Mortality, protoplasm decay rate, and reliability of staining techniques to recognize 'living' foraminifera: a review. *Micropaleontology*, **30**: 66–77.

Murray, J. W. and Hawkins, A. B. 1976. Sediment transport in the Severn estuary during the past 8000–9000 years. *Journal of the Geological Society, London*, **132**: 385–98.

Murray, J. W. and Pudsey, C. J. 2004. Living (stained) and dead foraminifera from the newly ice-free Larsen ice shelf, Weddell Sea, Antarctica: ecology and taphonomy. *Marine Micropaleontology*, **53**: 67–81.

Murray, J. W., Sturrock, S., and Weston, J. F. 1982. Suspended load transport of foraminiferal tests in a tide- and wave-swept sea. *Journal of Foraminiferal Research*, **12**: 51–65.

Murray, J. W., Weston, J. F. and Sturrock, S. 1983. Sedimentary indicators of water movement in the western approaches to the English Channel. *Continental Shelf Research*, **1**: 339–52.

Murray, J. W. and Wright, C. A. 1970. Surface textures of calcareous foraminiferids. *Palaeontology*, **13**: 184–7.

Murray, W. G. and Murray, J. W. 1987. A device for obtaining representative samples from the sediment-water interface. *Marine Geology*, **76**: 313–17.

Myers, E. H. 1935a. The life history of *Patellina corrugata* Williamson, a foraminifer. *Bulletin of the Scripps Institution of Oceanography of the University of California, Technical Series*, **3**: 355–92.

Myers, E. H. 1935b. Morphogenesis of the test and the biological significance of dimorphism in the foraminifer *Patellina corrugata* Williamson. *Bulletin of the Scripps Institution of Oceanography of the University of California, Technical Series*, **3**: 393–404.

Myers, E. H. 1936. The life cycle of *Spirillina vivipara* Ehrenberg, with notes on morphogenesis, systematics and the distribution of foraminifera. *Journal of the Royal Microscopical Society*, **56**: 120–46.

Myers, E. H. 1942. A quantitative study of the productivity of foraminifera in the sea. *Proceedings of the American Philosophical Society*, **85**: 325–42.

Myers, E. H. 1943. Life activities of foraminifera in relation to marine ecology. *Proceedings of the American Philosophical Society*, **86**: 439–58.

Naidu, P. D. and Malmgren, B. A. 1995. Do benthic foraminifer records represent a productivity index in oxygen minimum zone areas? An evaluation from the Oman margin, Arabian Sea. *Marine Micropaleontology*, **26**: 49–55.

Naidu, T. Y. and Subba Rao, M. 1988. Foraminiferal ecology of Bendi lagoon, east coast of India. *Revue de paléobiologie, volume spécial*, **2**: 851–8.

Nees, S. and Struck, U. 1999. Benthic foraminiferal response to major oceanographic changes. In Abrantes, F. and Mix, A., eds. *Reconstructing Ocean History: a Window into the Future*. New York: Kluwer, pp. 195–216.

Newton, A. C. and Rowe, G. T. 1995. The abundance of calcareous foraminifera and other meiofauna at a time series station in the Northeast Water Polynya, Greenland. *Journal of Geophysical Research*, **100**: 4423–38.

Nichols, M. and Norton, W. 1969. Foraminiferal populations in a coastal plain estuary. *Palaeogeography, Palaeoclimatology, Palaeoecology*, **6**: 197–213.

Nielsen, K. S. S. 1999. Foraminiferivory revisited: a preliminary investigation of holes in foraminifera. *Bulletin of the Geological Society of Denmark*, **45**: 139–42.

Nielsen, K. S. S., Collen, J. D. and Ferland, M. A. 2002. *Floresina*: a genus of predators, parasites or scavengers? *Journal of Foraminiferal Research*, **32**: 93–5.

Nigam, R. 1984. Living benthonic foraminifera in a tidal environment: Gulf of Khambhat (India). *Marine Geology*, **58**: 415–25.

Nixon, S. W. 1995. Coastal marine eutrophication: a definition, social causes, and future concerns. *Ophelia*, **41**: 199–219.

Nordberg, K., Gustafsson, M. and Krantz, A. L. 1999. Decreasing oxygen concentrations in the Gullmar Fjord, Sweden, as confirmed by benthic foraminifera, and the possible association with NAO. *Journal of Marine Systems*, **23**: 303–16.

Nordberg, K., Gustafsson, M. and Krantz, A. L. 2000. Decreasing oxygen concentrations in the Gullmar Fjord, Sweden, as confirmed by benthic foraminifera, and possible association with NAO. *Journal of Marine Systems*, **23**: 303–16.

Nyholm, K. G. 1961. Morphogenesis and biology of the foraminifer *Cibicides lobatulus*. *Zoologiska bidrag från Uppsala*, **33**: 157–96.

Ohga, T. and Kitazato, H. 1997. Seasonal changes in bathyal foraminiferal populations in response to the flux of organic matter (Sagami Bay, Japan). *Terra Nova*, **9**: 33–7.

Olsson, I. 1976. Distribution and ecology of the foraminiferan *Ammotium cassis* (Parker) and its ecological significance. *Zoon*, **5**: 11–14.

Osenberg, C. W. and Schmitt, R. J. 1996. Detecting ecological impacts caused by human activities. In Schmitt, R. J. and Osenberg, C. W., eds. *Detecting Ecological Impacts: Concepts and Applications in Coastal Habitats*. San Diego: Academic, pp. 3–16.

Ozarko, D. L., Patterson, R. T. and Williams, H. F. L. 1997. Marsh foraminifera from Nanaimo, British Columbia (Canada): implications of infaunal habitat and taphonomic biasing. *Journal of Foraminiferal Research*, **27**: 51–68.

Palmer, M. A. 1988. Epibenthic predators and marine meiofauna: separating predation, disturbance, and hydrodynamic effects. *Ecology*, **69**: 1251–9.

Parker, F. L. 1954. Distribution of the foraminifera in the northwest Gulf of Mexico. *Bulletin of the Museum of Comparative Zoology*, **111**: 453–588.

Parker, W. C. and Arnold, A. J. 1999. Quantitative methods of data analysis in foraminiferal ecology. In Sen Gupta, B. K., ed. *Modern Foraminifera*. Dordrecht: Kluwer, pp. 71–89.

Patterson, R. T. and Fishbein, E. 1989. Re-examination of the statistical methods used to determine the number of point counts needed for micropaleontological quantitative research. *Journal of Paleontology*, **63**: 245–8.

Patterson, R. T., Gehrels, W. R., Belknap, D. F. and Dalby, A. P. 2004. The distribution of salt marsh foraminifera at Little Dipper harbour New Brunswick, Canada: implications for development of widely applicable transfer functions in sea-level research. *Quaternary International*, **120**: 185–94.

Patterson, R. T., Guilbault, J. P. and Clague, J. J. 1999. Taphonomy of tidal marsh foraminifera: implications of surface sample thickness for high-resolution sea-level studies. *Palaeogeography, Palaeoclimatology, Palaeoecology*, **149**: 199–211.

Patterson, R. T., Guilbault, J. P. and Thomson, R. E. 2000. Oxygen level control on foraminiferal distribution in Effingham Inlet, Vancouver Island, British Columbia, Canada. *Journal of Foraminiferal Research*, **30**: 321–35.

Patterson, R. T. and McKillop, W. B. 1991. Distribution and possible paleoecological significance of *Annectina viriosa*, a new species of agglutinated foraminifera from nonmarine salt ponds in Manitoba. *Journal of Paleontology*, **65**: 33–7.

Patterson, R. T. and McKillop, W. B. 1997. Evidence for rapid avian-mediated foraminiferal colonization of Lake Winnipegosis, Manitoba, during the Holocene hypsithermal. *Journal of Paleolimnology*, **18**: 131–43.

Patterson, R. T., Scott, D. B. and McKillop, W. B. 1990. Recent marsh-type agglutinated foraminifera from inland salt springs, Manitoba, Canada. In Hemleben, C., Kaminski, M. A., Kuhnt, W. and Scott, D. B., eds. *Paleoecology, Biostratigraphy, Paleoceanography and Taxonomy of Agglutinated Foraminifera*. Dordrecht: Kluwer, pp. 765–81.

Pawlowski, J. 2000. Introduction to the molecular systematics of foraminifera. *Micropaleontology*, **46** (Supplement 1): 1–12.

Pawlowski, J., Bolivar, I., Farhni, J. and Zaninetti, L. 1995. DNA analysis of '*Ammonia beccarii*' morphotypes: one or more species? *Marine Micropaleontology*, **26**: 171–8.

Pawlowski, J., Holzmann, M., Fahrni, J. and Richardson, S. L. 2003. Small subunit ribosomal DNA suggests that the Xenophyophorean *Syringammina corbicula* is a foraminiferan. *Journal of Eukaryotic Microbiology*, **50**: 483–7.

Pawlowski, J., Vargas, C. de, Fahrni, J. and Bowser, S. 1998. *Reticulomyxa filoda* – an 'athalamiid' freshwater foraminifer. *Forams, '98, International Symposium on Foraminifera*, Sociedad Mexicana de Paleontologia, AC, Special Publication: 78–9.

Pecheux, M. J. F. 1994. Ecomorphology of a recent larger foraminifer, *Operculina ammonoides*. *Geobios*, **28**: 529–66.

Peebles, M. W., Hallock, P. and Hine, A. C. 1997. Benthic foraminiferal assemblages from current-swept carbonate platforms off the northern Nicaraguan Rise, Caribbean Sea. *Journal of Foraminiferal Research*, **27**: 42–50.

Peebles, M. W. and Lewis, R. D. 1991. Surface textures of benthic foraminifera from San Salvador, Bahamas. *Journal of Foraminiferal Research*, **21**: 285–92.

Petrucci, F., Medioli, F. S., Scott, D. B., Pianetti, F. A. and Cavazzini, R. 1983. Evalutation of the usefulness of foraminifera as sea level indicators in the Venice lagoon (N. Italy). *Acta naturalia de l'ateneo parmense*, **19**: 63–77.

Phleger, F. P. 1951. Ecology of foraminifera, northwest Gulf of Mexico. Part 1. Foraminiferal distribution. *Geological Society of America, Memoir*, **46**: 1–88.

Phleger, F. P. 1955. Ecology of foraminifera in southeastern Mississippi delta area. *Bulletin of the Association of Petroleum Geologists*, **39**: 712–52.

Phleger, F. P. 1956. Significance of living foraminiferal populations along the central Texas coast. *Contributions from the Cushman Foundation for Foraminiferal Research*, **7**: 106–51.

Phleger, F. P. 1960a. *Ecology and Distribution of Recent Foraminifera*. Baltimore: John Hopkins.

Phleger, F. P. 1960b. Foraminiferal populations of Laguna Madre, Texas. *Science Reports of the Tohoku University, Sendai, Japan, Series 2, Special Volume* 4: 83–91.

Phleger, F. P. 1960c. Sedimentary patterns of microfaunas in northern Gulf of Mexico. In Shepard, F. P., Phleger, F. B. and Andel, T. H. van, eds. *Recent Sediments, Northwest Gulf of Mexico*. Tulsa: American Association of Petroleum Geologists, pp. 267–301.

Phleger, F. P. 1964. Patterns of living benthonic foraminifera, Gulf of California. *Memoir of the American Association of Petroleum Geologists*, **3**: 377–94.

Phleger, F. P. 1965a. Patterns of marsh foraminifera, Galveston Bay, Texas. *Limnology and Oceanography*, **10** (Supplement): R169–84.

Phleger, F. P. 1965b. Living foraminifera from coastal marsh, southwestern Florida. *Boletin de la Sociedad Geologica Mexicana*, **28**: 45–60.

Phleger, F. P. 1965c. Sedimentology of Guerrero Negro Lagoon, Baja California, Mexico. In: Whittard, W. F. and Bradshaw, R., eds. *Submarine Geology and Geophysics. Colston Papers*, 205–35.

Phleger, F. P. 1966. Patterns of living marsh foraminifera in South Texas coastal lagoons. *Boletin de la Sociedad Geologica Mexicana*, **28**: 1–44.

Phleger, F. P. 1967. Marsh foraminiferal patterns, Pacific coast of North America. *Cienca del mar y limnologia Mexico*, **1**: 11–38.

Phleger, F. P. 1970. Foraminiferal populations and marsh processes. *Limnology and Oceanography*, **15**: 522–34.

Phleger, F. P. 1976a. Benthic foraminiferids as indicators of organic production in marginal marine areas. *Maritime Sediments Special Publication*, **1A**: 107–17.

Phleger, F. P. 1976b. Foraminiferal and ecological processes in St. Lucia lagoon, Zululand. *Maritime Sediments Special Publication*, **1A**: 195–204.

Phleger, F. P. and Ayala-Castañares, A. 1969. Marine geology of Topolobampo lagoons, Sinaloa, Mexico. In: Ayala-Castañares, A. and Phleger, F. P., eds. *Lagunas costeras, un simposio*. Mexico City: Universidad Nacional Autonoma de Mexico, pp. 101–36.

Phleger, F. P. and Bradshaw, J. S. 1966. Sedimentary environments in a marine marsh. *Science*, **154**: 1551–3.

Phleger, F. P. and Ewing, G. C. 1962. Sedimentology and oceanography of coastal lagoons in Baja California, Mexico. *Bulletin of the Geological Society of America*, **73**: 145–82.

Phleger, F. P. and Lankford, R. R. 1957. Seasonal occurrences of living benthonic foraminifera in some Texas bays. *Contributions from the Cushman Foundation for Foraminiferal Research*, **8**: 93–105.

Phleger, F. P. and Lankford, R. R. 1978. Foraminiferal and ecological processes in the Alvarado Lagoon area, Mexico. *Journal of Foraminiferal Research*, **8**: 127–31.

Phleger, F. P. and Soutar, A. 1973. Production of benthic foraminifera in three east Pacific oxygen minimum zones. *Micropaleontology*, **19**: 110–15.

Pielou, E. C. 1966. The measurement of diversity in different types of biological collections. *Journal of Theoretical Biology*, **13**: 131–44.

Pielou, E. C. 1974. *Population and Community Ecology: Principles and Methods*. New York: Gordon and Breach.

Pike, J., Bernhard, J. M, Moreton, S. and Butler, I. B. 2001. Microbioirrigation of marine sediments in dysoxic environments: implications for early sediment fabric formation and diagenetic processes. *Geology*, **29**: 923–6.

Platon, E. and Sen Gupta, B. K. 2001. Benthic foraminiferal communities in oxygen-depleted environments of the Lousiana continental shelf. In Rabalais, N. N. and Turner, R. E., eds. *Coastal Hypoxia: Consequences for Living Resources and Ecosystems*. American Geophysical Union, IV Series. V Series: Coastal and Estuarine Studies, 58: 147–64.

Poag, C. W. 1976. The foraminiferal community of San Antonia Bay. In Bouma, A. H., ed. *Shell Dredging and its Influence on Gulf Coast Environments*. Houston: Houston Publishing Co., pp. 304–36.

Poag, C. W. 1978. Paired foraminiferal ecophenotypes in Gulf Coast estuaries: ecological and palaeoecological implications. *Transactions of the Gulf Coast Association of Geological Societies*, **28**: 395–421.

Poag, C. W., Knebel, H. J. and Todd, R. 1980. Distribution of modern benthic foraminifers on the New Jersey outer continental shelf. *Marine Micropaleontology*, **5**: 43–69.

Poag, C. W. and Tresslar, R. C. 1979. Habitat associations among living foraminifers of West Flower Garden Bank, Texas continental shelf. *Transactions of the Gulf Coast Association of Geological Societies*, **29**: 347–51.

Poag, C. W. and Tresslar, R. C. 1981. Living foraminifers of West Flower Garden Bank, northernmost coral reef in the Gulf of Mexico. *Micropaleontology*, **27**: 31–70.

Polyak, L., Stanovoy, V. and Lubinski, D. J. 2003. Stable isotopes in benthic foraminiferal calcite from a river-influenced Arctic marine environment, Kara and Pechora seas. *Paleoceanography*, **18**: 3-1 to 3-17.

Pranovi, F. and Serandrei-Barbero, R. 1994. Benthic communities of Northern Adriatic area subject to anoxic conditions. *Memorie di scienze geologiche*, **46**: 79–92.

Rabalais, N. N. and Turner, R. E. 2001. Hypoxia in the northern Gulf of Mexico: description, causes and change. In Rabalais, N. N. and Turner, R. E., eds. *Coastal Hypoxia: Consequences for Living Resources and Ecosystems*. American Geophysical Union, IV Series V Series. Coastal and Estuarine Studies, 58: 1–36.

Radford, S. S. 1976a. Recent foraminifera from Tobago Island, West Indies. *Revista española de micropaleontología*, **8**: 193–218.

Radford, S. S. 1976b. Depth distribution of recent foraminifera in selected bays, Tobago Island, West Indies. *Revista española de micropaleontología*, **8**: 219–38.

Radford, S. S. 1995. Foraminifera from the southern Caribbean – Atlantic Province. *Transactions of the 3rd Geological Conference of the Geological Society of Trinidad and Tobago and the 14th Caribbean Geological Conference*, **1**: 163–78.

Rainer, S. F. 1992. Diet of prawns from the continental slope of North-Western Australia. *Bulletin of Marine Science*, **50**: 258–74.

Rao, D. C., Rao, M. S., Kaladhar, R. and Naidu, T. Y. 1982. Living foraminifera associated with algae from rock pools near Visakhapatnam, east coast of India. *Indian Journal of Marine Sciences*, **11**: 212–19.

Rasheed, D. A. and Ragothaman, V. 1977. Ecology and distribution of recent foraminifera from the Bay of Bengal off Porto-Novo, Tamil Nadu State, India. *Proceedings of the VII Indian Colloquium on Micropalaeontology and Stratigraphy*, Madras, India, pp. 263–98.

Rathburn, A. E. and Corliss, B. H. 1994. The ecology of living (stained) deep-sea benthic foraminifera from the Sulu Sea. *Paleoceanography*, **9**: 87–150.

Rathburn, A. E., Corliss, B. H., Tappa, K. D. and Lohmann, K. C. 1996. Comparisons of the ecology and stable isotopic compositions of living (stained) benthic foraminifera from the Sulu and South China seas. *Deep-Sea Research I*, **43**: 1617–46.

Rathburn, A. E., Levin, L. A., Held, Z. and Lohmann, K. C. 2000. Benthic foraminifera associated with cold methane seeps on the northern California margin: ecology and stable isotopic compositions. *Marine Micropaleontology*, **38**: 247–66.

Rathburn, A. E. and Miao, Q. 1995. The taphonomy of deep-sea benthic foraminifera: comparisons of living and dead assemblages from box and gravity cores taken in the Sulu Sea. *Marine Micropaleontology*, **25**: 127–49.

Rathburn, A. E., Pérez, M. E., Martin, J. B. *et al.* 2003. Relationships between the distribution and stable isotopic composition of living benthic foraminifera and cold methane seep biogeochemistry in Monterey Bay, California. *Geochemistry, Geophysics and Geosystems*, **4**: 1106, doi:10.1029/2003GC000595.

Reddy, K. R and Jagadiswara Rao, R. 1983. Diversity and dominance of living and total assemblages, Pennar estuary, Andhra Pradesh. *Journal of the Geological Society of India*, **24**: 594–603.

Reidenauer, J. A. 1989. Sand-dollar *Mellita quinquiesperforata* (Leske) burrow trails: sites of harpacticoid disturbance and nematode attraction. *Journal of Experimental Marine Biology and Ecology*, **130**: 223–36.

Reimers, C. E., Lange, C. B., Tabak, M. and Bernhard, J. M. 1990. Seasonal spillover and varve formation in the Santa Barbara Basin, California. *Limnology and Oceanography*, **35**: 1577–85.

Reinöhl-Kompa, S. 1985. Zur Populationsökologie der Foraminiferen im nordöstlichen Ems-Dollart-Ästuar, südliche Nordsee. *Senckenbergiana maritima*, **17**: 147–61.

Reiss, Z. and Hottinger, L. 1984. *The Gulf of Aqaba, Ecological Micropaleontology. Ecological Studies*, vol. 50. Berlin: Springer-Verlag.

Renema, W. 2002. Larger foraminifera as marine environmental indicators. *Scripta Geologica*, **124**: 1–263.

Renema, W. 2003. Larger foraminifera on reefs around Bali (Indonesia). *Zoologische Verhandelingen Leiden*, **345**: 337–66.

Renema, W., Hoeksema, B. W. and Hinte, J. E. van 2001. Larger benthic foraminifera and their distribution patterns on the Spermonde shelf, south Sulawesi. *Zoologische Verhandelingen Leiden*, **334**: 115–49.

Renema, W. and Troelstra, S. 2001. Larger foraminifera distribution on a mesotrophic carbonate shelf in SW Sulawesi (Indonesia). *Paleogeography, Palaeoclimatology, Palaeoecology*, **175**: 125–46.

Resig, J. M. 1958. Ecology of foraminifera of the Santa Cruz Basin, California. *Micropaleontology*, **4**: 287–308.

Resig, J. M. 1974. Recent foraminifera from a landlocked Hawaiian lake. *Journal of Foraminiferal Research*, **4**: 69–76.

Resig, J. M. and Glenn, C. R. 1997. Foraminifera encrusting phosphoritic hardgrounds of the Peruvian upwelling zone: taxonomy, geochemistry, and distribution. *Journal of Foraminiferal Research*, **27**: 133–50.

Rhumbler, L. 1935. Rhizopoden der Kieler Bucht, gesammelt durch R. Remane. I Teil. *Schriften des Naturwissenschaftlichen Vereins für Schleswig-Holstein*, **21**: 143–94.

Ribes, T., Salvado, H., Romero, J. and Gracia, M. P. 2000. Foraminiferal colonization on artificial seagrass leaves. *Journal of Foraminiferal Research*, **20**: 192–201.

Richardson, S. L. 2001. *Syringammina corbicula* sp. nov. (Xenophyophorea) from the Cape Verde Plateau, E. Atlantic. *Journal of Foraminiferal Research*, **31**: 201–9.

Richardson, S. L. and Rützler, K. 1999. Bacterial endosymbionts in the agglutinating foraminifera *Spiculodendron corallicolun* Rützler and Richardson, 1996. *Symbiosis*, **26**: 299–312.

Richter, G. 1961. Beobachtungen zur Ökologie einiger Foraminiferen des Jade-Gebietes. *Natur und Museum*, **91**: 163–70.

Richter, G. 1964a. Zur Ökologie der Foraminifera I: Die Foraminiferen-Gesellschaften der Jadesgebietes. *Natur und Museum*, **94**: 343–53.

Richter, G. 1964b. Zur Ökologie der Foraminifera II: Lebensraum und Lebenweise von *Nonion depressulum, Elphidium excavatum* und *Elphidium selseyense*. *Natur und Museum*, **94**: 421–30.

Richter, G. 1965. Zur Ökologie der Foraminifera III: Verdriftung und Transport in der Gezeitenzone. *Natur und Museum*, **95**: 51–62.

Riemann, F. 1983. Biological aspects of deep-sea manganese nodule formation. *Oceanlogica acta*, **6**: 303–11.

Rivkin, R. B. and DeLaca, T. E. 1990. Trophic dynamics in Antarctic benthic communities. I. *In situ* ingestion of microalgae by foraminifera and metazoan meiofauna. *Marine Ecology Progress Series*, **64**: 129–36.

Rohling, E. J. and Cooke, S. 1999. Stable oxygen and carbon isotopes in foraminiferal carbonate shells. In Sen Gupta, B. K., ed. *Modern Foraminifera*. Dordrecht: Kluwer, pp. 239–58.

Rosoff, D. B. and Corliss, B. H. 1992. An analysis of recent deep-sea benthic foraminiferal morphotypes from the Norwegian–Greenland seas. *Palaeogeography, Palaeoclimatology, Palaeoecology*, **91**: 13–20.

Ross, C. A. 1972. Biology and ecology of *Marginopora vertebralis* (Foraminiferida), Great Barrier Reef. *Journal of Protozoology*, **19**: 181–92.

Rosset-Moulinier, M. 1972. Etude des foraminifères des côtes nord et ouest de Bretagne. *Travaux du laboratoire de géologie, Ecole Normale Supérieur, Paris*, **6**: 1–225.

Rosset-Moulinier, M. 1976. Répartition des populations de foraminifères benthiques dans le golfe normand-breton at la baie de Seine. *Annales de l'Institut Océanographique, NS*, **54**: 107–26.

Rosset-Moulinier, M. 1986. Les populations de foraminifères benthiques de la Manche. *Cahiers de biologie marine*, **27**: 387–440.

Rottgardt, D. 1952. Mikropaläontologisch wichtige Bestandteile rezenter brackisher Sedimente an den Küsten Schleswig-Holstein. *Meyniana*, **1**: 169–228.

Röttger, R. 1974. Larger foraminifera: reproduction and early stages of development in *Heterostegina depressa*. *Marine Biology*, **26**: 5–12.

Röttger, R. 1976. Ecological observations on *Heterostegina depressa* (Foraminifera, Nummulitidae) in the laboratory and in its natural habitat. *Maritime Sediments, Special Publication*, **1**: 75–80.

Röttger, R. 1978. Unusual multiple fission in the gamont of the larger foraminifera *Heterostegina depressa*. *Journal of Protozoology*, **25**: 41–4.

Röttger, R. 1981. Vielteilung bei Großforaminiferen. *Mikrokosmos*, **70**: 137–8.

Röttger, R. 1990. Biology of larger foraminifera: present status of the hypothesis of trimorphism and ontogeny of the gamont of *Heterostegina depressa*. In Takayanagi, Y. and Saito, T., eds., *Proceedings of the Fourth International Symposium on Benthic Foraminifera Sendai, 1990*. Tokyo: Tokai University, pp. 43–54.

Röttger, R. and Berger, W. H. 1972. Benthic foraminifera: morphology and growth in clone culture of *Heterostegina depressa*. *Marine Biology*, **15**: 89–94.

Röttger, R., Irwan, A., Schmaljohann, R. and Franzisket, L. 1980. Growth of *Amphistegina lessonii* d'Orbigny and *Heterostegina depressa* d'Orbigny (Protozoa). In Schwemmler, W. and Schenk, H. E. A., eds. *Endocytobiology I*. Berlin: Walter der Gruyter, pp. 125–32.

Röttger, R. and Kruger, R. 1990. Observations on the biology of Calcarinidae (Foraminiferida). *Marine Biology*, **106**: 419–25.

Röttger, R., Kruger, R. and de Rijk, S. 1990a. Trimorphism in foraminifera (Protozoa): verification of an old hypothesis. *European Journal of Protistology*, **25**: 226–8.

Röttger, R., Krüger, R. and de Rijk, S. 1990b. Larger foraminifera: variation in outer morphology and prolocular size in *Calcarina gaudichaudii*. *Journal of Foraminiferal Research*, **20**: 170–4.

Röttger, R. and Schmaljohann, R. 1976. Foraminifera: Gamogonie, Teil des Entwicklungsgangs der rezenten Nummulitidae *Heterostegina depressa*. *Die Naturwissenschaften*, **10**: 486.

Röttger, R., Spindler, M., Schmaljohann, R., Richwien, M. and Fladanung, M. 1984. Functions of the canal system in the rotaliid foraminifer *Heterostegina depressa*. *Nature*, **309**: 789–91.

Rudnick, D. T. 1989. Time lags between the deposition and meiobenthic assimilation of phytodetritus. *Marine Ecology Progress Series*, **50**: 231–40.

Sabbatini, A., Morigi, C., Negri, A. and Gooday, A. J. 2002. Soft-shelled foraminifera from a hadal site (7800 m water depth) in the Atacama Trench (SE Pacific): preliminary observations. *Journal of Micropalaeontology*, **21**: 131–5.

Saffert, H. and Thomas, E. 1998. Living foraminifera and total populations in salt marsh peat cores: Kelsey marsh (Clinton, CT) and the Great Marshes (Barnstable, MA). *Marine Micropaleontology*, **33**: 175–202.

Sakai, K. and Nishihira, M. 1981. Population study of the benthic foraminifer *Baculogypsina sphaerulata* on the Okinawan reef flat and preliminary estimation of its annual production. *Proceedings of the Fourth International Coral Reef Symposium, Manila*, **2**: 763–6.

Salvat, B. and Vénec-Peyré, M. T. 1981. The living foraminifera in the Scilly atoll lagoon (Society Island). *Proceedings of the Fourth International Coral Reef Symposium, Manila*, **2**: 767–74.

Schaaf, A., Hoffert, M., Karpoff, A. M. and Wirrmann, D. 1977. Associations de structures stromatolithiques et de foraminifères sessiles dans un encroûtment ferromanganésifère à coeur granitique en provenance de l'Atlantique nord. *Comptes rendus de l'Academie de Science, Paris*, **284**: 1705–8.

Schafer, C. T. 1967. Preliminary survey of the distribution of living benthonic foraminifera in Northumberland Strait. *Maritime Sediments*, **3**: 105–8.

Schafer, C. T. 1971. Sampling and spatial distribution of benthonic foraminifera. *Limnology and Oceanography*, **16**: 944–51.

Schafer, C. T. 1976. In situ environmental response of benthonic foraminifera. *Geological Survey of Canada Paper*, **76–1C**: 27–32.

Schafer, C. T. 1982. Foraminiferal colonization of an offshore dump site in Chaleur Bay, New Brunswick, Canada. *Journal of Foraminiferal Research*, **12**: 317–26.

Schafer, C. T. and Cole, F. E. 1982. Living benthic foraminifera distributions on the continental slope and rise east of Newfoundland, Canada. *Bulletin of the Geological Society of America*, **932**: 208–17.

Schafer, C. T., Cole, F. E., Frobel, D., Rice, N. and Buzas, M. A. 1996. An *in situ* experiment on temperature sensitivity of nearshore temperate benthic foraminifera. *Journal of Foraminiferal Research*, **26**: 53–63.

Schafer, C. T., Cole, F. E. and Syvitski, J. P. M. 1989. Bio- and lithofacies of modern sediments in Knight and Bute inlets, British Columbia. *Palaios*, **4**: 107–26.

Schafer, C. T. and Prakash, A. 1968. Current transport and deposition of foraminiferal tests, planktonic organisms and other lithogenic particles in Bedford Basin, Nova Scotia. *Maritime Sediments*, **4**: 100–3.

Schafer, C. T., Winters, G. V., Scott, D. B. *et al.* 1995. Survey of living foraminifera and polychaete populations at some Canadian aquaculture sites: potential for impact mapping and monitoring. *Journal of Foraminiferal Research*, **25**: 236–59.

Schafer, C. T. and Young, J. A. 1977. Experiments on mobility and transportability of some nearshore benthonic foraminifera species. *Geological Survey of Canada Paper*, **99-1C**: 27–31.

Schaudinn, F. 1895. Über den Dimorphismus bei Foraminiferen. *Sitzungsberichte der Gessellschaft naturforschender Freunde zu Berlin*, **5**: 87–97.

Schewe, I. 2001. Small-sized benthic organisms of the Alpha Ridge, central Arctic Ocean. *Internationale Revue der gesamten Hydrobiologie*, **86**: 317–35.

Schiebel, R. 1992. Rezente benthische Foraminiferen in Sedimenten des Schelfes und oberen Kontintentalhanges im Golf von Guinea (Westafrika). *Berichte-Reports, Geologisch-Paläontologisches Institut und Museum, Universität Kiel*, **51**: 1–179.

Schlitzer, R., Usbeck, R. and Fisher, G. 2003. Inverse modelling of particulate organic carbon fluxes in the South Atlantic. In Wefer, G., Mulitza, S. and Ratmeyer, V., eds. *The South Atlantic in the Late Quaternary: Reconstructions of Material Budgets and Current Systems*. Berlin: Springer-Verlag, pp. 1–19.

Schmiedl, G., de Bovée, F., Buscail, R. *et al.* 2000. Trophic control of benthic foraminiferal abundance and microhabitat in the bathyal Gulf of Lions, western Mediterranean. *Marine Micropaleontology*, **40**: 167–88.

Schmiedl, G. and Mackensen, A. 1993. *Cornuspiroides striolatus* (Brady) and *C. rotundus* nov. spec.: large miliolid foraminifera from Arctic and Antarctic oceans. *Journal of Foraminiferal Research*, **23**: 221–30.

Schmiedl, G., Mackensen, A. and Müller, P. J. 1997. Recent benthic foraminifera from the eastern South Atlantic Ocean: dependence on food supply and water masses. *Marine Micropaleontology*, **32**: 249–87.

Schnitker, D. 1974. Ecophenotypic variation in *Ammonia beccarii* (Linné). *Journal of Foraminiferal Research*, **4**: 217–23.

Schönfeld, J. 1997. The impact of the Mediterranean Outflow Water (MOW) on benthic foraminiferal assemblages and surface sediments at the southern Portuguese continental margin. *Marine Micropaleontology*, **29**: 211–36.

Schönfeld, J. 2001. Benthic foraminifera and pore-water oxygen profiles: a re-assessment of species boundary conditions at the western Iberian margin. *Journal of Foraminiferal Research*, **31**: 86–107.

Schönfeld, J. 2002. Recent benthic foraminiferal assemblages in deep high-energy environments from the Gulf of Cadiz (Spain). *Marine Micropaleontology*, **44**: 141–62.

Schönfeld, J. and Zahn, R. 2000. Late Glacial to Holocene history of the Mediterranean Outflow. Evidence from benthic foraminiferal assemblages and stable isotopes at the Portuguese margin. *Palaeogeography, Palaeoclimatology, Palaeoecology*, **159**: 85–111.

Schrader, H., Cheng, G. and Mahood, R. 1983. Preservation and dissolution of foraminiferal carbonate in an anoxic slope environment, southern Gulf of California. *Utrecht Micropaleotological Bulletins*, **30**: 205–27.

Schröder, C. J. 1988. Subsurface preservation of agglutinated foraminifera in the northwest Atlantic Ocean. *Abhandlungen der geologischen Bundesanstalt*, **41**: 325–36.

Schröder, C. J., Scott, D. B. and Medioli, F. S. 1987. Can smaller foraminifera be ignored in paleoenvironmental analyses? *Journal of Foraminiferal Research*, **17**: 101–5.

Schröder, C. J., Scott, D. D., Medioli, F. S., Bernstein, B. B. and Hessler, R. R. 1988. Larger agglutinated foraminifera: comparisons of assemblages from central North Pacific and western North Atlantic (Nares Abyssal Plain). *Journal of Foraminiferal Research*, **18**: 25–41.

Schröder-Adams, C. J. 1990. High latitude agglutinated foraminifera: Prydz Bay (Antarctica) vs. Lancaster Sound (Canadian Arctic). In Hemleben, C., Kaminski, M. A., Kuhnt, W. and Scott, D. B., eds. *Paleoecology, Biostratigraphy, Paleoceanography and Taxonomy of Agglutinated Foraminifera*. Dordrecht: Kluwer, pp. 315–43.

Schröder-Adams, C. J., Cole, F. E., Medioli, F. S. *et al*. 1990. Recent Arctic shelf foraminifera: seasonally ice covered vs. perennially ice covered areas. *Journal of Foraminiferal Research*, **20**: 8–36.

Schultz, E. 1915. Die Hyle des Lebens. I. Beobachtungen und Experimente an *Astrorhiza limicola*. *Archiv für Entwicklungsmechanik der Organismen*, **41**: 215–36.

Scott, D. B. 1976a. Brackish water foraminifera from Southern California and description of *Polysaccammina ipohalina* n. gen., n. sp. *Journal of Foraminiferal Research*, **6**: 312–21.

Scott, D. B. 1976b. Quantitative studies of marsh foraminiferal patterns in southern California and their application to Holocene stratigraphical problems. *Maritime Sediments, Special Publication*, **1**: 153–70.

Scott, D. B., Collins, E. S., Duggan, J. *et al*. 1996. Pacific rim marsh foraminiferal distributions: implications for sea-level studies. *Journal of Coastal Research*, **12**: 850–61.

Scott, D. B., Hasegawa, S., Saito, T., Ito, K and Collins, E. 1995. Marsh foraminifera and vegetation distributions in Nemuro Bay wetland areas, east Hokkaido. *Transactions of the Palaeontological Society of Japan, NS*, **180**: 282–95.

Scott, D. B. and Hermelin, J. O. R. 1993. A device for precision splitting of micropaleontological samples in liquid suspension. *Journal of Paleontology*, **67**: 151–4.

Scott, D. B. and Martini, I. P. 1982. Marsh foraminifera zonations in western James and Hudson bays. *Naturaliste canadien*, **109**: 399–414.

Scott, D. B. and Medioli, F. 1978. Vertical zonations of marsh foraminifera as accurate indicators of former sea levels. *Nature*, **272**: 528–31.

Scott, D. B. and Medioli, F. 1980a. Living vs. total foraminiferal populations; their relative usefulness in palaeoecology. *Journal of Paleontology*, **54**: 814–31.

Scott, D. B. and Medioli, F. 1980b. Quantitative studies of marsh foraminiferal distribution in Nova Scotia: implications for sea level studies. *Contributions from the Cushman Foundation for Foraminiferal Research, Special Publication*, **17**: 1–58.

Scott, D. B., Medioli, F. S. and Schafer, C. T. 1977. Temporal changes in foraminiferal distributions in Miramichi River estuary, New Brunswick. *Canadian Journal of Earth Sciences*, **14**: 1566–87.

Scott, D. B., Medioli, F. S. and Schafer, C. T. 2001. *Monitoring in Coastal Environments using Foraminifera and Thecamoebian Indicators.* Cambridge: Cambridge University Press.

Scott, D. B., Mudie, P. J. and Bradshaw, J. S. 1976. Benthonic foraminifera of three southern Californian lagoons: ecology and recent stratigraphy. *Journal of Foraminiferal Research,* **6**: 59–75.

Scott, D. B., Piper, D. J. W. and Panagos, A. G. 1979. Recent salt marsh and intertidal mudflat foraminifera from the western coast of Greece. *Rivista italiana de paleontologia,* **85**: 243–66.

Scott, D. B., Schafer, C. T. and Medioli, F. S. 1980. Eastern Canadian estuarine foraminifera: a framework for comparison. *Journal of Foraminiferal Research,* **10**: 205–34.

Scott, D. B., Schnack, E. J., Ferrero, L., Espinosa, M. and Barbosa, C. F. 1990. Recent marsh foraminifera from the east coast of South America: comparison to the northern hemisphere. In: Hemleben, C., Kaminski, M. A., Kuhnt, W. and Scott, D. B., eds., *Paleoecology, Biostratigraphy, Paleoceanography and Taxonomy of Agglutinated Foraminifera.* Dordrecht: Kluwer, pp. 717–37.

Scott, D. B., Suter, J. R. and Klosters, E. 1991. Marsh foraminifera and arcellaceans of the lower Mississippi delta: controls on spatial distribution. *Micropaleontology,* **37**: 373–92.

Scott, D. B., Williamson, M. A. and Duffett, T. E. 1981. Marsh foraminifera of Prince Edward Island: their recent distribution and application for former sea level studies. *Maritime Sediments and Atlantic Geology,* **17**: 98–129.

Scott, G. A., Scourse, J. D. and Austin, W. E. N. 2003. The distribution of benthic foraminifera in the Celtic Sea: the significance of seasonal stratification. *Journal of Foraminiferal Research,* **33**: 32–61.

Scourse, J. D., Austin, W. E. N., Long, B. T., Assinder, D. J. and Huws, D. 2002. Holocene evolution of seasonal stratification in the Celtic Sea: refined age model, mixing depths and foraminiferal stratigraphy. *Marine Geology,* **191**: 119–45.

Scourse, J. D., Kennedy, H., Scott, G. A. and Austin, W. E. N. 2004. Stable isotopic analyses of modern benthic foraminifera from seasonally stratified shelf seas: disequilibria and the 'seasonal effect'. *The Holocene,* **14**: 747–58.

Seibold, I. and Seibold, E. 1981. Offshore and lagoonal benthic foraminifera near Cochin (southwest India) – distribution, transport, ecological aspects. *Neues Jahrbuch für Geologie und Paläntologie Abhandlungen,* **162**: 1–56.

Seiglie, G. A. 1974. Foraminifers of Mayagüez and Añasco Bays and its surroundings. *Caribbean Journal of Science,* **14**: 1–58.

Seiglie, G. A. 1975. Foraminifers of Guayanilla Bay and their use as environmental indicators. *Revista española de micropaleontología,* **7**: 453–87.

Seiler, W. C. 1975. Tiefenverteilung benthischer Foraminiferen am portugiesichen Kontinentalhang. *"Meteor" Forschungsergebnisse, Reihe C,* **23**: 47–94.

Sejrup, H. P., Birks, H. J. B., Klitgaard Kristensen, D. and Madsen, H. 2004. Benthonic foraminiferal distributions and quantitative transfer functions for the northwest European continental margin. *Marine Micropaleontology,* **53**: 197–226.

Semeniuk, T. A. 2000. Variability in epiphytic foraminifera from micro- to regional scale. *Journal of Foraminiferal Research*, **30**: 99–109.

Semeniuk, T. A. 2001. Epiphytic foraminifera along a climatic gradient, Western Australia. *Journal of Foraminiferal Research*, **31**: 191–200.

Sen Gupta, B. K. 1971. The benthonic foraminifera of the tail of the Grand Banks. *Micropaleontology*, **17**: 69–98.

Sen Gupta, B. K., ed. 1999. *Modern Foraminifera*. Dordrecht: Kluwer.

Sen Gupta, B. K. and Aharon, P. 1994. Benthic foraminifera of bathyal hydrocarbon vents of the Gulf of Mexico: initial report on communities and stable isotopes. *Geo-Marine Letters*, **14**: 88–96.

Sen Gupta, B. K. and Hayes, W. B. 1979. Recognition of Holocene benthic foraminiferal facies by recurrent group analysis. *Journal of Foraminiferal Research*, **9**: 233–45.

Sen Gupta, B. K., Lee, R. F. and May, M. S. 1981. Upwelling and an unusual assemblage of benthic foraminifera on the northern Florida continental slope. *Journal of Paleontology*, **55**: 853–7.

Sen Gupta, B. K. and Machain-Castillo, M. L. 1993. Benthic foraminifera in oxygen-poor habitats. *Marine Micropaleontology*, **20**: 183–202.

Sen Gupta, B. K., Platon, E., Bernhard, J. M. and Aharon, P. 1997. Foraminiferal colonization of hydrocarbon-seep bacterial mats and underlying sediment, Gulf of Mexico slope. *Journal of Foraminiferal Research*, **27**: 292–300.

Sen Gupta, B. K. and Schafer, C. T. 1973. Holocene benthonic foraminifera in leeward bays of St. Lucia, West Indies. *Micropaleontology*, **19**: 341–65.

Sen Gupta, B. K. and Strickert, D. P. 1982. Living benthic foraminifera of the Florida–Hatteras slope: distribution and trends. *Bulletin of the Geological Society of America*, **93**: 218–24.

Sen Gupta, B. K., Turner, R. E. and Rabalais, N. N. 1996. Seasonal oxygen depletion in continental shelf waters of Louisiana: historical record of benthic foraminifers. *Geology*, **24**: 227–30.

Serandrei-Barbero, R., Morisieri, M. R., Carbognin, L. and Donnici, S. 2003. An inner shelf foraminiferal fauna and its response to environmental processes (Adriatic Sea, Italy). *Revista española de micropaleontología*, **35**: 241–64.

Severin, K. P. 1987. Laboratory observations on the rate of subsurface movement of a small miliolids foraminifer. *Journal of Foraminiferal Research*, **17**: 110–16.

Severin, K. P., Culver, S. J. and Blandpied, C. 1982. Burrows and trails produced by *Quinqueloculina impressa* Reuss, a benthic foraminifer, in fine-grained sediment. *Sedimentology*, **26**: 897–901.

Severin, K. P. and Erskian, M. G. 1981. Laboratory experiments on vertical movement of *Quinqueloculina impressa* Reuss through sand. *Journal of Foraminiferal Research*, **11**: 133–6.

Severin, K. P. and Lipps, J. J. 1989. The weight–volume relationship of the test of *Alveolinella quoyi*: implications for the taphonomy of large fusiform foraminifera. *Lethaia*, **22**: 1–12.

Shannon, C. E. 1948. A mathematical theory of communication. *Bell System Technical Journal*, **27**: 379–423; 623–56.

Sharifi, A. R., Croudace, I. W. and Austin, T. L. 1991. Benthic foraminiferids as pollution indicators in Southampton Water, southern England, United Kingdom. *Journal of Micropalaeontology*, **10**: 109–13.

Shirayama, Y. 1984a. The abundance of deep sea meiobenthos in the western Pacific in relation to environmental factors. *Oceanologica acta*, **7**: 113–21.

Shirayama, Y. 1984b. Vertical distribution of meiobenthos in the sediment profile in bathyal and hadal deep-sea systems of the western Pacific. *Oceanologica acta*, **7**: 123–9.

Shires, R., Gooday, A. J. and Jones, A. R. 1994a. The morphology and ecology of an abundant new komokiacean mudball (Komokiacea, Foraminiferida) from the bathyal and abyssal NE Atlantic. *Journal of Foraminiferal Research*, **24**: 214–25.

Shires, R., Gooday, A. J. and Jones, A. R. 1994b. A new large agglutinated foraminifer (Arboramminidae n. fam.) from an oligotrophic site in the abyssal northeast Atlantic. *Journal of Foraminiferal Research*, **24**: 149–57.

Sibuet, M. and Olu-Le Roy, K. 2002. Cold seep communities on continental margins: structure and quantitative distribution relative to geological and fluid venting patterns. In Wefer, G., Billett, D., Hebbeln, D. et al., eds. *Ocean Margin Systems*. Berlin: Springer-Verlag, pp. 235–51.

Siddall, J. D. 1878. The foraminifera of the River Dee. *Proceedings of the Chester Society for Natural Science*, **2**: 42–56.

Silva, K. A., Corliss, B. H., Rathburn, A. E. and Thunell, R. 1996. Seasonality of living benthic foraminifera from the San Pedro Basin, Californian borderland. *Journal of Foraminiferal Research*, **26**: 71–93.

Sjoerdsma, P. G. and Zwaan, G. J. van der 1992. Simulating the effect of changing organic flux and oxygen content on the distribution of benthic foramnifera. *Marine Micropaleontology*, **19**: 163–80.

Sliter, W. V. 1965. Laboratory experiments on the life cycle and ecological controls of *Rosalina globularis* d'Orbigny. *Journal of Protozoology*, **12**: 210–15.

Sliter, W. V. 1971. Predation on benthic foraminifers. *Journal of Foraminiferal Research*, **1**: 20–9.

Smith, C. R. and Hamilton, C. 1983. Epibenthic megafauna of a bathyal basin off southern California; patterns of abundance, biomass, and dispersion. *Deep-Sea Research*, **30**: 907–28.

Smith, D. A., Scott, D. B. and Medioli, F. S. 1984. Marsh foraminifera in the Bay of Fundy: modern distribution and application to sea-level determinations. *Maritime Sediments and Atlantic Geology*, **20**: 127–42.

Smith, P. B. 1963. Quantitative and qualitative analysis of the Family Bolivinidae. Recent foraminifera off central America. *United States Geological Survey, Professional Paper*, **429**: A1–39.

Smith, P. B. 1964. Ecology of benthonic species. Recent foraminifera off central America. *United States Geological Survey, Professional Paper*, **429**: B55.

Smith, R. K. 1968. An intertidal *Marginopora* colony in Suva harbour, Fiji. *Contributions from the Cushman Foundation for Foraminiferal Research*, **19**: 12–7.

Snyder, S.W., Hale, W.R. and Kontrovitz, M. 1990. Assessment of postmortem transportation of modern benthic foraminifera of the Washington continental shelf. *Micropaleontology*, **36**: 259–82.

Sournia, A. 1976. Primary production of sands in the lagoon of an atoll and the role of foraminiferan symbionts. *Marine Biology*, **37**: 29–32.

Sousa, W.P. 2001. Natural disturbance and the dynamics of marine benthic communities. In Bertness, M.D., Gaines, S.D. and Hay, M.E., eds. *Marine Community Ecology*. Sunderland: Sinauer Associates, pp. 85–130.

Spindler, M. 1980. The pelagic gulfweed *Sargassum natans* as a habitat for the benthic foraminifera *Planorbulina acervalis* and *Rosalina globularis*. *Neues Jahrbuch für Geologie und Paläntologie Monatheft*, **9**: 569–80.

Stackelberg, U. von. 1984. Significance of benthic organisms for the growth and movement of manganese nodules, equatorial north Pacific. *Geo-Marine Letters*, **4**: 37–42.

Stanley, D.J., Culver, S.J. and Stubblefield, W.L. 1986. Petrologic and foraminiferal evidence for active downslope transport in Wilmington Canyon. *Marine Geology*, **69**: 207–18.

Stein, R. and Macdonald, R., eds. 2004. *The Organic Carbon Cycle in the Arctic Ocean*. Berlin: Springer.

Steineck, P.L. and Bergstein, J. 1979. Foraminifera from Hommocks salt-marsh, Larchmont harbor, New York. *Journal of Foraminiferal Research*, **9**: 147–58.

Steinker, D.C. and Clem, K.V. 1984. Some nearshore foraminiferal assemblages from phytal substrates and bottom sediments, Bermuda. *The Compass*, **61**: 98–115.

Steinker, D.C. and Rayner, A.L. 1981. Some habitats of nearshore foraminifera, St. Croix, US Virgin Islands. *The Compass*, **59**: 15–26.

Steinker, P.J. and Steinker, D.C. 1976. Shallow-water foraminifera, Jewfish Cay, Bahamas. *Maritime Sediments, Special Publication*, **1**: 171–80.

Stouff, V., Debenay, J.P. and Lesourd, M. 1999a. Origin of double and multiple tests in benthic foraminifera: observations in laboratory cultures. *Marine Micropaleontology*, **36**: 189–204.

Stouff, V., Geslin, E., Debenay, J.P. and Lesourd, M. 1999b. Origin of morphological abnormalities in *Ammonia* (Foraminifera): studies in laboratory and natural environments. *Journal of Foraminiferal Research*, **29**: 152–70.

Stouff, V., Lesourd, M. and Debenay, J.P. 1999c. Laboratory observations on asexual reproduction (schizogony) and ontogeny of *Ammonia tepida* with comments on the life cycle. *Journal of Foraminiferal Research*, **29**: 75–84.

Streeter, S.S. 1972. Living benthonic foraminifera of the Gulf of California, a factor analysis of Phleger's (1964) data. *Micropaleontology*, **18**: 64–73.

Streeter, S.S. 1973. Bottom water and benthonic foraminifera in the North Atlantic – glacial–interglacial contrasts. *Quaternary Research*, **3**: 131–41.

Sturrock, S. and Murray, J.W. 1981. Comparison of low energy middle shelf foraminiferal faunas: Celtic Sea and western English Channel. In Neale,

J. W. and Brasier, M. D., eds. *Microfossils from Recent and Fossil Shelf Seas*. Chichester: Ellis Horwood, pp. 250-60.

Swallow, J. E. 2000. Intra-annual variability and patchiness in living assemblages of salt-marsh foraminifera from Mill Rythe Creek, Chichester harbour, England. *Journal of Micropalaeontology*, **19**: 9-22.

Swallow, J. E. and Culver, S. J. 1999. Living (rose Bengal) stained benthic foraminifera from New Jersey continental margin canyons. *Journal of Foraminiferal Research*, **29**: 104-16.

Tachikawa, K. and Elderfield, H. 2002. Microhabitat effects on Cd/Ca and δ^{13}C of benthic foraminifera. *Earth and Planetary Science Letters*, **202**: 607-24.

Talge, H. K. and Hallock, P. 1995. Cytological examination of symbiont loss in a benthic foraminifera, *Amphistegina gibbosa*. *Marine Micropaleontology*, **26**: 107-13.

Talge, H. K., Williams, D. E., Hallock, P. and Harney, J. N. 1997. Symbiont loss in foraminifera: consequences for affected populations. *Proceedings of the Eighth Coral Reef Symposium*, **1**: 589-94.

Temnikov, N. K. 1976. US Antarctic Research Program, 1975-1976. *Antarctic Journal of the United States*, **11**: 220-2.

Tendal, O. S. 1979. Aspects of the biology of Komokiacea and Xenophyophorea. *Sarsia*, **64**: 13-17.

Tendal, O. S. and Gooday, A. J. 1981. Xenophyophoria (Rhizopoda, Protozoa) in bottom photographs from the bathyal and abyssal NE Atlantic. *Oceanologica acta*, **4**: 415-22.

Tendal, O. S. and Thomsen, E. 1988. Observations on the life position and size of the large foraminifer *Astrorhiza arenaria* Norman, 1876 from the shelf off northern Norway. *Sarsia*, **73**: 39-42.

Ter Kuile, B. 1991. Mechanisms for calcification and carbon cycling in algal symbionts-bearing foraminifera. In Lee, J. J. and Anderson, O. R., eds. *Biology of Foraminifera*. London: Academic, pp. 73-89.

Ter Kuile, B. and Erez, J. 1984. *In situ* growth rate experiments on the symbionts-bearing foraminifera *Amphistegina lobifera* and *Amphisorus hemprichii*. *Journal of Foraminiferal Research*, **14**: 262-76.

Ter Kuile, B. and Erez, J. 1991. Carbon budgets for two species of benthonic symbiont-bearing foraminifera. *Biological Bulletin*, **180**: 489-95.

Ter Kuile, B., Erez, J. and Lee, J. J. 1987. The role feeding in the metabolism of larger symbiont bearing foraminifera. *Symbiosis*, **4**: 335-50.

Thiel, H., Pfannkuche, O., Schriever, G. *et al.*, 1988. Phytodetritus on the deep-sea floor in a central oceanic region of the northeast Atlantic. *Biological Oceanography*, **6**: 203-39.

Thiel, H. and Schneider, J. 1988. Manganese nodule–organism interactions. In Halbach, P., Friedrich, G. and Stackelberg, U. von, eds. *The Manganse Nodule Belt of the Pacific Ocean. Geological Environment, Nodule formation, and Mining Aspects*. Stuttgart: Ferdinand Enke Verlag, pp. 102-10.

Thies, A. 1990. The ecology, distribution and taxonomy of *Crithionina hispida* Flint, 1899. In Hemleben, C., Kaminski, M. A., Kuhnt, W. and Scott, D. B., eds.

Paleoecology, Biostratigraphy, Paleoceanography and Taxonomy of Agglutinated Foraminifera. Dordrecht: Kluwer, pp. 305–13.

Thomas, E., Gapotchenko, T., Varekamp, J. C., Mecray, E. L. and Buchholtz ten Brink, M. R. 2000. Benthic foraminifera and environmental changes in Long Island Sound. *Journal of Coastal Research,* **16**: 641–55.

Thomas, E. and Gooday, A. J. 1996. Cenozoic deep-sea benthic foraminifers: tracers for changes in oceanic productivity? *Geology,* **24**: 355–8.

Thompson, L. B. 1978. Distribution of living benthic foraminifera, Isla de los Estados, Tierra del Fuego, Argentina. *Journal of Foraminiferal Research,* **8**: 241–57.

Thompson, P. R. 1992. Foraminiferal evidence for the source and timing of mass-flow deposits south of Baltimore canyon. In Ishizaki, K. and Saito, T., eds. *Centenary of Japanese Micropaleontology.* Tokyo: Terra Scientific Publishing Company, pp. 97–128.

Thomsen, L. and Altenbach, A. V. 1993. Vertical and areal distribution of foraminiferal abundance and biomass in microhabitats around inhabited tubes of marine echiurids. *Marine Micropaleontology,* **20**: 303–9.

Timm, S. 1992. Rezente Tiefsee-Benthosforaminiferen aus Oberflächensedimenten des Golfes von Guinea (Westafrika) – Taxonomie, Verbreitung, Ökologie und Korngrößenfraktionen. *Berichte-Reports, Geologisch-Paläontologisches Institut und Museum, Universität Kiel,* **519**: 1–192.

Todo, Y., Kitazato, H., Hashimoto, J. and Gooday, A. J. 2005. Simple foraminifera flourish at the ocean's deepest point. *Science,* **307**: 689.

Travis, J. L. and Bowser, S. S. 1991. The motility of foraminifera. In Lee, J. J. and Anderson, O. R., eds. *Biology of Foraminifera.* London: Academic, pp. 91–155.

Travis, J. L., Welnhofer, E. A. and Orokos, D. D. 2002. Autonomous reorganization of foraminiferan reticulopodia. *Journal of Foraminiferal Research,* **32**: 425–33.

Tsuchiya, M., Kitazato, H. and Pawlowski, J. 2000. Phylogenetic relationships among species of Glabratellidae (Foraminifera) inferred from ribosomal DNA sequences: comparison with morphological and reproductive data. *Micropaleontology,* **46** (Supplement 1): 13–20.

Tuckwell, G. W., Allen, K., Roberts, S. and Murray, J. W. 1999. Simple models of agglutinated foraminifera test construction. *Journal of Eukaryotic Microbiology,* **46**: 248–53.

Uchio, T. 1960. Ecology of living benthic foraminifera from the San Diego, California, area. *Cushman Foundation for Foraminiferal Research, Special Publication,* **5**: 1–72.

Ujiié, H. and Kusukawa, T. 1969. Analysis of foraminiferal assemblages from Miyako and Yamada bays, northeastern Japan. *Bulletin of the National Science Museum Tokyo,* **12**: 735–72.

Valiela, I. 1995. *Marine Ecological Processes,* 2nd edn. New York: Springer Verlag.

Vangerow, E. F. 1972. Ökologische Beobachtungen an Foraminiferen des Brackwasserbereiches der Rhônemündung (Sudfrankreich). *Journal of Experimental Marine Biology and Ecology,* **8**: 145–65.

Vangerow, E. F. 1974. Récentes observations écologiques des foraminifères dans la zone saumâtre de l'embouchure du Rhône. *Revista española de micropaleontología*, **17**: 95–106.

Vangerow, E. F. 1977. Häufigkeitsverteilung der Foraminiferen im Flaschen Wasser der Rhônemundung. *Paläontologische Zeitschrift*, **51**: 145–51.

Vénec-Peyré, M. T. 1984. Etude de la distribution des foraminifères vivant dans la Baie de Banyuls-sur-Mer. In Bizon, J. J. and Burolet, P. F., eds. *Ecologie des microorganismes en Méditerranée occidentale "Ecomed"*. Paris: Association Française des Techniciens du Pétrole, pp. 60–80.

Vénec-Peyré, M. T. 1985a. La role de certains foraminifères dans la bioérosion et la sédimentogenèse. *Comptes rendus de l'Académie des Sciences, Paris*, séries 2, **300**: 83–8.

Vénec-Peyré, M. T. 1985b. Etude de la distribution des foraminifères vivants dans le lagon de l'île haute volcanique de Moorea (Polynésie française). *Proceedings of the Fifth International Coral Reef Congress, Tahiti*, **5**: 227–32.

Vénec-Peyré, M. T. 1988. Two new species of bioeroding Trochamminidae (Foraminiferida) from French Polynesia. *Journal of Foraminiferal Research*, **18**: 1–5.

Vénec-Peyré, M. T. 1991. Distribution of living benthic foraminifera on the back-reef and outer slopes of a high island (Moorea, French Polynesia). *Coral Reefs*, **9**: 193–203.

Vénec-Peyré, M. T. 1993. Mise en evidence d'un mode de vie endolithe chez les foraminifères *Gypsina globulus* (R.) et *Cribrobaggina reniformis* (H. A. et E.) dans les récifs de Polynésie. Révision taxinomique de *G. globulus*. *Revue de micropaléontologie*, **36**: 67–75.

Vénec-Peyré, M. T. 1996. Bioeroding foraminifera: a review. *Marine Micropaleontology*, **28**: 19–30.

Vénec-Peyré, M. T. and Le Calvez, Y. 1981. Etude des foraminifères de l'herbier á Posidonies de Banyuls-sur-Mer. *106ᵉ Congrès national des sociétés savants, Perpignan, 1981*, **1**: 191–203.

Vénec-Peyré, M. T. and Le Calvez, Y. 1986. Foraminifères benthiques et phénomènes de transfert; importance des études comparatives de la biocénose et de la thanatocénose. *Annales de l'Institut Océanographique NS*, **57**: 79–110.

Vénec-Peyré, M. T. and Le Calvez, Y. 1988. Les foraminifères epiphytes de l'herbier de Posidonies de Banyuls-sur-Mer (Méditerranée occidentale): étude des variations spatiotemporelles du peuplement. *Cahiers de micropaléontologie, NS*, **3**: 21–40.

Vénec-Peyré, M. T. and Salvat, B. 1981. Les foraminifères de l'atoll de Scilly (Archipel de la Société): étude comparée de la biocénose et de la thanatocénose. *Annales de l'Institut Océanographique, NS*, **57**: 79–110.

Violanti, D. 1996. Taxonomy and distribution of recent benthic foraminifera from Terra Nova Bay (Ross Sea, Antarctica), Oceanographic Campaign 1987/1988. *Paleontographia Italica*, **83**: 25–71.

Violanti, D. 2000. Morphogroup analysis of recent agglutinated foraminifers off Terra Nova Bay, Antarctica (Expedition 1987–1988). In Faranda,

F. M., Gugliemo, L., Ionora, A. *et al.* eds. *Ross Sea Ecology, Italiantartide Expeditions* (1987–1995). Berlin: Springer, pp. 479–92.

Voigt, E. and Bromley, R. G. 1974. Foraminifera as commensals around clionid sponge papillae: cretaceous and recent. *Senckenbergiana maritima*, **6**: 33–45.

Voorthuysen, J. H. van 1960. Die Foraminiferen des Dollart-Ems-Estuarium. *Verhandelingen van het koninck nederlandsch geologisch-mijnbouwkundig genootschap (Geology Series)*, **19**: 237–69.

Wakefield, M. I. 2003. Bio-sequence stratigraphic utility of SHE diversity analysis. *Society of Economic Paleontologists and Mineralogists Special Publication*, **75**: 81–7.

Walker, D. A., Linton, A. E. and Schafer, C. T. 1974. Sudan Black B: a superior stain to rose Bengal for distinguishing living from non-living foraminifera. *Journal of Foraminiferal Research*, **4**: 205–15.

Walker, D. A. 1971. Etching of the test surface of benthic foraminifera due to ingestion by the gastropod *Littorina littorea* Linné. *Canadian Journal of Earth Science*, **8**: 1487–91.

Walton, W. R. 1952. Techniques for recognition of living foraminifera. *Contributions from the Cushman Foundation for Foraminiferal Research*, **3**: 56–60.

Walton, W. R. 1955. Ecology of living benthonic foraminifera, Todos Santos Bay, Baja California. *Journal of Paleontology*, **29**: 952–1018.

Walton, W. R. 1964. Ecology of benthonic foraminifera in the Tampa-Sarasota Bay area, Florida. In Miller, R. L., ed. *Papers in Marine Geology*. New York: Macmillan, pp. 429–54.

Walton, W. R. and Sloan, B. J. 1990. The genus *Ammonia* Brünnich, 1772: its geographic distribution and morphologic variability. *Journal of Foraminiferal Research*, **20**: 128–56.

Wang, P. 1983. Verbreitung der Benthos-Foraminiferen im Elbe-Ästuar. *Meyniana*, **35**: 67–83.

Wang, P., Hong, X. and Zhao, Q. 1985. Living foraminifera and ostracoda: distribution in the coastal area of the East China Sea and Huanghai Sea. In Wang, P., ed. *Marine Micropaleontology of China*. Beijing: China Ocean Press, pp. 243–55.

Wang, P. and Murray, J. W. 1983. The use of foraminifera as indicators of tidal effects in estuarine deposits. *Marine Geology*, **51**: 239–50.

Ward, B. L., Barrett, P. J. and Vella, P. 1987. Distribution and ecology of benthic foraminifera in McMurdo Sound, Antarctica. *Palaeogeography, Palaeoclimatology, Palaeoecology*, **58**: 139–53.

Ward, J. N., Pond, D. W. and Murray, J. W. 2003. Feeding of benthic foraminifera on diatoms and sewage-derived organic matter: an experimental application of lipid biomarker techniques. *Marine Environmental Research*, **56**: 515–30.

Wassenberg, J. and Hill, B. J. 1993. Diet and feeding behavior of juvenile and adult banana prawns *Penaeus merguiensis* in the Gulf of Carpenteria, Australia. *Marine Ecology Progress Series*, **94**: 287–95.

Wefer, G. 1976a. Umwelt, Produktion und Sedimentation benthischer Foraminiferen in der Westlichen Ostsee. *Report Sondersforschungsbereich 95, Kiel*, **14**: 1–103.

Wefer, G. 1976b. Environmental effects on growth rates of benthic foraminifera (shallow water, Baltic Sea). *Maritime Sediments, Special Publication*, **1**: 39–50.

Wefer, G. and Berger, W. 1991. Isotope paleontology, growth and composition of extant calcareous species. *Marine Geology*, **100**: 207–48.

Wefer, G. and Lutze, G. F. 1976. Benthic foraminifera biomass production in the western Baltic. *Kieler Meeresforschungen*, **3**: 76–81.

Wefer, G. and Lutze, G. F. 1978. Carbonate production by benthic foraminifera and accumulation in the western Baltic. *Limnology and Oceanography*, **23**: 992–6.

Wefer, G. and Richter, W. 1976. Colonization of artificial substrates by foraminifera. *Kieler Meeresforschungen*, **3**: 72–75.

Weinberg, J. R. 1991. Rates of movement and sedimentary traces of deep-sea foraminifera and Mollusca in the laboratory. *Journal of Foraminiferal Research*, **21**: 213–17.

Wetmore, K. L. 1987. Correlations between test strength, morphology and habitat in some benthic foraminifera from the coast of Washington. *Journal of Foraminiferal Research*, **17**: 1–13.

Wetmore, K. L. 1988. Burrowing and sediment movement by benthic foraminifera, as shown by time-lapse cinematography. *Revue de paléobiologie, Volume Spécial*, **2**: 921–7.

Wetmore, K. L. and Plotnick, R. E. 1992. Correlations between test morphology, crushing strength, and habitat in *Amphistegina gibbosa*, *Archaias angulatus*, and *Laevipeneroplis proteus* from Bermuda. *Journal of Foraminiferal Research*, **22**: 1–12.

Whittaker, R. H. 1975. *Communities and Ecosystems*, 2nd edn. MacMillan, New York.

Williams, C. B. 1964. *Patterns in the Balance of Nature*. London: Academic.

Williams, D. E. and Hallock, P. 2004. Bleaching in *Amphistegina gibbosa* d'Orbigny (Class Foraminifera): observations from laboratory experiments using visible and ultraviolet light. *Marine Biology*, **145**: 641–9.

Williams, D. E., Hallock, P., Talge, H. K., Harney, J. N. and McRae, G. 1997. Response of *Amphistegina gibbosa* populations in the Florida Keys (USA) to a multi-year stress event (1991–1996). *Journal of Foraminiferal Research*, **27**: 264–9.

Winter, F. W. 1907. Zur Kenntnis der Thalamophoren. I, Untersuchung über *Peneroplis pertusus* (Forskål). *Archiv für Protistenkunde*, **10**: 1–113.

Witte, L. J. 1994. Man-made dispersal of microbenthos. *Marine Micropaleontology*, **23**: 87–8.

Wollenburg, J. 1992. Taxonomie von rezenten benthischen Foraminiferen aus dem Nansen Becken, Arktischer Ozean. *Berichte zur Polarforschung*, **112**: 1–137.

Wollenburg, J. 1995. Benthische Foraminiferenfaunen als Wassermassen-, Produktions- und Eisdriftanzeiger im Arktischen Ozean. *Berichte zur Polarforschung*, **179**: 1–227.

Wollenburg, J. E. and Kuhnt, W. 2000. The response of benthic foraminifers to carbon flux and primary production in the Arctic Ocean. *Marine Micropaleontology*, **40**: 189–231.

Wollenburg, J. E., Kuhnt, W. and Mackensen, A. 2001. Changes in Arctic Ocean paleoproductivity and hydrography during the last 145 kyr: the benthic foraminiferal record. *Paleoceanography*, **16**: 65–77.

Wollenburg, J. E. and Mackensen, A. 1998a. On the vertical distribution of the living (rose Bengal stained) benthic foraminifers in the Arctic Ocean. *Journal of Foraminiferal Research*, **28**: 268–85.

Wollenburg, J. E. and Mackensen, A. 1998b. Living benthic foraminifers from the central Arctic Ocean: faunal composition, standing stock and diversity. *Marine Micropaleontology*, **34**: 153–85.

Woo, H. J., Culver, S. J. and Oertel, G. F. 1997. Benthic foraminiferal communities of a barrier–lagoon system, Virginia, USA. *Journal of Coastal Research*, **13**: 1192–200.

Wright, R. C. and Hay, W. W. 1971. The abundance and distribution of foraminifers in a back-reef environment, Molasses Reef, Florida. *Memoir Miami Geological Society*, **1**: 121–74.

Wynn, R. B., Weaver, P. P. E., Masson, D. G. and Stow, D. A. V. 2002. Turbidite depositional architecture across three interconnected deep-water basins on the north-west African margin. *Sedimentology*, **49**: 669–95.

Yamamoto, S., Tokuyama, H., Fujioka, K., Takeuchi, A. and Ujiié, H. 1988. Carbonate turbidites deposited on the floor of the Palau trench. *Marine Geology*, **82**: 217–33.

Zampi, M., Benocei, S. and Focardi, S. 1997. Epibiont foraminifera of *Sertella frigida* (Waters) (Bryozoa, Cheilostomata) from Terranova Bay, Ross Sea, Antarctica. *Polar Biology*, **17**: 363–70.

Zaninetti, L. 1979. L'étude des foraminifères des mangroves actuelles: réflexion sur les objectifs et sur l'état des connaissances. *Archives des sciences genève*, **32**: 151–61.

Zaninetti, L. 1982. Les foraminifères des marais salants de Salin-de-Giraud (sud de la France): milieu de vie et transports dans le salin, comparaison avec les microfaunes marines. *Géologie méditerranéene*, **9**: 447–70.

Zaninetti, L. 1984. Les foraminifers du salin de Bras del Port (Santa Pola, Espagne) avec remarques sur le distribution des ostracodes. *Revista d'investigacions geologiques*, **38/39**: 123–38.

Zebrowski, G. 1937. New genera of Cladochytriaceae. *Annals of the Missouri Botanical Gardens*, **33**: 553–64.

Zieman, J. C. 1975. Seasonal variation of turtle grass *Thalassia testudinum* König, with reference to temperature and salinity effects. *Aquatic Botany*, **1**: 107–23.

Zmiri, A., Kaha, D., Hochstein, S. and Reiss, Z. 1974. Phototaxis and thermotaxis in some species of *Amphistegina* (Foraminifera). *Journal of Protozoology*, **21**: 133–8.

Zobel, B. 1973. Biostratigraphische Untersuchungen an Sedimenten des indisch-pakistanischen Kontinentalrandes (Arabisches Meer). *"Meteor" Forschungs-Ergebnisse, Reihe C*, **12**: 9–73.

Zumwalt, G. S. and Delaca, T. E. 1980. Utilization of brachiopod feeding currents by epizoic foraminifera. *Journal of Paleontology*, **54**: 477–84.

Zwaan, G. J. van der 2000. Variation in natural vs. anthropogenic eutrophication of shelf areas in front of major rivers. In Martin, R. E., ed. *Environmental Micropaleontology*, New York: Kluwer, pp. 385–404.

Zwaan, G. J. van der, Duinstee, I. A. P., Dulk, M. den *et al.* 1999. Benthic foraminifers: proxies or problems? A review of paleoecological concepts. *Earth-Science Reviews*, **46**: 213–36.

Taxonomic index

Page numbers refer to the start of the section in which the species is mentioned.

412

General index

Page numbers refer to the start of the section in which the species is mentioned.
Page numbers in **bold** refer to main entries.